Electricity, Electronics, and Control Systems for HVAC

Thomas E. Kissell

Terra Community College
Fremont, Ohio

Prentice Hall
Upper Saddle River, New Jersey   Columbus, Ohio

To Chris and Amy
and Jeff and Kelly

Library of Congress Cataloging-in-Publication Data

Kissell, Thomas E.
  Electricity, electronics, and control systems for HVAC / Thomas E. Kissell.
      p. cm.
    Includes bibliographical references and index.
    ISBN 0-13-328659-2
    1. Buildings—Electric equipment. 2. Heating—Control.
  3. Ventilation—Control. 4. Air conditioning—Control. 5. Electric controllers. I. Title.
  TK4035.A35K568 1997
  697—dc20   96-42083
                                                              CIP

Cover art/photo: Courtesy of United Technologies Carrier Corporation
Editor: Ed Francis
Production Editor: Rex Davidson
Production Coordination: bookworks
Design Coordinator: Julia Zonneveld Van Hook
Cover Designer: Brian Deep
Production Manager: Deidra M. Schwartz
Marketing Manager: Danny Hoyt

This book was set in Times Roman by The Clarinda Company and was printed and bound by R.R. Donnelley & Sons Company. The cover was printed by Phoenix Color Corp.

 © 1997 by Prentice-Hall, Inc.
Simon & Schuster/A Viacom Company
Upper Saddle River, New Jersey 07458

Printed in the United States of America

10 9 8 7 6 5 4 3 2 1

ISBN: 0-13-328659-2

Prentice-Hall International (UK) Limited, *London*
Prentice-Hall of Australia Pty. Limited, *Sydney*
Prentice-Hall Canada, Inc., *Toronto*
Prentice-Hall Hispanoamericana, S. A., *Mexico*
Prentice-Hall of India Private Limited, *New Delhi*
Prentice-Hall of Japan, Inc., *Tokyo*
Simon & Schuster Asia Pte. Ltd., *Singapore*
Editora Prentice-Hall do Brasil, Ltda., *Rio de Janeiro*

Preface

Electricity, Electronics, and Control Systems for HVAC is written to help students understand the electrical circuits and controls of air-conditioning, heating, and refrigeration systems. Electricity tends to be very difficult to comprehend for a beginning student, and this book makes learning this subject much easier. The author has over 20 years of experience in installation, troubleshooting, and repair of motors and controls as well as 20 years of teaching electricity to beginning students of all ages. The author incorporates his experience in this textbook to help students learn and comprehend electricity. An extensive number of electrical diagrams are presented in each chapter to explain the theory of operation and show test points for troubleshooting. An appendix is also provided to help students to learn how to read and write electrical ladder diagrams. This text is designed to be simple enough for the beginner, yet comprehensive enough to be a complete reference when the student finishes school and needs to review a troubleshooting technique or theory of operation for a control when on the job.

The book is basically split in two parts. Chapters 1–16 provide an introduction to the basic theory and operation of all the motors and controls that are found in air-conditioning, heating, and refrigeration systems. Chapters 17–20 present detailed diagrams, theory of operation, and troubleshooting of complete systems such as air conditioners, heating systems, (electric, gas, and oil), heat pumps, and refrigeration systems such as reach-in coolers, freezers, and ice makers. This approach allows students to understand the basic parts, and then study the complete electrical system of each of these types of equipment as they would see them when they are on the job. The text always presents information in terms of what the student is expected to do on the job. All mathematics are applied to practical problems students would use to install or troubleshoot electrical equipment, rather than to theoretical analyses that are never used.

The first chapter of the text introduces the basic terminology and theory of electricity. The second chapter explains how a technician must use voltmeters, ammeters, and ohmmeters to troubleshoot circuits and controls. Chapters 3 and 4 explain basic series, parallel, and series-parallel circuits that are the basis of all electrical control circuits in air conditioning and refrigeration.

Chapter 5 introduces the fundamentals of magnetics, and Chapter 6 incorporates this information to explain the fundamentals of AC electricity and the theory of operation of components such as transformers and motors. The explanation of reactance is limited to how it affects AC motors and transformers. Chapter 7 continues this information to explain where single-phase and three-phase voltage comes from. Detailed pictures and diagrams are provided to show the types of circuit breaker panels, disconnects, and terminal points where students will find single-phase and three-phase voltage.

Chapter 8 provides information about symbols and diagrams that are commonly used in installation and troubleshooting. This information is usually presented in very early chapters of most books, well before students can comprehend names and terminology. The author has learned over the years to present information only when it is needed and when students have had the foundation to add more in-depth material. In the first seven chapters a limited number of symbols and diagrams are introduced. After the students have completed the first seven chapters, they are ready to learn more about symbols and diagrams.

Chapters 9–15 present information about the theory of operation and methods of troubleshooting relays, motors, and controls. These chapters provide detailed information to help beginning learners and provide a reference that will be usable when the students reach the job. Every type of open and hermetic motor the students will encounter is explained in a format that is easy to understand. Extensive diagrams will help students when they must identify motors and their terminals to test their windings.

Chapter 16 provides the theory of operation of basic electronics and a complete explanation of the electronic components found in modern equipment. This chapter presents only the electronic fundamentals that students will actually encounter in air-conditioning, heating, and refrigeration equipment.

Chapters 17–20 are perhaps the most important chapters of this text since they provide examples of complete air-conditioning, heating, and refrigeration systems. Many texts present electrical theory, motors, and controls without ever discussing complete systems as students will find them when they are on the job. These chapters provide detailed pictures and diagrams that include cut-away views of the equipment to show where the motors and controls are actually located in the equipment. As veteran teachers, we often forget that the most important question the student has is: Where will I find a compressor motor or fan motor on this particular type of equipment?

This text is designed to be used in a single quarter or semester course, or it can be used where basic electricity and electrical motors and controls are presented as separate courses. At the end of each chapter, a variety of multiple choice, true or false, and open-ended questions is provided to be used as homework assignments or as end-of-chapter tests.

Acknowledgments

I would like to thank my wife, Kathleen Kissell, for her help as a graphic artist and, for keeping all of the figures straight for this book. I would also like to thank her for the support she has provided every time we write a book. I would like to thank Ed Francis, the editor for this book for being very patient and providing help at all stages.

I would like to thank the following people for reviewing this book at various stages: William E. Whitman, Triton College; Gregory E. Jourdan, Wenatchee Valley College; Joseph Zagrobski, Massasoit Community College.

I would also like to thank the following people and companies for their help with this book. This book would not be possible without their help and support in providing pictures, diagrams, and information about their products.

Mr. Tom Falconi
Acme Electric
Acme Transformer Division

Mr. Joe Daleo
Alco Controls Division, Emerson
Electric Co.

Ms. Linda Sorenson
Allen-Bradley, a Rockwell
Automation Business

Ms. Debbie Beihl
Armstrong Manufacturing

Mr. Bill Steel
Bard Manufacturing Co.

Ms. Susan Carlin
Carlyle Compressor Company,
Division of Carrier Corp.

Mr. Gary Forcey
Challenger Electrical Equipment
Corp.

Ms. Lil Nichols
Copeland Corporation

Mr. Bob Roderique
Copper Industries, Bussman
Division

Mr. Max Robinson
Danfoss Automatic Controls

Ms. Patricia Pruis
GE Motors and Industrial
Systems, Fort Wayne, Indiana

Mr. John Rhoads Jr.
Heat Controller, Inc.

Ms. Chari Paul
Honeywell

Mr. John Bernaden and
Ms. Donyel Smith
Johnson Controls

Ms. Cathrine Burkhardt
Magnatek, Century Motors

Mr. Larry Hagman
Manitowac Equipment Works

Ms. Linda Capcara
Motorola Semiconductor

Ms. Terri Johnston
North American Capacitor Co.

Mr. Bob Rank
Paragon Electric Company, Inc.

Mr. Donald Faulhaber
R.W. Beckett Corporation

Mr. Ron Smith
Reliance Electric

Mr. Joe Clark
Scotsman

Ms. Jennifer Toffler
Square D Groupe Schneider

Mr. John Mullen
Supco Sealed Unit Parts Co. Inc.

Mr. Charlie Dunham
SymCom, Inc.

Mr. Kenneth Goldman
Tecumseh Products Company

Mr. Jim Holbrook
THERM-O-DISC, Incorporated
Subsidiary of Emerson Electric

Mr. Ted Suever
Tripplet Corporation

Mr. Wayne Gray
Tyler Refrigeration Corp.

Ms. Lisa Abbott and
Ms. Christine Domorat
United Technologies Carrier Corporation
and Gibbs & Soell Inc.

Ms. Yissell Fernandez
Watsco Components, Inc.

Ms. Cindy Nuspl
White Rogers Division,
Emerson Electric Co.

Mr. Elwood Spangler
York International Corporation

Contents

14   THERMOSTATS AND HEATING CONTROLS   317

Fundamentals of Electricity

OBJECTIVES:

After reading this chapter, you will be able to:
1. Explain the term *electricity*.
2. Describe the term *voltage*.
3. Explain the term *current*.
4. Describe the term *resistance*.
5. Understand the sources of electricity.
6. Understand Ohm's law.

1.0   Overview of Electricity

As a heating, air-conditioning, and refrigeration technician, you must thoroughly understand electricity, so that you are able to troubleshoot the electrical components of systems such as motors and controls. This chapter will provide you with the basics of electricity which will be the building blocks for more complex circuits that you will encounter in the field. You do not need any previous knowledge of electricity to understand this material. The simplest form of electricity to understand is *direct current* (DC) electricity, so many of the examples in this chapter will use DC electricity. It is also important to understand that the majority of electrical components and circuits in field equipment use *alternating cur-*

1

rent (AC) electricity, so some of the examples will use AC electricity. This chapter will begin by defining basic terminology such as voltage, current, and resistance, and showing simple relationships between them.

1.1   Example of a Simple Electrical Circuit in a Heating System

It is easier to understand the basic terms of electricity if they are connected with a working circuit that you can identify. We will use the fan that moves warm air in a furnace as an example. The fan must have a source of voltage and current. *Current* is the flow of electrons, and *voltage* is the force that makes the electrons flow. Since this is just an introduction, much more detail about the terms will be provided later in this chapter.

In the diagram in Fig. 1–1 the source voltage is 110 volts, which is identified by the letters L1 and N. The thermostat is a temperature-actuated switch for this circuit, which provides a way to turn the fan motor on and off. When the temperature in the furnace rises, the thermostat switch closes. This allows voltage and current to move through the wires to the fan motor. When voltage and current reach the motor, its shaft begins to turn. The wires in the circuit are called conductors, which provide a path for voltage and current and they provide a small amount of opposition to the current flow. This opposition is called *resistance*.

After the fan has run for a while, the temperature in the furnace cools down, and the thermostat switch opens. This causes the voltage and current to stop at the switch and the motor is turned off. When the temperature in the furnace increases, the thermostat will close and allow voltage and current to reach the motor again. Now that you have an idea of the basic terms, *voltage, current,* and *resistance,* you are ready to learn more about them.

The source of energy that makes the shaft in the motor turn is called *electricity,* which is defined as the flow of electrons. You should remember from the previous discussion that the definition of *current* is also the flow of electrons. In many cases the terms *electricity* and *current* will be used interchangeably. Electrons are the negative part of an atom, and atoms are the basic building blocks of

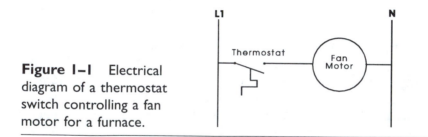

Figure 1–1   Electrical diagram of a thermostat switch controlling a fan motor for a furnace.

all matter. All matter can be broken into one of 103 basic materials called elements. Two of the elements that are commonly used in electrical components and circuits are copper, which is used for wire, and iron, which is used to make many of the parts such as motors.

To understand electricity at this point, we only need to understand three parts of the atom: electrons that have a negative charge, protons that have a positive charge, and neutrons that have no charge, which is also called a neutral charge. Fig. 1–2 shows a diagram of the simplest atom that has one electron, one proton, and one neutron. From this diagram you can see that the proton and neutron are located in the center of the atom. The center of the atom is called the *nucleus*. The electron moves around the nucleus in a path called an orbit, in much the same way the earth revolves around the sun.

Since the electron is negatively charged and the proton is positively charged, an attraction between these two charges holds the electron in orbit around the nucleus until sufficient energy, such as heat, light, or magnetism, is used to break it loose so that it can become free to flow. Anytime an electron breaks out of its orbit around the proton and becomes free to flow, it is called electrical current. The energy to create electricity can come from burning coal, nuclear power reactions, or hydroelectric dams.

The atom shown in Fig. 1–2 is a hydrogen atom and it has only one electron. In the circuit presented in Fig. 1–1, copper wire is used to provide a path for the electricity to reach the fan motor. Since copper is the element most often used for the conductors in circuits, it is important to study the copper atom. The copper atom is different from the hydrogen atom in that the copper atom has 29 electrons, and the hydrogen atom only has one. The copper wire that provides a path for the electricity for the fan motor contains millions of other copper atoms.

Fig. 1–3 shows a diagram of the copper atom with its 29 electrons that move about the nucleus in a series of orbits. Since the atom has so many electrons, they will not all fit in the same orbit. Instead the copper atom will have four orbits. These orbits are referred to as *shells*. Since there are also 29 protons in

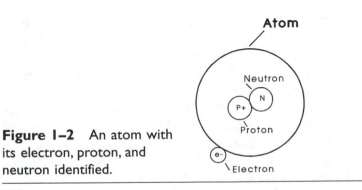

Figure 1–2 An atom with its electron, proton, and neutron identified.

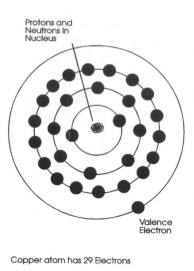

Protons and
Neutrons In
Nucleus

Valence
Electron

Copper atom has 29 Electrons

Figure 1–3 A copper atom has 29 electrons. This atom has one electron in its valence shell that can break free to become current flow.

the nucleus of the copper atom, the positive charges of the protons will attract the negative charges of the electrons and keep the electrons orbiting in their proper shell. The attraction for the electron in the three shells that is closest to the nucleus is so strong that these electrons cannot break free to become current flow. The only electrons that can break free from the atom are the electrons in the outer shell. The outer shell is called the *valence shell* and the electrons in the valence shell are called *valence electrons.* You should notice that the copper atom only has one electron in the valence shell. All good conductors have a low number of valence electrons. For example, gold is one of the best conductors and has only one valence electron and silver, which is a good conductor, has three valence electrons.

In the fan circuit we studied earlier, a large amount of electrical energy called *voltage* is applied to the copper atoms in the copper wire. The voltage provides a force to cause a valence electron in each copper atom to break loose and begin to move freely. The free electrons actually move from one atom to the next. This movement is called *electron flow* and, more precisely, it is called *electrical current.* Since the copper wire has millions of copper atoms, it will be easier to understand if we study the electron flow between three of them. Fig. 1–4 shows a diagram of the electrons breaking free from three separate copper atoms.

When the electron breaks loose from one atom, it leaves a space called a *hole* where the electron was located. This hole will attract a free electron from the next nearest atom. As voltage causes the electrons to move, each electron only moves to the *hole* that was created in the adjacent atom. This means that electrons move through the conductor by moving from hole to hole in each atom.

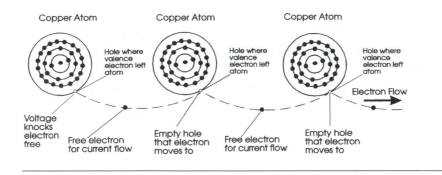

Figure 1–4   Voltage is the force that knocks one electron out of its orbit. In this diagram the valence electron in the atom on the far right breaks free from its orbit and leaves a hole. The electrons in the atoms to the left of the first atom move from hole to hole to create current flow.

The movement will begin to look like cars that are bumper to bumper on a freeway.

1.2   Example of Voltage in a Circuit

Voltage is the force that moves electrons through a circuit. Fig. 1–5 shows the effects of voltage in a circuit where two identical light bulbs are shown connected to separate voltage supplies. You should notice that the bulb that is connected to the 12-volt battery is much brighter than the bulb that is connected to the 6-volt battery. The bulb connected to the 12-volt battery is brighter because 12 volts can exert more force on the electrons in the circuit than 6 volts and higher voltage will also cause more electrons to flow.

In the original diagram of the fan motor, the amount of voltage is 110 volts. The energy that produces this force comes from a generator. The shaft of the

Figure 1–5   Two identical light bulbs have different voltages applied to them. The light bulb with 12 volts applied to it is brighter than the one with 6 volts applied to it.

generator is turned by a source of energy such as steam. Steam is created from heating water to its boiling point by burning coal or from a nuclear reaction. The energy to turn the generator shaft can also come from water moving past a dam at a hydroelectric facility. In the light bulb example the voltage comes from a battery. The voltage in a battery has been previously stored in a chemical form when the battery was originally charged. The larger the amount of voltage is in a circuit, the larger the force that can be exerted on the electrons, which makes more of them flow. The scientific name for voltage is *electromotive force* (EMF).

1.3   Example of Current in a Circuit

The definition of *current* is the flow of electrons. The actual number of electrons that flow at anytime in a circuit can be counted, and their number will be very large. The unit of current flow is called the *ampere*. The word *ampere* is usually shortened to *amp*. One amp is equal to 6.24×10^{18} electrons flowing past a point in 1 second. Fig. 1–6 shows an example of a meter measuring the number of electrons flowing. Each electron that is flowing in the circuit is measured and when 6,240,000,000,000,000,000 electrons pass the meter in 1 second, it is called an *ampere*. The word *ampere* originated from the unit of measure called a *coulomb* and when the International System (SI) of Units was adopted in 1948, the term *ampere* was selected to replace the term *coulomb*. The term *ampere* is used in all reference to heating, air-conditioning, and refrigeration equipment as the unit of electrical current.

1.4   Example of Resistance in a Circuit

Resistance is defined as the opposition to current flow. Resistance is present in the wire that is used for conductors and in the insulation that covers the wires.

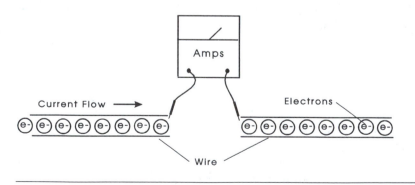

Figure 1–6   A number of electrons passing a point where they are measured. When the number reaches 6.24×10^{18}, 1 ampere has been measured.

The unit of resistance is called the ohm and its symbol is Ω. The Ω symbol comes from the Greek word *omega,* which is the Greek character for the letter O. When the resistance of a material is very high, it is considered to be an insulator; if its resistance is very low, it is considered to be a conductor.

It is important for conductors to have very low resistance so that electrons do not require a lot of force to move through them. It is also important for insulation that is used to cover wire to have very high resistance so that the electrons cannot move out of the wire into another wire that is nearby. The insulation also ensures that a wire on a hand tool such as a drill or saw can be touched without electrons traveling through it to shock the person using it. Examples of materials that have very high resistance and can be used as insulators are rubber, plastic, and air.

The amount of resistance in an electrical circuit can also be adjusted to provide a useful function. For example, the heating element for a furnace is manufactured to have a specific amount of resistance so that the electrons will heat it up when they try to flow through it. The same concept is used in determining the amount of resistance in the filament of a light bulb. The light bulb filament of a 100-watt light bulb has approximately 100 ohms of resistance, which is sufficient resistance to cause the electrons to work harder as they move through the filament. This causes it to heat up to a point where it glows. When the amount of resistance in a circuit is designed to convert energy such as a heating element or motor, it is referred to as a *load.*

1.5　Identifying the Basic Parts of a Circuit

It is important to understand that each electrical circuit must have a *supply of voltage, conductors* to provide a path for the electrons to flow, and at least one *load.* The circuit may also have one or more controls. Fig. 1–7 shows an example of a typical electrical circuit with the voltage supply, conductors, control, and load identified. In this circuit, the voltage source is supplied through terminals L1 and N, the load is a fan motor, and the two wires connecting the fan

Figure 1–7 The basic parts of a circuit are identified in this example. They are the power source, conductors, control, and load.

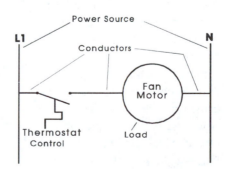

motor to the voltage source are called the conductors. The thermostat switch is called a control, since it controls the current flow to the motor. In a large system such as a residential air-conditioning system, it is possible to have more than one load. For example, the compressor motor, the condenser fan motor, and the evaporator fan motor are all loads in an air-conditioning system. You will learn more about these components later.

1.6   Equating Electricity to a Water System

It may be easier to understand electricity if you can equate it to a system like a water system that may be more familiar to you. If you would examine water flowing through a hose, for example, the flow of water would be equivalent to the flow of current (amps), and the water pressure would be equal to voltage. If you stepped on the hose or bent the end over, it would create a resistance that would slow the flow, just like resistance in an electrical circuit slows the flow of current. Fig. 1–8 provides a diagram that shows the similarities between the water system and an electrical circuit.

By definition, 1 *volt* is the amount of force required to push the number of electrons equal to 1 amp through a circuit that has 1 ohm of resistance. Likewise, 1 *amp* is the number of electrons that flow through a circuit that has 1 ohm of resistance when a force of 1 volt is applied, and 1 *ohm* is the amount of resistance that causes 1 amp of current to flow when a force of 1 volt is applied. You can see that voltage, current, and resistance can all be defined in reference to each other.

Figure 1–8 A diagram of a water system used to indicate the similarities to an electrical system. The flow of water is similar to the flow of current (amps), the pressure on the water is similar to voltage, and the resistance from the valve causes the flow of water to slow just like resistance in an electrical system causes current flow to slow.

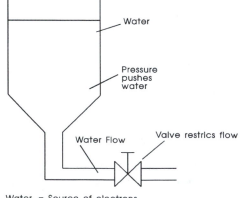

Water = Source of electrons
Water pressure = Voltage that pushes electrons
Water flow = Current flow
Valve restrics flow = Electrical resistance

1.7   Using Ohm's Law to Calculate Volts, Amps, and Ohms

There are times when you are working in a circuit that you need to calculate or estimate the amount of volts, amps, or ohms the circuit should have. In most cases when you are troubleshooting an electrical problem, you will also need to make measurements of the volts, amps, and ohms to determine if the circuit is operating correctly. The problem with taking a measurement is that you will not be sure if the values are too high or too low, or if they are just about right. A calculation can tell you if the measurements are within the estimated values the circuit should have.

A relationship exists between the volts, amps, and ohms in every DC circuit that can be identified by Ohm's law. Ohm's law simply states that the amount of voltage in a DC circuit is always equal to the amps multiplied by the resistance in that circuit. If a circuit has 2 amps of current and 4 ohms of resistance, the total voltage will be 8 volts.

The formula can be changed to determine the amount of amps or ohms. Since we know that $2 \times 4 = 8$, it stands to reason that 8 divided by 2 equals 4, and 8 divided by 4 equals 2.

If you measure the volts in a circuit and find that you have 8 volts, and measure the amps in a circuit and find that you have 2 amps, you could calculate that you have 4 ohms by dividing 8 by 2. You could also determine the amount of amps you had in the circuit by measuring the volts and ohms and calculating: 8 volts divided by 4 ohms is 2 amps.

1.8   Ohm's Law Formulas

When scientists, engineers, and technicians use calculations, letters of the alphabet are used to represent units such as volts, amps, and ohms. The letters are abbreviations that are accepted by everyone so that when a calculation is passed from one person to another, it will be understood by everyone.

In Ohm's law formulas, the letter *E* is used to represent voltage. The *E* is derived from the first letter of the words *electromotive force.* The letter *R* is used to represent resistance. The letter *I* is used to represent current (amps) and it has been derived from the first letter of the word *intensity.*

Ohm's law can be represented by the formula $E = IR$. When you use Ohm's law the basic rule of thumb is that you should use *E, I,* and *R* when the values are *unknown,* and you should use V for volts, A for amps, and Ω for ohms when you have determined an answer or anytime the values are *known* or measured.

1.9   Using Ohm's Law to Calculate Voltage

When you are solving a calculation for voltage, you need to know the amount of current (amps) and the amount of resistance (ohms). If you do not know these two values, you cannot calculate the voltage. For example, if you are asked to calculate the amount of voltage when the current is 20 amps and the resistance is 40 Ω, you can multiply 20 $\times$ 40 to get an answer of 800 volts. When the problem is stated, you should start with the formula and continue by substituting the values for I and R. Then you can continue the calculation. Notice that the units A for amps and Ω for ohms are added when values are known.

$$E = IR \qquad E = 20 \text{ A} \times 40 \text{ } \Omega \qquad E = 800 \text{ volts}$$

1.10   Using Ohm's Law to Calculate Current

The formula for calculating an unknown amount of current in a circuit is $I = \frac{E}{R}$. When you are calculating the current, you must divide the voltage by the resistance. For example, if you have measured the circuit voltage and found that it is 50 volts, and you have measured the resistance and found that it is 10 ohms, the current is found by dividing 50 V by 10 Ω, which equals 5 amps.

$$I = \frac{E}{R} \qquad \frac{50 \text{ V}}{10 \text{ } \Omega} = 5\text{A}$$

1.11   Using Ohm's Law to Calculate Resistance

The formula for calculating an unknown amount of resistance in a circuit is $R = \frac{E}{I}$. When you are calculating the resistance of a circuit, you must divide the voltage by the current. For example, if you have measured 60 volts and you have measured the current and found that it is 12 amps, the resistance is found by dividing the voltage by the current. Again you should start with the formula and substitute values to get an answer.

$$R = \frac{E}{I} \qquad R = \frac{60 \text{ V}}{12 \text{ A}} \qquad R = 5 \text{ } \Omega$$

1.12   Using Ohm's Law Wheel to Remember Ohm's Law Formulas

A learning aid has been developed to help you remember all of the Ohm's law formulas. Fig. 1–9 shows this aid, and you can see that is in the shape of a wheel or a pie. The top of the pie has the letter E to represent voltage, the bottom left has the letter I to represent current (amps), and the bottom right has the letter R to represent resistance (ohms).

    The way the pie functions is that it helps you remember the formulas for Ohm's

Figure 1–9 Ohm's law pie
that shows a memory tool
for remembering all of the
Ohm's law formulas.

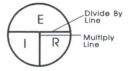

law. For example, if you want to remember the formula for voltage, you would put your finger over the E and the letters I and R show up side by side. The vertical line separating the I and R is called the *multiply line* because that is the math function you are supposed to use with I and R. This represents the formula $E = IR$.

If you want to know the formula for current (I), you should put your finger over the I and the remaining letters will be the E over the R. The horizontal line that separates the E and R is called the *divide by line* because you are supposed to divide the E by R. This represents the formula $I = \dfrac{E}{R}$.

If you want to know the formula for resistance (R), you should put your finger over the R and the remaining letters will be the E over the I. This represents the formula $R = \dfrac{E}{I}$.

1.13   Calculating Electrical Power

Electrical power is work that electricity can do. The units for electrical power are *watts*. Watts are calculated by multiplying amps times volts ($P = IE$). Most electrical loads that are resistive in nature such as heating elements will be rated in watts. For example, if a heating element is rated for 5 amps and 110 volts, it would use 550 watts of power. You could calculate the amount of power by using the formula $P = IE$: 5 A × 110 V = 550 W.

This basic formula can be rearranged like the Ohm's law formula. You can calculate current by dividing the power (550 W) by the voltage (110 V), and you can calculate voltage by dividing the power (550 W) by the current (5 A). Fig. 1–10 shows a pie that is a memory aid for remembering all of the formulas for Watt's law. From this pie you can see that the original formula, $P = IE$, can be found by placing your finger over the P. The formula to solve for current

Figure 1–10 The formula
pie for Watt's law formulas
that are used to calculate
electrical power.

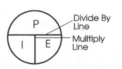

$\left(I = \dfrac{P}{E}\right)$ can be found by placing your finger over the *I*, and the formula to solve

for voltage $\left(E = \dfrac{P}{I}\right)$ can be found by placing your finger over the *E*.

1.14   Presenting All of the Formulas

When you are troubleshooting, you may be able to get a voltage reading easily. If you can find a data plate to provide the wattage for a heating element or other load, you will be able to use one of the formulas that was presented to calculate (estimate) the current or the resistance. This estimate will give you an idea of what the values should be so that you can determine if the system is working correctly or not. When you are in the field making these measurements, it may be difficult to remember all of the formulas, so a chart has been developed that provides all of the formulas. Fig. 1–11 shows this chart. From the figure you can see that the chart is in the shape of a wheel with spokes emanating from an inner ring. A cross is shown in the middle of the inner wheel and it separates the inner wheel into four sections. These sections are represented by the letters *P* (power or watts), *I* (amps), *E* (volts), and *R* (ohms). (You should notice that power can be identified as either *W* for watts or *P* for power.)

The formulas in each section of the outer ring correspond to either *P, I, E,* or *R*. The formulas for wattage are shown emanating from the inner section marked with the *P*. All of the formulas for current are shown emanating from the inner section marked with an *I*. All the formulas for voltage are shown emanating from the inner section marked with an *E*, and all the formulas for resistance are shown emanating from the inner section marked with an *R*.

If you are trying to calculate the power in a circuit or the power used by any

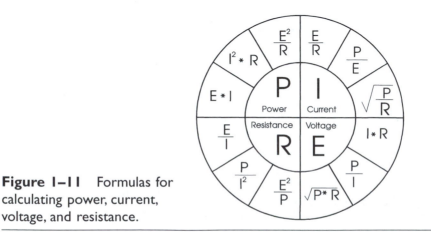

Figure 1–11 Formulas for calculating power, current, voltage, and resistance.

individual component, you could use any of the three formulas that are shown on the wheel emanating from the section identified by the letter P. This means that you could use the formula $P = EI$, $P = I^2/R$, $P = E^2/R$. Your choice of formula would depend on the two values that are given.

Questions for This Chapter

1. Define the term *electricity*.
2. What is voltage?
3. What is current?
4. Explain what a valence electron is and why it is important to current flow.
5. What is resistance?
6. Define the term *power*.

True or False

1. The insulation coating on wire has high resistance.
2. Voltage is the flow of electrons.
3. Resistance is the opposition to current flow.
4. Watts are the units for current.
5. A good conductor has low resistance.
6. The negative part of an atom is an electron.

Multiple Choice

1. A valence electron is _____

   a. the electron closest to the nucleus.
   b. the electron in the outer shell.
   c. an electron with a positive charge.
2. Wattage is _____

   a. the unit for power.
   b. the unit for current.
   c. the unit for voltage.
3. Resistance is _____

   a. the force that moves electrons.
   b. wattage.
   c. the opposition to current flow.

4. Current will _____ when voltage increases and resistance stays the same.

   a.   increase
   b.   decrease
   c.   stay the same

5. Wattage will _____ when voltage increases and current stays the same.

   a.   increase
   b.   decrease
   c.   stay the same

Problems

1. Use Ohm's law to calculate the amount of voltage if current in a DC circuit is 5 A and resistance is 8 Ω.
2. Use Ohm's law to calculate the amount of current in a DC circuit if voltage is 25 V and resistance is 50 Ω.
3. Use Ohm's law to calculate the amount of resistance in a DC circuit if voltage is 40 V and current is 10 A.
4. Use the power formula to determine the amount of watts in Problem 1.
5. Use the power formula to determine the amount of watts in Problem 2.

Voltmeters, Ammeters, and Ohmmeters

OBJECTIVES:

After reading this chapter, you will be able to:
1. Use a VOM meter to measure voltage.
2. Use a VOM meter to measure resistance.
3. Use a VOM meter to measure milliamps.
4. Understand safety warnings for using ohmmeters and ammeters.
5. Explain the operation of a clamp-on ammeter.

2.0   Measuring Volts, Amps, and Ohms

It is important to be able to measure the amount of voltage (volts), current (amps), and resistance (ohms) in a heating, air-conditioning, or refrigeration system to be able to determine when it is working correctly, or if something is faulty. If you are checking a tire on your automobile, you can see if it is flat by simply looking at it. Since the flow of electrons through a wire is invisible, it is not possible to simply look at the wire and determine if it has voltage applied to it or if it has current flowing through it, so a meter must be used to make voltage, current, and resistance measurements.

2.1   **Measuring Voltage**

The meter that is used to measure voltage is called a voltmeter. Fig. 2–1 and Fig. 2–2 show pictures of two types of voltmeters. The meter in Fig. 2–2 is called a digital voltmeter because the display uses numbers to indicate the amount of voltage that is being measured. The meter in Fig. 2–1 is called an analog voltmeter because it uses a needle and a scale to indicate the amount of voltage being measured.

The voltmeter has very high internal resistance, which is approximately 20,000Ω per volt, so its probes can be safely placed directly across the terminals of the power source without damaging the meter. The amount of voltage that is measured by the voltmeter is shown on the display. The range selector switch adjusts the internal resistance of the meter to set the maximum amount of voltage the meter can measure. The voltmeter can be used to make many different voltage measurements. For example, it is used to measure the amount of supply voltage in a system as well as measuring the amount of voltage that is provided to each load. Fig. 2–3 shows three examples where the probes of a voltmeter should be placed to safely make voltage measurements in a circuit. In the diagram in Fig. 2–3a, the voltmeter probes are shown across the terminals of a battery to measure the amount of battery voltage supplied. Fig. 2–3b shows the proper location to place the voltmeter probes to measure the voltage available to lamp 1, and Fig. 2–3c shows the proper location to place the voltmeter probes to measure the

Figure 2–1   An analog voltmeter. *(Courtesy of Triplett Corporation)*

Figure 2–2 A digital
voltmeter. *(Courtesy of Triplett
Corporation)*

voltage available to lamp 2. In each case the voltmeter probes are placed *across* the terminals where the voltage is being measured.

SAFETY NOTICE!!

It is important to remember that the voltage to a circuit must be turned on to make a voltage measurement. This presents an electrical shock hazard. You must take extreme caution when you are working around a circuit that has voltage applied so that you do not come into contact with exposed terminals or wires when you are making voltage measurements.

Typical meter ranges for digital voltmeters are 0–3 volts, 0–30 volts, 0–300 volts, and 0–600 volts. Special high-voltage probes must be used to measure voltages above 600 volts. Typical voltages that are found in HVAC systems are 24 volts, 120 volts, 208 volts, 240 volts, and 480 volts. The actual amount of voltage in your area may be slightly different; for example, the 120 volts may actually measure 110 volts, and the 480 volts may be closer to 440 volts.

The ranges for analog-type voltmeters are typically 0–3 volts, 0–12 volts, 0–60 volts 0–300 volts, and 0–600 volts. If you try to measure 120 volts when the selector switch is set to 60 volts, the needle will try to go past the maximum reading and the meter will be damaged. If you are using a digital voltmeter and you try to read a voltage that exceeds the range selector switch setting, the display will indicate you are over the range but the meter will not be damaged.

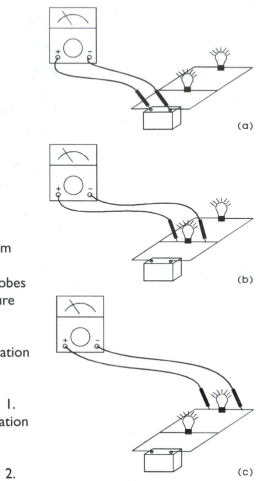

Figure 2–3a A diagram that shows the proper location to place the probes of a voltmeter to measure the amount of voltage available at the battery.
b Shows the proper location to place the voltmeter probes to measure the voltage available at lamp 1.
c Shows the proper location to place the voltmeter probes to measure the voltage available at lamp 2.

2.2   Measuring Electrical Current

An ammeter is used to measure electrical currents. Fig. 2–4 shows an example of two types of digital ammeters. (It should be noted that ammeters can also be a digital or an analog meter.) The ammeter in Fig. 2–4a is used to measure current that is less than 10 amps. In many cases the ammeter is actually built into the same meter with the voltmeter, and the selector switch allows you to set the meter to be a voltmeter, ohmmeter, or ammeter. When a meter can measure volts, ohms, or amps, it is called a *VOM meter* (volt, ohm, and milliamp meter). The ammeter is actually designed to read very small amounts of current such as 1/1000 of an amp, which is called a *milliamp,* and it can also be written as 0.001 amp. The prefix *milli-* means one thousandth. The VOM meter has a single meter move-

Figure 2–4a   A digital volt, ohm, milliamp meter that is called a VOM meter. *(Courtesy of Triplett Corporation)*

ment that causes a needle to deflect when it senses current flow. This means that when a voltage is applied to the VOM meter, it must be routed through resistors that limit the amount of current that actually flows through the meter movement. When the VOM meter is set to measure current or resistance, the same meter movement is used. The only part of the meter that changes as it is switched from a voltmeter, ammeter, or ohmmeter is the arrangement of resistors to limit the amount of current flow through the meter movement.

The meter shown in Fig. 2–4b is called a clamp-on ammeter and it is used to measure current in larger AC circuits. From the figure you can see this type of meter has two *claws* at the top that form a circle when they are closed. A button on the side of the meter is depressed to cause the claws to open so that they can fit around a wire without having to disconnect the wire at one end. The claws of the clamp-on ammeter are actually a transformer that measures the amount of current flowing through a wire by *induction*. When current flows through a wire, it creates a magnetic field that has flux lines. Since AC voltage reverses polarity once every 1/60 of a second, the flux lines collapse and cause a small current to flow through the transformer in the claw of the ammeter. This small current is measured by the meter movement and the display indicates the actual current flowing in the conductor that the meter is clamped around. Induction will be explained in more detail in Chapter 5.

Figure 2–4b   A clamp-on
ammeter that is used to
read high currents. *(Courtesy
of Triplett Corporation)*

2.3   Measuring Milliamps

At times you will need to measure the amount of milliamps in a circuit. For
example, you will need to measure the amount of current that a gas valve is
using so that you can set the heat anticipator on a heating thermostat. You will
also need to make milliamp measurements on the solid-state control boards found
in many newer furnaces and air conditioners. Fig. 2–5 shows that an open should
be made in the circuit when a milliamp measurement is made and each of the
terminals of the ammeter should be placed on the wires where the hole has been
made in the circuit. This places the meter in series with the circuit and all of the
electrons flowing in the circuit must also go through the meter because the ammeter has become part of the circuit.

Figure 2–5   Diagram that
shows the proper way to
place the meter probes to
make a current
measurement.

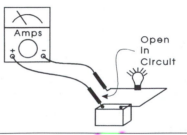

Since the ammeter must become part of the circuit to measure the current, it must have very low internal resistance so that it does not change the total resistance of the circuit. When the VOM meter is set to read milliamps, it is vulnerable to damage because it has very small internal resistance.

SAFETY NOTICE!

If you mistakenly placed the meter leads across a power source as you would to take a voltage reading when the meter is set to read current, the meter would be severely damaged. You could also be severely burned because the meter could explode, since it is a very low-resistance component when it is set to read current.

This would be similar to taking a piece of bare wire and bending it so it could be inserted into the two holes of an electrical outlet. You must always be sure to check how you have the meter set when you are making voltage, current, or resistance measurements.

As a technician you will be expected to use both the milliamp meter and the clamp-on ammeter. Milliamp meters are used to measure small amounts of current and clamp-on meters are used to measure larger currents such as the amount of current a motor is using. If a motor is using (pulling) too much current, you will be able to determine that it is overloaded and that it will soon fail. Later chapters in this book will explain how to use the clamp-on ammeter to measure motor current.

Digital VOM meters and some analog VOM meters have circuits to measure both DC and AC current. These types of meters have several ranges for measuring very small current in the range of milliamps up to 10 amps.

2.4   Creating a Multiplier with a Clamp-On Ammeter

Sometimes the amount of current that is measured with the clamp-on ammeter is very small, and the needle will not deflect sufficiently to create an accurate measurement. Since the claws of the clamp-on ammeter are actually a transformer, the amount of current in the reading can be multiplied by wrapping the wire that you are measuring the current in around the claw several times to cause the reading to be multiplied. For example, if you wrapped the wire around the claw two times, and the wire has 1.25 amps flowing through it, the meter reading would indicate 2.5 amps, which is double the amount. You would need to take the meter reading and divide by 2 to determine the exact current flowing in the wire. If you wrapped the wire around the claw four times, the meter reading would be 5 amps, and you would need to divide the meter reading by 4 to get the true current flow in the wire. Creating a multiplier will make very small current measurements easier and more accurate to read. Fig. 2–6 shows an example of a wire twisted around the claws of a clamp-on ammeter to create a multiplier.

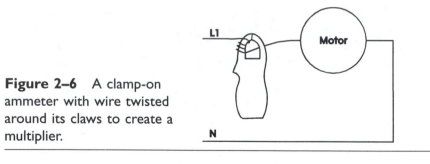

Figure 2–6 A clamp-on ammeter with wire twisted around its claws to create a multiplier.

2.5   **Measuring Electrical Resistance**

The VOM meter is also able to measure resistance when the selector switch is moved to one of the ohms positions such as R × 1, R × 10, or R × 1000. When the VOM meter is set in the ohm configuration, it uses an *internal battery* to provide voltage to make the meter movement deflect the needle. When this voltage is applied to a resistance, a small amount of current will flow which the meter detects.

SAFETY NOTICE!

Since the battery in the meter provides the power source for the current that the meter movement will read, it is important the components being tested are not connected to any other source of voltage. It is also a good practice to remove the component from the circuit when resistance measurements are made. When a component is disconnected from a circuit and all sources of power, it is said to be *isolated.*

The ohmmeter can be used to measure the resistance of a wire or fuse to ensure that the resistance is near zero ohms (0 Ω). A wire is said to have *continuity* and it is considered to be a good wire when it has near zero ohms. If the wire has extremely high resistance (several million ohms), it is considered to be bad since it must have an *open* that is causing the high resistance. The same terms apply to fuses. If a fuse is good, it will be described as having continuity, and if it is bad, it will be described as an *open fuse.* When a wire or fuse is open, it is also said to have *infinite resistance.* The word *infinite* indicates the amount of resistance is very high due to the fact the wire or fuse has an open. Since a typical VOM meter cannot accurately measure resistance that is over 2 million ohms, its scale will be marked with the sign for infinity (∞) at the highest point on its scale. Digital VOM meters flash an overrange display when an infinite resistance reading is made.

Fig. 2–7a shows an example of the ohmmeter indicating 0 Ω when a good wire is tested, and Fig. 2–7b shows the ohmmeter indicating (∞) Ω when the meter is testing a wire with an open in it.

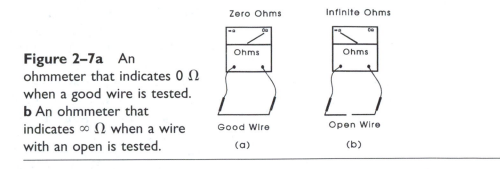

Figure 2–7a An ohmmeter that indicates 0 Ω when a good wire is tested. **b** An ohmmeter that indicates ∞ Ω when a wire with an open is tested.

The electrical loads in an air-conditioning, heating, and refrigeration system, such as motors or heating elements, may have some amount of resistance that falls between zero ohms and infinite ohms. For example, a fan motor may have 12 Ω of resistance in its run winding and 30 Ω of resistance in its start winding. The ohmmeter can be used to accurately measure each winding and determine whether it has 12 Ω or 30 Ω. The ohmmeter can be used to test motor leads and determine if they are open, and if they are not open, the run and start windings can be distinguished by the amount of resistance each has. You will see in later chapters that the start winding of an AC induction motor will always have more resistance than its run winding and the ohmmeter allows you to easily tell the run winding from the start winding.

2.6   Reading the Scales of the VOM Meter

The scale of a typical VOM meter is shown in Fig. 2–8. From this figure you can see that the face has four separate scales (lines that are shown in the shape of an arc). The bottom scale is used to measure decibels (dB) and it will not be used at this time. The second scale from the bottom is used to measure AC amps,

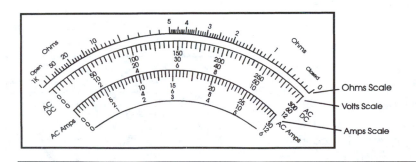

Figure 2–8 The face of a typical VOM meter that shows the voltage, current (amps), and resistance (ohms) scales.

and the third scale from the bottom that is identified as AC DC is used to measure AC and DC voltages and DC milliamps. The top scale identified by the word Ohms is used to measure resistance. You should notice that the needle that is attached to the meter movement should rest at the far left corner of the meter face when voltage, current, or resistance is not being measured.

The way the VOM meter is able to measure volts, ohms, and milliamps with the same meter movement is to switch a number of resistors into and out of the circuit that connects the two probes to the meter movement. The switch is moved by a large knob on the front of the meter. Fig. 2–9 shows the knob and the range selections on the face of the meter. If we use the hands of a clock to explain the range selections, you can see that the AC voltage circuit is located between 1 and 3 o'clock. The AC voltage circuit has five ranges: 0–3 volts, 0–12 volts, 0–60 volts, 0–300 volts, and 0–600 volts.

The ohms circuits used to measure resistance are located between the 3 and 6 o'clock positions. You should notice that an ✕ precedes each of the ranges because you will need to multiply the reading on the scale of the meter times the multiplier factor that is identified by the switch. You should get into the habit of saying the word *times* when you encounter the ✕. For example, the first range of the ohms scales is identified as ✕1. You will call this range the *times 1 range*. The other ohms ranges are: ✕10 (times 10), ✕100 (times 100), ✕1000 or ✕1k (times 1000), and ✕100000 or 100k (times 100,000).

You should notice that when the range reaches 1000 and 100,000, the letter k is substituted. The letter k is the unit for kilo, which represents 1000. The k is used many times where there is not enough space to write in all of the zeros.

The DC milliamp circuits are located between the 6 to 9 o'clock positions. The ranges for milliamps are: 0 to 0.06, 0 to 1.2, 0 to 12, and 0 to 120. The DC voltage ranges are located between the 9 and 12 o'clock positions. The ranges for DC voltage are: 0–3 volts, 0–12 volts, 0–60 volts, 0–300 volts, and 0–600 volts.

Figure 2–9 The face of a VOM meter that shows the ranges and the knob that is attached to the switch that connects the meter movement to different circuits of resistors inside the meter.

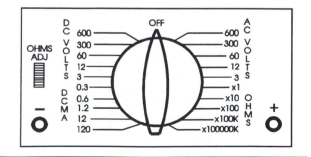

2.7   Measuring DC Volts and Reading the VOM Scales

When you are making a DC voltage measurement, you will place the meter leads (probes) in the sockets identified as + and − and you should set the selector switch on the face of the meter to the highest range. For this meter the selector switch would be set to 600 volts. Setting the selector switch to 600 V will protect the meter, since you are not sure of how large a voltage you will be measuring. When you touch the tips of the probes to a voltage source, the needle that is attached to the meter movement will begin to move from the position at the far left where it rests when no voltage is being sensed toward the middle. If you have the positive meter probe touching the negative voltage source, the meter movement will try to deflect farther to the left. If this occurs, simply switch the two probes so that the positive meter probe is touching the positive voltage source. This is called *changing the polarity* of the meter.

If the voltage that is being measured is small, the needle will not move very far, and you can change the switch on the meter face to the next lower scale. If the needle still does not move sufficiently, you can continue to change the switch to the next lower scale. Fig. 2–10 shows an example of the meter that is being used to measure the amount of DC voltage in a circuit that has approximately 4 volts. Since the needle would not move sufficiently when the switch was set on the 0–600 V scale, the switch was changed to the next lower range until it was

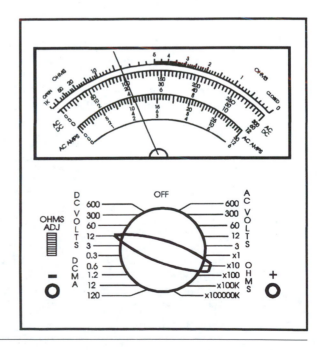

Figure 2–10 Voltmeter with its selector switch set on the 0–12 V range and its needle indicating 4 volts are being measured.

switched to the 0–12 V range and the needle moved to the position as shown in the example.

When you are ready to read the position of the needle on the face of the meter, you would start by looking at the position of the switch. Since the switch in this example is at the 12 V DC setting, you must use the scale marked AC DC for this reading. Fig. 2–11 shows an enlarged diagram of only the face of the meter. The AC DC scales are located directly below the ohms scale, which is the top scale. The voltage scales are highlighted in this figure so they are easier to identify. You should notice that the scale marked AC DC has three numbers shown at the far right of the scale: 300, 60, 12. Since the selector switch is set on 12 volts, we will use the scale that has the numbers 0–12 marked on it. These numbers are also highlighted in this example.

When you look closely at the meter face, you can see that the needle is pointing to the 4 on this scale. This means that the meter is measuring 4 volts DC. You may be confused at first because the numbers 100 and 20 are also marked at this same position. You can be sure that the reading is 4 volts by starting at the far right side of the scale and locating the number 12. Since the number 12 is the bottom number, you should use the bottom set of numbers for this scale. (A good habit to get into is to start with the number on the right side of the scale and count backward as you locate the exact position the needle is pointing to. For this example you would start at 12 and count backward 8, 6, 4 as you moved to the left where the needle is pointing.)

2.8   Measuring and Reading the Milliamp Scales on the VOM

When you are making a current reading with a milliamp meter, you must open the circuit and connect the meter probes so that the meter becomes part of the circuit. You should set the selector switch to the highest milliamp setting (120 DC mA range). If the needle does not deflect enough so that you can read it easily, you can switch the meter to the next lower range. When the selector switch

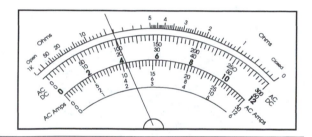

Figure 2–11 Example of the face of a VOM meter with the 0–12 V DC scale shown in bold.

is set on this setting, you will need to read the amount of current from the same AC DC scale that you used for the DC voltage reading.

NOTICE!

The scale marked AC Amps is only used when measuring AC current and it is not usable for DC current readings.

In the example shown in Fig. 2–12 the selector switch is set on the 120 DC mA setting. Since there is not a scale that ends with 120, the 0–12 scale will be used and you will have to move the decimal point or multiply the reading by 10 to determine the exact amount of current flowing through the wire.

For example, in the figure you can see that the needle is pointing to the number 2. You can determine this by finding the 0–12 scale and begin counting backward (12, 10, 8, 6, 4, 2) until you reach the place where the needle is resting. Since the selector switch is set to the 120 mA setting, you must treat the 12 as 120. You can do this by multiplying the original reading by 10 or by counting backwards from 120, to 20.

If the needle was pointing to the same position and the selector switch was pointing to 1.2 mA, the value the meter is measuring would be 0.2 mA. The major point to remember when reading the milliamp scales is that all three ranges

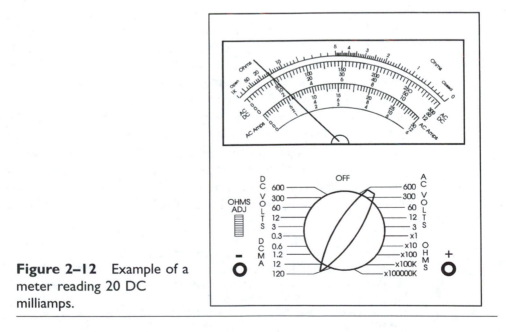

Figure 2–12 Example of a meter reading 20 DC milliamps.

use the numbers "one-two" (12) and the difference between each range is the placement of the decimal point.

2.9   Measuring and Reading Resistance Scales on the VOM

When you are making a resistance measurement with the VOM, you must ensure that the device or component that you are measuring does not have power connected to it and that it is isolated from other components. You should remember that this is a necessary step because the VOM meter uses an internal battery as the power source when it is used to measure resistance. When the meter is set to read resistance, the voltage from the battery is rather small: 3 V, 9 V, or 30 V depending on the brand name of meter. The meter would be seriously damaged if you touched its probes to a power source such as 110 V by mistake when the meter is set for resistance.

Another point to remember when using the VOM to read resistance is to zero the meter. Since the meter uses a battery for the power source, the voltage in the battery must be able to move the needle to the 0 Ω mark on the scale when the two probes are touched together. This process is called *zeroing the meter* and it should be done prior to all resistance readings. If the battery is weak, the needle will not be able to move to the zero point and all resistance readings will be faulty. This step is necessary because the actual voltage in the battery will constantly change as the battery is used and it starts to drain down.

The first step in the process to zero the meter is to place the selector knob on the ohms range you want to use. Next you need to touch the two meter probes together, and the meter needle should swing toward the right side of the scale. Since the VOM meter is being used to measure resistance, the very top scale that is identified with the word *Ohms* will be used. If the meter is zeroed correctly, the needle will rest directly on the 0 ohm setting. If the needle does not move to this position, you can adjust the needle position by changing the OHMS ADJ knob that is located to the left of the selector switch, just above the terminal for the negative probe. The OHMS ADJ knob is actually a potentiometer, and when you adjust it, you are changing the amount of resistance in the meter movement circuit, which will change the amount of current the battery is sending to the meter movement. If you change the OHMS ADJ knob and the meter needle does not reach all the way to the 0 ohm position, the meter has a weak battery and it should be replaced.

NOTICE!

You must zero the VOM meter anytime you switch resistance scales or anytime you change from one resistance scale to another because a different amount of internal resistance is used for each circuit. You do not need to zero the VOM meter when it is used to make voltage or current measurements because these types of readings do not use the meter's internal battery.

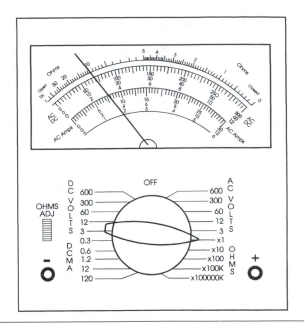

Figure 2–13 VOM used to measure resistance. The scale indicates the needle is resting on the 10 Ω value and the selector is set to the X1 (times 1) range so the meter is measuring 10 ohms.

Fig. 2–13 shows an example of the VOM meter measuring resistance. In this example, the meter selector switch is set to the ×1 (times 1) setting. Since the meter is being used to measure resistance, the top scale is used. The needle is resting on the value 10 on the top scale. Since the selector knob is setting on the ×1 (times 1) scale, you will need to multiply the reading by 1. This means that the VOM meter is measuring 10 Ω.

If the selector switch was setting on the ×10 (times 10) range and the needle was resting in the same position, the meter would be measuring 100 Ω because you would need to multiply the scale reading by 10.

2.10   Making Measurements with Digital VOM Meters

When you are making a measurement with a digital VOM, you should follow the same rules as when you use an analog VOM. The major difference is that the measurement is presented as a digital number on the meter's display. All you need to do is observe the range selector switch to determine the units that the meter is measuring, and use the value that is displayed as the measurement. Be sure to observe polarity (+/−) if the voltage or current is DC. Some digital meters also provide an audible signal for resistance measurements, which is useful when you are locating wires in multiconductor cables.

Questions for This Chapter

1. Explain the term *infinite resistance.*
2. Use the diagrams in Fig. 2–3 to explain where you would place the probes of a voltmeter when making voltage measurements.
3. Use the diagram in Fig. 2–5 to explain where you would place the probes of an ammeter when making a current measurement.
4. List two things that you should be aware of when making resistance measurements.
5. Explain why you should zero an ohmmeter every time you change scales or make a measurement.
6. Explain why you should set the voltmeter and ammeter to the highest setting when you first make a measurement of an unknown voltage or current.

True or False

1. All voltage should be turned off when making a resistance measurement.
2. Infinity is the highest resistance reading for the ohmmeter.
3. The VOM meter must be zeroed prior to making voltage, current, or resistance measurements.
4. The voltage should be turned off for safety reasons when making a voltage measurement.
5. The ohmmeter uses an internal battery to supply the power for all resistance readings.

Multiple Choice

1. Milliamps are _____
   a. one thousandth of an ampere.
   b. one millionth of an ampere.
   c. one million amps.
2. When a current measurement is made with a VOM meter, you should _____
   a. place the meter probes across the load.
   b. turn off all power to the circuit during the current reading so you don't get shocked.
   c. create an open in the circuit and place the meter probes so that the meter is in series with the load.

3. When a voltage measurement is made, you should _____
    a. place the meter probes across the load.
    b. turn off all power to the circuit during the voltage reading so you don't get shocked.
    c. create an open in the circuit and place the meter probes so that the meter is in series with the load.
4. When a resistance measurement is made, you should _____.
    a. keep all voltage applied to the component you are measuring so you can determine the true resistance.
    b. turn off all power to the circuit and isolate the component you want to measure.
    c. create an open in the circuit and place the meter probes so that the meter is in series with the load.
5. When you are measuring current with a clamp-on ammeter, you should _____.
    a. open the claws of the meter and place them around the wire where you want to measure the current.
    b. create an open in the circuit and place the meter probes so that the meter is in series with the load.
    c. turn off all power to the circuit so that you don't get shocked.

Problems

1. Use the diagram in Fig. 2–11 to determine the voltage reading for the meter if the selector switch is set on 60 V.
2. Use the diagram in Fig. 2–11 to determine the current reading for the meter if the selector switch is set on 12 mA.
3. Use the diagram in Fig. 2–13 to determine the resistance reading for the meter if the selector switch is set for ×1K.
4. Determine the amount of current flowing in a wire if a clamp-on ammeter has four turns of the wire twisted around its claw and the reading on the scale is 2 amps.
5. List the kind of problems that could be present in the circuit if a milli-amp meter reading indicates 0 milliamps are present in the circuit.

3 Series Circuits

OBJECTIVES:

After reading this chapter, you will be able to:
1. Explain the operation of switches that are connected in series.
2. Explain the operation of electrical loads that are connected in series.
3. Utilize Ohm's law for series circuit calculations.
4. Calculate the voltage, current, resistance, and power in a series circuit.

3.0   Introduction

Simple circuits consisting of resistors will be used to explain the relationship between voltage, current, and resistance in a series circuit. It should be pointed out that resistors are not common loads in air-conditioning and refrigeration systems, but they are common in modern electronic circuit boards that are used to control furnaces, air conditioners, and refrigeration systems. The motors and other loads in air-conditioning and refrigeration systems will react in some part as resistances, so it is important for you to understand the relationship between multiple resistances in series circuits. Understanding resistances in series circuits and in parallel circuits will also provide the initial theory that you will require to understand solid-state components that are commonly used in solid-state controls found in furnaces, air conditioners, and refrigeration systems.

33

As a technician you must understand the effects of switches in series circuits and electrical loads in series circuits. The first part of this chapter will explain the relationships between switches that are connected in series. The second part of this chapter will explain the effects of electrical loads that are connected in series.

The relationship between voltage, current, and resistance in series circuits will be introduced in the last part of this chapter. It will be easier to understand these relationships when a known value of resistance is used. A component called a resistor is used to provide the resistance for these simple series circuits. Each resistor has a color code consisting of colored bands that are painted on the resistor to identify the amount of resistance the resistor has. Ohm's law will also be introduced to help you understand how voltage and current should react in all types of series circuits. The basic rules provided in this chapter will be used to help you troubleshoot larger circuits that are commonly found in heating, air-conditioning, and refrigeration systems.

3.1   Examples of Series Circuits

All electrical circuits have at least a power source, one load, and conductors to provide a path for current. As circuits become more complex, switches are added for control. These switches can provide a variety of functions depending on the way they are connected in the circuit. In Chapter 1 (Fig. 1–1) we saw a diagram of a simple circuit that included a power source, a thermostatic switch for control, and a motor as the load. This circuit is called a *series* circuit because the current has only one path to travel to the load and back to the power source.

When additional switches are added to the circuit to control the load, they can be connected so that all current continues to have only one path to travel around the circuit. Fig. 3–1 shows an example of a typical circuit that controls power to a compressor for a refrigeration case. This circuit has a thermostat that controls the temperature where the compressor turns on and off, a high-pressure switch that protects the compressor against excessive pressure if the condenser coils should get dirty, and an oil pressure switch that protects the compressor against loss of oil pressure.

This circuit is called a series circuit because the current has only one path to move from L1 through the switches to the load and back to L2. If any of the three switches are opened, current is interrupted and the motor will be turned off. When all of the switches are closed, the compressor motor will run. When the refrigeration case becomes too cold, the thermostat (temperature switch) will open and turn the compressor motor off. If the case becomes warm again, the thermostat will close and reenergize the compressor. This means the thermostat is an *operational control,* since its job is to operate the system at the setpoint

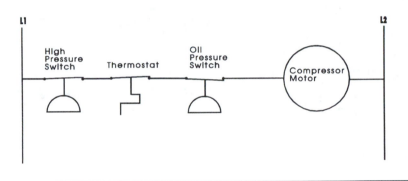

Figure 3–1   Electrical diagram of a power source, high-pressure switch, thermostat, oil pressure switch, and compressor motor. All of the switches are in series with the compressor.

temperature. If the condenser becomes dirty and causes the high side pressure (head pressure) to become excessive, the high-pressure switch will open and turn the compressor motor off. The same is true of the oil pressure switch. If the oil pressure becomes too low, the oil pressure switch will open and turn the motor off. The high-pressure switch and oil pressure switches are in the circuit for safety purposes.

Since all three of these switches are connected in series, they each are capable of opening and causing current to stop flowing to the compressor motor. If additional operational or safety switches need to be added to this circuit, they would be connected in series with the original switches. It is important to remember that when switches are connected in series, any one of them can open and stop current flow in the entire circuit.

3.2   Adding Loads in Series

In air-conditioning and refrigeration systems, loads such as motors are not connected in series with other loads. Instead they are connected in parallel with each other so that they all receive the same voltage. Even though loads such as motors are not connected in series, you should understand the problems that exist if motor loads were connected in series. This section will explain the problems associated with connecting motor-type loads in series, and Section 3.3 will explain how resistors are connected in series to create useful voltage drop circuits for electronic circuit boards and components used in HVAC electronic controls. Parallel circuits will be explained later in Chapter 4.

The reason loads are not generally connected in series is that they will split

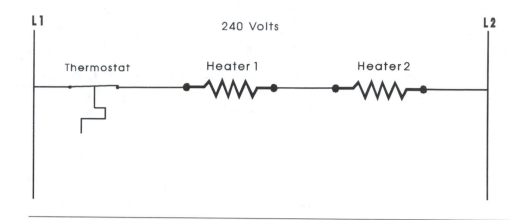

Figure 3–2   Electrical diagram of two 120-volt heaters connected in series to a 240-volt power supply.

the amount of applied voltage. For example, if you connected two light bulbs that have the same amount of resistance in their filaments in series with each other, and supplied the circuit with 110 volts, each light bulb would receive 55 volts. Each bulb would be glowing half as brightly as when compared to having only one light connected so that it would receive the entire 110 volts.

There are some exceptions to connecting loads in series such as heating circuits where two heating elements are connected in series on purpose. Fig. 3–2 shows an example of two heating elements connected in series. If the amount of resistance for each heater is equal, the supply voltage will be split equally between them. In this example the supply voltage to the heaters is 240 volts, and each heater will receive 120 volts. This means that each heater is rated for 120 volts. This type of circuit allows smaller heaters to be used on larger voltage sources.

Another problem with connecting loads in series is that if one of the loads has a defect and develops an open circuit, the current to all the loads in the circuit will be interrupted. An example of this type of circuit is the strings of lights that are used to decorate a Christmas tree. If all of the lights are connected in series, they will all go out if one of the lights burns out. The advantage of connecting 50 lights in series is that each of the light bulbs will each receive approximately 2 volts when the circuit is plugged into a 110-volt power source.

3.3   Using Resistors as Loads

Some circuits that you encounter will have resistors that are used to change the amount of voltage and current in a circuit. These types of circuits are commonly found in the solid-state control boards that are now used in many heating, air-

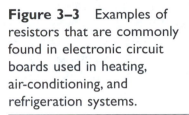

Figure 3–3 Examples of resistors that are commonly found in electronic circuit boards used in heating, air-conditioning, and refrigeration systems.

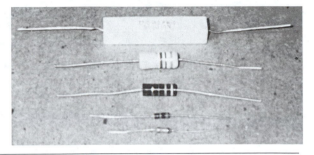

conditioning, and refrigeration systems. These solid-state boards are used to control motor speeds, time-delay functions, and to convert small signal voltages from sensors to larger voltages that can energize motor controls.

Fig. 3–3 shows a picture of several resistors. The electrical symbol for a resistor is the same as the heaters shown in Fig 3–2. A resistor is a component that is specifically manufactured with a precise amount of resistance. The amount of resistance will be identified on each resistor by color stripes or bands. Since most resistors are very small, a color code system has been designed to show the amount of resistance each resistor has.

3.4   Resistor Color Codes

The resistor color bands are shown in their respective positions in Fig. 3–4. From this figure you can see that the resistor has four basic color bands that are grouped to one end of the resistor. (Note: A few resistors will have fifth and sixth bands that represent other manufacturing information.) The color bands are used to represent numbers that indicate the amount of resistance each resistor has. The first color band represents the first digit of the number, and the second color band represents the second digit of the number. Each digit can be a value 0–9. The third band is a multiplier band. This will basically be the number of zeros that you place with the two-digit numbers to show numbers that have values above 99. The fourth band is called the tolerance band. The actual amount of resistance in the resistor will be close to the amount identified by the color bands, but it

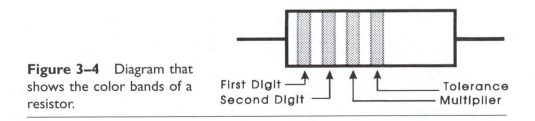

Figure 3–4 Diagram that shows the color bands of a resistor.

First Digit

Second Digit

Tolerance

Multiplier

may be slightly more or less depending on the amount of tolerance that is used. Typical tolerances are 5% and 10%. This means that the actual amount of resistance can be plus or minus the percentage identified by the tolerance band.

The colors that are assigned to the color code are shown in Fig. 3–5 and you can see each number 0–9 is represented by a color. The tolerances are also represented by the colors silver and gold. A silver band means the resistor has a tolerance of 10% and a gold band means the resistor has a tolerance of 5%.

3.5   Decoding a Resistor

A resistor with a brown first band, blue second band, and a red third band would be decoded as follows. The first two colors, brown and blue, mean that the first two numbers are 16. The third band, which is the multiplier band, is red so you need to add two zeros (0s) to the first two digits so the total resistance is 1600 Ω. If you want to use the multiplier for red, you would multiply the first two digits by 100: (16 $\times$ 100 = 1600 Ω).

A second resistor has a violet first band, a white second band, and a green third band. This resistor would be decoded as follows. The first two colors, violet and white, mean the first two numbers are 79. The third band, which is

Color	Digit Value	Multiplier	Tolerance
Black[1]	0	1	
Brown	1	10	
Red	2	100	
Orange	3	1,000	
Yellow	4	10,000	
Green	5	100,000	
Blue	6	1,000,000	
Violet[2]	7		
Gray[2]	8		
White[2]	9		
Silver		0.01	10%
Gold		0.1	5%

[1]Cannot be used in the first band (first digit).
[2]Maximum resistance is 22 M. Therefore, these colors are not used as multipliers.

Figure 3–5   The numerical value assigned to each of the colors for the resistor color code.

the multiplier band, is green so you need to add five zeros to the first two digits. This makes the total resistance 7,900,000 Ω. If you want to use the multiplier for green, you would multiply 79 by 100,000: (79 $\times$ 100,000 = 7,900,000 Ω).

PROBLEM 3–1

Determine the total resistance of a resistor that has a color code of yellow, brown, and orange.

Solution:

Since the first and second colors are yellow and brown, the first two digits are 41. The multiplier is orange so three zeros must be added to 41 to get a total of 41,000 Ω. If you use the multiplier for orange, you would multiply 41 $\times$ 1000 = 41,000 Ω.   ♦

3.6   Calculating the Resistor Tolerance

The first resistor that was decoded had color bands of brown, blue, and red and a total resistance of 1600 Ω. If this resistor had a fourth band of silver, the resistor tolerance would be calculated as +10% and −10%. The first step in calculating this tolerance is to multiply the total resistance by 10%.

$$1600 \ \Omega \times 0.10 = 160 \ \Omega$$

The second step in calculating the tolerance is to add the 160 Ω and subtract the 160 Ω from 1600 Ω.

$$
\begin{array}{cc}
1600 \ \Omega & 1600 \ \Omega \\
+160 \ \Omega & -160 \ \Omega \\
\hline
1760 \ \Omega & 1440 \ \Omega
\end{array}
$$

These calculations show that the tolerance of the resistor with the color bands of brown, blue, red, and silver would be 1760 Ω to 1440 Ω. If you are measuring this resistor with an ohmmeter and its total resistance falls between 1440 Ω and 1760 Ω, it would be considered a good resistor. If the total resistance is less than 1440 Ω or greater than 1760 Ω, the resistor would be considered out of tolerance and it would not be usable. If the fourth band was gold instead of silver, the multiplier would be 5% instead of 10%. The remainder of the calculation would be completed the same way in that you would add 5% and subtract 5% from the total to get the upper and lower limits.

PROBLEM 3–2

Determine the total resistance with the tolerance of a resistor whose color code is yellow, red, orange, and gold.

Solution:

Since the first two colors are yellow and red, the first two digits are 42. The multiplier is orange so you need to add three zeros to make the total resistance 42,000 Ω.   ◆

The tolerance band is gold so the tolerance is +5% and −5%; 5% of 42,000 Ω is found by multiplying 42,000 Ω × 0.05.

$$42,000 \times 0.05 = 2,100$$

The next step requires that you add 2,100 Ω to 42,000 Ω to get the upper tolerance, and subtract 2,100 Ω from 42,000 Ω to get the lower tolerance.

42,000 Ω	42,000 Ω
+2,100 Ω	−2,100 Ω
44,100 Ω	39,900 Ω

The upper tolerance of this resistor is 44,100 Ω and the lower tolerance is 39,900 Ω.

3.7   Using Ohm's Law to Calculate Ohms, Volts, and Amps for Resistors in Series

In some electronic circuits all of the loads are resistors. These resistors are sized so that the supply voltage will be dropped into smaller increments. These types of circuits are widely used in electronic circuits for air-conditioning, heating, and refrigeration equipment, and they are also useful to learn the concepts of how voltage and current are affected by changes in resistance. Several circuits that have resistors connected in series will be used to explain these concepts. Fig. 3–6 shows an example of three resistors connected in series with a voltage source (battery). Since the resistors are connected in series, there is only one path for current. The resistors are numbered R_1, R_2, and R_3 and the supply voltage is identified as E_T. The formula for calculating total resistance in a series circuit is $R_T = R_1 + R_2 + R_3$. If more than three resistors are used in the circuit, the

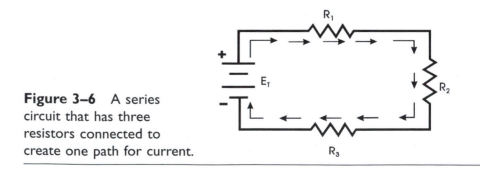

Figure 3–6   A series circuit that has three resistors connected to create one path for current.

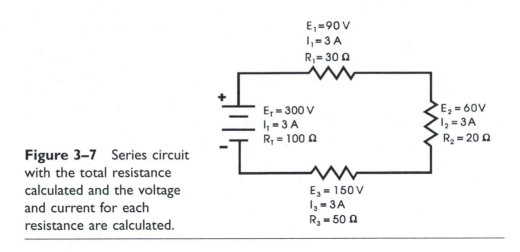

Figure 3–7 Series circuit with the total resistance calculated and the voltage and current for each resistance are calculated.

additional resistors are added to the first three to get a total. The arrows show conventional current flow. [Electron flow would show the flow of electrons (current) moving from the negative battery terminal to the positive terminal. This text will use conventional current flow in its explanations and diagrams.]

Fig. 3–7 shows a series circuit that has the size of each resistor identified. Resistor R_1 is 30 ohms, R_2 is 20 ohms, and R_3 is 50 ohms. The total resistance for this circuit can be calculated by the formula:

$$R_T = R_1 + R_2 + R_3 \qquad R_T = 30 \ \Omega + 20 \ \Omega + 50 \ \Omega \qquad R_T = 100 \ \Omega$$

3.8   Solving for Current in a Series Circuit

The total current can be calculated by the Ohm's law formula, $I = \dfrac{300 \text{ V}}{100 \ \Omega}$, or $I = $ 3 A. The total current in a series circuit is the same everywhere in the circuit because there is only one path for the current. This means that once you find the

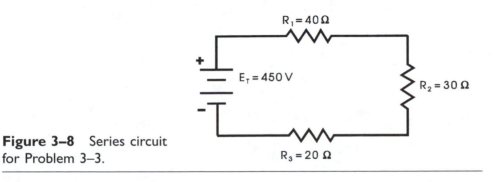

Figure 3–8 Series circuit for Problem 3–3.

total resistance of a series circuit, you also have determined the current that flows through each resistor. The current (3 A) has been listed with each resistor in the diagram in Fig. 3–8. Since the current is the same everywhere in the series circuit, it may be identified as I_T or I_1 where it is shown at resistor R_1.

3.9   Calculating the Voltage Drop Across Each Resistor

The voltage drop across each resistor can be calculated by the Ohm's law formula ($E = IR$). It is important to remember that the current in the series circuit is the same in all places so I_T is equal to 3 A and I_1, I_2, and I_3 are also equal to 3 A. In this case the voltage drop across resistor R_1 will be calculated by the formula:

$$E_1 = I_1R_1 \qquad E_1 = 3\text{ A} \times 30\ \Omega \qquad E_1 = 90\text{ V}$$

If you placed the probes from a voltmeter on each side of resistor R_1, you would measure 90 V. This is the actual voltage that the resistor is causing to *drop* when current is flowing through it. The voltage drop across resistor R_2 can be calculated by the formula:

$$E_2 = I_2 \times R_2 \qquad E_2 = 3\text{ A} \times 20\ \Omega \qquad E_2 = 60\text{ V}$$

The voltage drop across resistor R_3 can be calculated by the formula:

$$E_3 = I_3 \times R_3 \qquad E_3 = 3\text{ A} \times 50\ \Omega \qquad E_3 = 150\text{ V}$$

From these calculations you can see that the voltage dropped across R_1 is 90 V, across R_2 is 60 V, and across R_3 is 150 V. If you added all of these voltage drops, you would find that they are equal to the supply voltage. This means that $E_1 + E_2 + E_3 = E_T$ (90 V + 60 V + 150 V = 300 V).

PROBLEM 3–3

Fig. 3–8 shows three resistors connected in series. Resistor R_1 is 40 ohms, R_2 is 30 ohms, and R_3 is 20 ohms. Calculate the total resistance for this circuit. After you have calculated the total resistance, calculate the total current. When you have determined the total current for this circuit, calculate the voltage drop that would be measured across each resistor.

Solution:

R_T can be calculated by the formula: $R_T = R_1 + R_2 + R_3$.

$$R_T = 40\ \Omega + 30\ \Omega + 20\ \Omega \qquad R_T = 90\ \Omega$$

I_T is calculated by the formula:

$$I_T = \frac{E_T}{R_T} \qquad I_T = \frac{450\text{ V}}{90\ \Omega} \qquad I_T = 5\text{ A}$$

Since current is the same in all parts of the circuit:

$$I_T = 5 \text{ A} \qquad I_1 = 5 \text{ A} \qquad I_2 = 5 \text{ A} \qquad I_3 = 5 \text{ A}$$

The voltage that is dropped across each resistor from the current flowing through it is calculated by the formulas:

$$
\begin{aligned}
E_1 &= R_1 \times I_1 & E_1 &= 40 \ \Omega \times 5 \text{ A} & E_1 &= 200 \text{ V} \\
E_2 &= R_2 \times I_2 & E_2 &= 30 \ \Omega \times 5 \text{ A} & E_2 &= 150 \text{ V} \\
E_3 &= R_3 \times I_3 & E_3 &= 20 \ \Omega \times 5 \text{ A} & E_3 &= 100 \text{ V}
\end{aligned}
$$

You can check your answers by adding all of the drops to see that they equal the total supply voltage.

$$E_T = E_1 + E_2 + E_3 \qquad E_T = 200 \text{ V} + 150 \text{ V} + 100 \text{ V} \qquad E_T = 450 \text{ V} \qquad \blacklozenge$$

3.10   Calculating the Power Consumption of Each Resistor

The power consumed by each resistor or the power consumed by the total circuit can be calculated by the formulas for power (wattage): $P = EI$, $P = I^2R$, and $P = E^2/R$. In the circuit for Problem 3–3, we have previously calculated the voltage, current, and resistance for each resistor. This means that we can use any of the three formulas to calculate the power for any resistor. For this example we will use all three formulas to show that they all work. We will use the values for R_1: $E_{R1} = 200 \text{ V}$, $I_{R1} = 5 \text{ A}$, and $R_1 = 40 \ \Omega$.

$$
\begin{aligned}
P &= EI & P &= 200 \text{ V} \times 5 \text{ A} & P &= 1000 \text{ watts} \\
P &= I^2R & P &= (5 \text{ A})^2 \times 40 \ \Omega & P &= 1000 \text{ watts} \\
P &= E^2/R & P &= (200 \text{ V})^2 / 40 \ \Omega & P &= 1000 \text{ watts}
\end{aligned}
$$

Since all three formulas work the same, you can use any of the three that you choose. The major factor in deciding which formula to use will be the values that you have been given or that you can determine by measuring.

3.11   Calculating the Power Consumption of an Electrical Heating Element

The power formula can also be used to calculate the power consumption of any other type of resistance used in a circuit. The electric heating elements used in an electric furnace have large resistances, so if you know the amount of resistance in the element and the amount of voltage applied to the circuit, you can calculate the current the element uses and the amount of power it consumes. For example, if the heating element has 10 ohms of resistance and it is connected to 220 volts, it would draw 22 amps. The power this element would consume would be 4,840 watts.

PROBLEM 3–4

A heating element in an electric furnace has a resistance of 2.5 Ω. Calculate the current and the power consumption of this heating element if the furnace is connected to 240 V AC.

Solution:

The amount of current is calculated by 240 V/2.5 Ω = 96 A. The amount of power can be calculated by $P = I^2R$ or $P = IE$: Using $P = I^2R$, $(96)^2 \times 2.5$ Ω = 23,040 watts.

Using $P = IE$, $96 \times 240 = 23{,}040$ watts.

(Note! This answer could be expressed as 23.040 kilowatts or 23.040 kW).   ◆

3.12   Review of Series Circuits

In a series circuit all of the loads will be connected in such a way that there is only one path for current to flow. If one or more switches are used to control current flow to the loads in the circuit, they must also be connected so that only one path exists. The total resistance in a circuit can be calculated by adding all of the resistor values together ($R_T = R_1 + R_2 + R_3$ . . .). (Notice that the . . . means that any number of resistors can be used in this formula since they are all added.)

The total current of a series circuit can be calculated by the formula $I = E/R$, and it is important to remember that the current will be the same in all parts of the circuit. This means that once you measure or calculate current at any point or through any component, it will be the same at all other components in the circuit.

The voltage that is measured across each component in the series circuit can be different. If the resistance values of any two resistors in a circuit are the same, the amount of voltage measured across them will be the same. The amount of voltage that is measured across each resistor is caused by the value of the resistor and the amount of current flowing through it. The formula for calculating voltage is $E = IR$. The amount of voltage measured across each resistor in a series circuit is also called the voltage drop for that resistor. The total of all the voltage drops in a series circuit will always be equal to the supply voltage for that circuit.

The power consumed by each component in a series circuit can be calculated by the formula $P = IE$. The total power consumed by the components in a series circuit can be calculated by the same formula or by adding the amount of power consumed by each component. For example, $P_T = P_1 + P_2 + P_3$.

Questions for This Chapter

1. Discuss the advantage of connecting switches in series with a load.
2. Explain why loads are typically not connected in series with each other.

3. In some cases loads may be connected in series with each other. Explain when this would be advantageous and what type of loads would be used.
4. Explain what a resistor is and where you would find it in HVAC circuits.
5. Explain what each of the four color bands on a resistor is used to indicate.

True or False

1. When a series circuit has an open, current flow stops in all parts of the circuit.
2. The main reason the switches in Fig. 3–1 are connected in series is so that if any one of the switches opens, current to the load will be interrupted.
3. It is a good practice to connect the fan motor and compressor motor for an air conditioner in series so that when one quits the other will stop.
4. Two heating elements should be connected in series if you want them to split the amount of voltage applied to them.
5. All resistors have the same amount of resistance if they are the same physical size.

Multiple Choice

1. In a series circuit the current in each part of the circuit is _____
   a. always zero.
   b. the same.
   c. equal to the voltage.
2. The amount of resistance a resistor has _____
   a. is the same if the resistors are the same size.
   b. can be determined by color codes.
   c. changes as the amount of current that flows through it changes.
3. When three resistors are connected in series, the amount of voltage that is measured across each one is _____
   a. determined by the amount of current flowing through it and the amount of resistance it has.
   b. determined by the wattage rating of each resistor.
   c. impossible to determine by calculations.

4. If a thermostat, high-pressure switch, and oil pressure switch are con-
    nected in series with a compressor motor and the oil pressure switch is
    opened because of low oil pressure, the compressor motor will
    _____

    a. still run because two of the other switches are still closed.
    b. will not be affected because no other loads are connected to it in
        series.
    c. will stop running because current flow will be zero.

5. If the resistance of a heating element increases from 2 Ω to 5 Ω, the
    current it draws will _____

    a. increase if the voltage stays the same.
    b. decrease if the voltage stays the same.
    c. not change if the voltage does not change.

Problems

1. Decode the following resistors and indicate their tolerances. The first
    color band is listed on the left.

	FIRST BAND	SECOND BAND	THIRD BAND	FOURTH BAND	
a.	red	blue	brown	gold	_____
b.	green	gray	red	silver	_____
c.	orange	black	blue	gold	_____
d.	brown	yellow	violet	gold	_____

2. Calculate the amount of voltage for a circuit that has 6 amps and 20
    ohms.

3. Calculate the amount of current for a circuit that has 100 volts and 25
    ohms.

4. Calculate the wattage for the circuit that has 120 volts and 8 ohms.

5. Calculate the wattage for the circuit that has 15 ohms total and draws
    25 amps.

Parallel and Series-Parallel Circuits

OBJECTIVES:

After reading this chapter, you will be able to:
1. Explain the operation of components connected in parallel.
2. Calculate the voltage, current, resistance, and power in a parallel circuit.
3. Explain Ohm's law as it applies to parallel circuits.
4. Calculate the voltage, current, resistance, and power in a series-parallel circuit.

4.0   Introduction

Parallel circuits are used frequently in air-conditioning, heating, and refrigeration systems. Fig. 4–1 shows an example circuit and you can see the difference between the series circuit and the parallel circuit is that all of the loads in a parallel circuit will have the same voltage supplied, whereas in a series circuit, each load has a different amount of voltage. The load components in air-conditioning, heating, and refrigeration systems, such as compressor motors and fan motors, must all be provided with the same voltage, so they must be connected in parallel when they are in the same circuit. Each point where a resistor is connected in parallel in this circuit is called a branch circuit. The parallel circuit can have any number of branch circuits. The current in a parallel circuit allows current more

Figure 4–1 Example of a parallel circuit. Notice in this circuit that all of the resistors will have the same amount of voltage supplied to each.

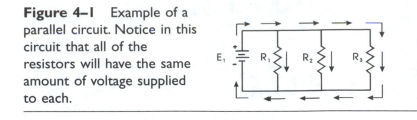

than one path to return to the power supply. These paths are identified by the arrows in the diagram.

The current in a parallel circuit is additive and the formula to calculate total resistance is: $I_T = I_1 + I_2 + I_3$ . . . (remember the . . . means that additional currents would be added to the total). Another point to understand is that the current in a parallel circuit gets larger as more loads are added to the circuit. The parallel circuit also provides a means so that any load can be disconnected from the power source and voltage will still be supplied to the remaining loads. This is accomplished by placing a switch in each branch circuit just ahead of each resistor.

Series-parallel circuits are used in air-conditioning, heating, and refrigeration systems where some parts of the circuit are series in nature, and other parts of the circuit are parallel. For example, the fuse and disconnect for the circuit must be able to interrupt all power that is supplied to the loads in a circuit, so the fuse and disconnect must be connected in series with the power supply. The condenser fan motor and compressor must have the same amount of voltage supplied to each of them, so they must be connected in parallel with each other and the power supply. The important point to remember when working with series-parallel circuits is that you treat the part of the circuit that is in series with the formulas for series circuits, and the part of the circuit that is in parallel uses the formulas for parallel circuits. It is also possible to combine parts of the circuit to make it a simplified series or simplified parallel circuit.

4.1   Calculating Voltage, Current, and Resistance in a Parallel Circuit

Voltage, current, and resistance can be calculated in a parallel circuit just as in a series circuit by using Ohm's law. Fig. 4–2 shows an example circuit with the voltage, current, and resistance calculated at each point in the circuit. From this diagram you can see that the supply voltage is 300 V. Since the supply voltage is 300 V, you can determine that the voltage of each branch circuit (across each resistor) is also 300 V because the voltage at each branch circuit in a parallel circuit is the same as the supply voltage.

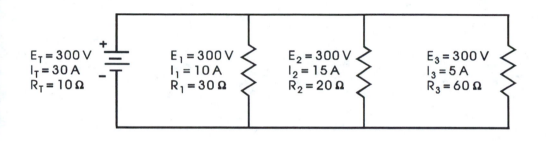

Figure 4–2   Example of a parallel circuit with the voltage, resistance, and current shown at each load.

The current that is flowing through each resistor can be calculated by using Ohm's law formula for current: $I = E/R$. The current in each branch circuit is calculated from this formula and it is placed in the diagram beside each resistor.

$$
\begin{array}{lll}
I_1 = E/R_1 & I_1 = 300 \text{ V}/30 \ \Omega & I_1 = 10 \text{ A} \\
I_2 = E/R_2 & I_2 = 300 \text{ V}/20 \ \Omega & I_2 = 15 \text{ A} \\
I_3 = E/R_3 & I_3 = 300 \text{ V}/60 \ \Omega & I_3 = 5 \text{ A} \\
I_T = E_T/R_T & I_T = 300 \text{ V}/10 \ \Omega & I_T = 30 \text{ A} \\
I_T = I_1 + I_2 + I_3 & I_T = 10 \text{ A} + 15 \text{ A} + 5 \text{ A} & I_T = 30 \text{ A}
\end{array}
$$

If the voltage and current were given and the resistance needed to be calculated, the Ohm's law formula for resistance could be used: $R = E/I$.

$$
\begin{array}{lll}
R_1 = E/I_1 & R_1 = 300 \text{ V}/10 \text{ A} & R_1 = 30 \ \Omega \\
R_2 = E/I_2 & R_2 = 300 \text{ V}/15 \text{ A} & R_2 = 20 \ \Omega \\
R_3 = E/I_3 & R_3 = 300 \text{ V}/5 \text{ A} & R_3 = 30 \ \Omega \\
R_T = E/I_T & R_T = 300 \text{ V}/30 \text{ A} & R_T = 10 \ \Omega
\end{array}
$$

If the total resistance and the total current in a circuit are known, you can calculate the voltage for the circuit. If you know the current and resistance at any branch circuit, you could also calculate the voltage using the Ohm's law formula: $E = IR$. The nice part about calculating voltage is that once you determine the voltage at any branch circuit, the same amount of voltage is at every other branch circuit and at the supply. The same is true if you calculate the voltage at the supply; you do not need to calculate the voltage at any other branch because it will be the same.

$$
\begin{array}{lll}
E_T = I_T R_T & E_T = 30 \text{ A} \times 10 \ \Omega & E_T = 300 \text{ V} \\
E_1 = I_1 R_1 & E_1 = 10 \text{ A} \times 30 \ \Omega & E_1 = 300 \text{ V} \\
E_2 = I_2 R_2 & E_2 = 15 \text{ A} \times 20 \ \Omega & E_2 = 300 \text{ V} \\
E_3 = I_3 R_3 & E_3 = 5 \text{ A} \times 60 \ \Omega & E_3 = 300 \text{ V}
\end{array}
$$

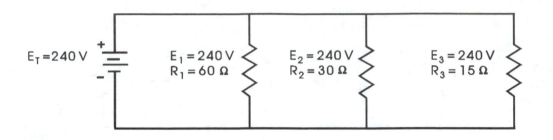

Figure 4–3   Parallel circuit for Problem 4–1.

PROBLEM 4–1

Use the circuit in Fig. 4–3 to calculate the individual branch current, total current, and total resistance.

Solution:

Use the following formulas to calculate the current for each branch.

$$
\begin{aligned}
I_1 &= E/R_1 & I_1 &= 240\ \text{V}/60\ \Omega & I_1 &= 4\ \text{A} \\
I_2 &= E/R_2 & I_2 &= 240\ \text{V}/30\ \Omega & I_2 &= 8\ \text{A} \\
I_3 &= E/R_3 & I_3 &= 240\ \text{V}/15\ \Omega & I_3 &= 16\ \text{A}
\end{aligned}
$$

Use the following formula to calculate the total current I_T.

$$I_T = I_1 + I_2 + I_3 \qquad I_T = 4\ \text{A} + 8\ \text{A} + 16\ \text{A} \qquad I_T = 28\ \text{A}$$

Use the following formula to calculate total resistance R_T.

$$R_T = E_T/I_T \qquad R_T = 240\ \text{V}/28\ \text{A} \qquad R_T = 8.57\ \Omega \qquad\qquad \blacklozenge$$

4.2   Calculating Resistance in a Parallel Circuit

At times you will need to calculate the total resistance of a parallel circuit when only the branch resistance and supply voltage are provided. You could calculate the individual currents at each branch circuit and then calculate the total with the formula $I_T = I_1 + I_2 + I_3$. After you have determined the total current, you can use the formula $R_T = E_T/I_T$.

Another method called the *product over the sum* method could be used, which would require you to use the formula for calculating total resistance in a parallel circuit. This method is called the product over the sum method because you multiply the two resistors to get the product, and then you add the two resistors to get the sum. The third step in the calculation includes dividing the product by the sum (product over sum). The formula is presented as follows: $R_T = \dfrac{R_1 \times R_2}{R_1 + R_2}.$

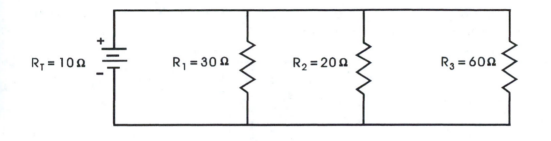

Figure 4–4   Parallel circuit for calculating total resistance.

You should notice that with this formula you can only calculate the total resistance of two resistors at a time. Since this circuit has three resistors, you would need to find the total of the first two resistors in the branch and then use that total and use the formula again to find the *grand* total resistance. We will use this method to calculate the total resistance for the resistors shown in the circuit in Fig. 4–4.

$$R_{T_{R1R2}} = \frac{R_1 \times R_2}{R_1 + R_2} \qquad R_{T_{R1R2}} = \frac{30\,\Omega \times 20\,\Omega}{30\,\Omega + 20\,\Omega} \qquad R_{T_{R1R2}} = \frac{600\,\Omega}{50\,\Omega} \qquad R_{T_{R1R2}} = 12\,\Omega$$

$$R_{T_{R1R2}} = \frac{R_{T_{R1R2}} \times R_3}{R_{T_{R1R2}} + R_3} \qquad R_{T_{R1R2}} = \frac{12\,\Omega \times 60\,\Omega}{12\,\Omega + 60\,\Omega} \qquad R_{T_{R1R2}} = \frac{720\,\Omega}{72\,\Omega} \qquad R_{T_{R1R2}} = 10\,\Omega$$

From these calculations you can see that the total parallel resistance is 10 Ω. *It is important to understand that in all parallel circuits, the total resistance will always be smaller than the smallest branch circuit resistance.* You can see that in this circuit, the smallest resistance in the branch circuits is 20 Ω, and the total resistance is 10 Ω.

The next calculation shows the total resistance calculated from the formula:

$R_T = \dfrac{1}{\dfrac{1}{R_1} + \dfrac{1}{R_2} + \dfrac{1}{R_3}}$. This formula is designed to be used with a calculator. If

you do not have a calculator, it is recommended that you use the previous method. If you use a calculator, you should use the following keystrokes to get an answer from this formula:

$$R_T = \frac{1}{\dfrac{1}{R_1} + \dfrac{1}{R_2} + \dfrac{1}{R_3}} \qquad R_T = \frac{1}{\dfrac{1}{20\,\Omega} + \dfrac{1}{10\,\Omega} + \dfrac{1}{30\,\Omega}}$$

Keystrokes   (Notice the boxes indicate keys on the calculator to use and numbers indicate number keys to use.)

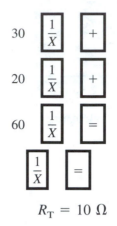

$$R_\text{T} = 10 \ \Omega$$

PROBLEM 4–2

Use the parallel circuit in Fig. 4–5 to calculate the current through each resistor, the total current, and the total resistance for this circuit. The supply voltage is 100 V.

Solution:

$$I_1 = E/R_1 \qquad\qquad I_1 = 100 \ \text{V}/20 \ \Omega \qquad\qquad I_1 = 5 \ \text{A}$$
$$I_2 = E/R_2 \qquad\qquad I_2 = 100 \ \text{V}/10 \ \Omega \qquad\qquad I_2 = 10 \ \text{A}$$
$$I_3 = E/R_3 \qquad\qquad I_3 = 100 \ \text{V}/5 \ \Omega \qquad\qquad I_3 = 20 \ \text{A}$$
$$I_\text{T} = I_1 + I_2 + I_3 \qquad I_\text{T} = 5 \ \text{A} + 10 \ \text{A} + 20 \ \text{A} \qquad I_\text{T} = 35 \ \text{A}$$
$$R_\text{T} = E_\text{T}/I_\text{T} \qquad\qquad R_\text{T} = 100 \ \text{V}/35 \ \text{A} \qquad\qquad R_\text{T} = 2.86 \ \Omega$$

$$R_\text{T} = \cfrac{1}{\dfrac{1}{R_1} + \dfrac{1}{R_2} + \dfrac{1}{R_3}} \qquad R_\text{T} = \cfrac{1}{\dfrac{1}{20 \ \Omega} + \dfrac{1}{10 \ \Omega} + \dfrac{1}{5 \ \Omega}} \qquad R_\text{T} = 2.86 \ \Omega$$

◆

$E_\text{T} = 100 \ \text{V}$ $E_1 = 100 \ \text{V}$ $E_2 = 100 \ \text{V}$ $E_3 = 100 \ \text{V}$
                            $R_1 = 20 \ \Omega$ $R_2 = 10 \ \Omega$ $R_3 = 5 \ \Omega$

Figure 4–5   Parallel circuit for Problem 4–2.

4.3 Calculating Power in a Parallel Circuit

The formula for calculating power in a parallel circuit is the same as the formula for a series circuit: $P_T = P_1 + P_2 + P_3$. The formula for power at each individual resistor is found from the original Watt's law formula $P = IE$. This means that you must calculate the power consumed by each branch resistor and then add them all together to get the total power consumed. For example, in the parallel circuit that was previously shown in Fig. 4–2, the voltage at R_1 is 300 V and the current is 10 A, the voltage at R_2 is 300 V and the current is 15 A, and the voltage at R_3 is 300 V and the current is 5 A. The following calculations will be used to determine the power consumed by each individual resistor and the total power used by the whole circuit.

$$P_1 = I_1E_1 \qquad\qquad P_1 = 10\ \text{A} \times 300\ \text{V} \qquad\qquad P_1 = 3000\ \text{watts}$$
$$P_2 = I_2E_2 \qquad\qquad P_2 = 15\ \text{A} \times 300\ \text{V} \qquad\qquad P_2 = 4500\ \text{watts}$$
$$P_3 = I_3E_3 \qquad\qquad P_3 = 5\ \text{A} \times 300\ \text{V} \qquad\qquad P_3 = 1500\ \text{watts}$$
$$P_T = P_1 + P_2 + P_3 \quad P_T = 3000\ \text{W} + 4500\ \text{W} + 1500\ \text{W} \quad P_T = 9000\ \text{watts}$$

You could also calculate the total power consumed by this circuit by using the formula:

$$P_T = I_TE_T \qquad P_T = 30\ \text{A} \times 300\ \text{V} \qquad P_T = 9000\ \text{watts}$$

4.4 Reviewing the Principles of Parallel Circuits

Parallel circuit are used in air-conditioning, heating, and refrigeration systems because this type of circuit provides the same amount of voltage to each motor in the circuit. For example, if an air-conditioning system has a compressor that requires 240 volts, and a condenser fan that requires 240 volts, a parallel circuit will provide both motors this voltage if they are connected in parallel. A parallel circuit provides more than one path for current to return to the power source. Each path is called a branch circuit. The amount of current in each branch of the parallel circuit can be determined by dividing the amount of voltage supplied to that branch by the amount of resistance for the load in that branch. The amount of total resistance in a parallel circuit will always be smaller than the smallest branch resistance. This also means that as additional loads (resistances) are added to a parallel circuit, the total amount of resistance will become smaller. Since the amount of resistance is reduced as loads are added, the total amount of current will increase.

4.5 Using Prefixes with Units of Measure

At times in previous examples you have seen the amount of resistance exceed 1000 ohms or the amount of power exceed 1000 watts. In these cases it would be useful to substitute the large number of zeros with a letter. For example, we learned that the word for 1000 is *kilo,* and that the letter k could be substituted for the value 1000. This means that if the answer to a problem is 1000 Ω, we could use the term 1 kΩ to indicate the same amount. The letter k in this case replaces three zeros.

When the resistor values were calculated, the amount of resistance sometimes was as large as 300,000 Ω. This amount could be represented by the term 300 kΩ, since the letter k represents 1000. If the value of a resistor is determined to be 5,000,000 (5 million), the six zeros could be replaced by the letter M. The letter M is the abbreviation for the word *mega,* which is the word for *million.* This means that the value of 5,000,000 Ω could be represented by the value 5 MΩ. When you see the value 5 MΩ, you would say that you have measured 5 megohms.

If the amount of current was one thousandth of an amp, you could show this value as 0.001 A or 1/1000th of an amp. The prefix *milli-* is used to represent the value of one thousandth of an amp. This prefix is represented by the lowercase letter m. This means that the value 0.001 A could be written as 1 mA. (Notice that the lowercase letter m is used to specify *milli,* and the capital letter A is used to represent amps.)

If the amount of current was measured as one millionth of an amp, it could be shown as 0.000001 A or 1/1,000,000 A. The prefix *micro-* is used to represent this amount, and the Greek letter μ. This means that the value 0.000001 A can be represented as 1 μA. Fig. 4–6 shows a table listing the common prefixes you will use most often in electricity.

The table in this figure shows the four common prefixes that you will use in electricity. The exponent value for each prefix is also shown. The exponent is another way of expressing the number without showing a large number of zeros. For example, the prefix *mega-,* indicates the value of 1 million. If you want to show the value 6 kilo-ohms you could express this value as 6 kΩ. If you want to

Prefix	Symbol	Number	Exponent
mega	M	1,000,000	$\times 10^6$
kilo	k	1,000	$\times 10^3$
milli	m	0.001	$\times 10^{-3}$
micro	μ	0.000001	$\times 10^{-6}$

Figure 4–6   Table of common prefixes with the numerical value and exponential value.

multiply this value times the value of 20 A, you could simply multiply 6,000 × 20 to get the answer 120,000 volts. The exponential value for 6,000 is 6×10^3. The exponent $\times10^3$ means that you have taken 10 and multiplied it by itself three times (10 × 10 × 10). As you know, 10 × 10 × 10 is equal to 1000.

The exponential value for 1 million is $\times10^6$, which means that you would multiply 10 times itself six times (10 × 10 × 10 × 10 × 10 × 10), which equals 1 million. The representation of the value 1,000,000 as 1×10^6 makes it easier to see that the value has six zeros. The other advantage to exponents is that when you multiply numbers that have exponents in them, you can simply add the exponents to get the correct answer. For example, if you needed to multiply 5,000,000 times 4,000 you could do this longhand as shown:

$$\begin{array}{r} 5,000,000 \\ \times\ 4,000 \\ \hline 20,000,000,000 \end{array}$$

You could also do it with exponents and simply multiply 5 × 4 to get 20, and then add the exponents $1 \times 10^6 + 1 \times 10^3$ to get the value 1×10^9. This means the answer will be 20 followed by nine zeros (20,000,000,000).

$$\begin{array}{r} 5 \times 10^6 \\ \times\ 4 \times 10^3 \\ \hline 20 \times 10^9 \end{array}$$

Most calculators have an exponent button on them. This button is identified as

$$\boxed{EE}$$

The *EE* on the label for this button stands for *enter exponent*. If you used a calculator to solve the previous problem, you would use the following keystrokes:

5 $\boxed{EE}$ 6 $\boxed{\times}$

4 $\boxed{EE}$ 3 $\boxed{=}$ and the answer would show 20^9.

(It is important to remember that the exponent will be shown in the upper right corner of the calculator's display as a number only; there is not room to display the $\times10$.)

If the value 2 mA (milliamps) is shown as an exponent, the exponential value will be displayed as a negative number (2×10^{-3}). The negative number (-3) that is used to indicate the exponent means that the decimal point should be moved three places to the left. Since the original value of the number was 2, you will need to write this value as 2.0 to show where the decimal point resides origi-nally. When the exponent -3 is used, the decimal point should be moved three places to the left as shown: 0.002. (It is also important to understand that the first zero that is shown to the left of the decimal point does not add any value to the number. It is used so that you do not confuse the decimal point with the punctuation mark called a period.)

When the value for 7 μA (microamps) is shown, it will use the same scheme, and the exponent for this value is shown as the number -6: (7×10^{-6}). If you need to multiply the value 15 mA times 4 MΩ, you could use the calculator and use exponents instead of trying to remember how many zeros to use, and where to place the decimal point. The keystrokes would be as follows:

$$15 \quad \boxed{EE} \quad -3 \quad \boxed{\times}$$

$$4 \quad \boxed{EE} \quad 6 \quad \boxed{=} \quad \text{the answer would show } 60^3.$$

The answer would show 60^3 to indicate 60,000 volts. Notice that since the exponent is 3, you could also give the answer as 60 k volts.

When you are working with electrical circuits, you will find a prefix is com-monly used where it saves space to keep from displaying large numbers of zeros. You should notice that on most meters, the kilo-ohms scale is shown as kΩ, and the mega-ohm scale is shown as MΩ. The milliamp scale is shown as mA, and the microamp scale is shown as μA. Some companies that manufacture meters use the capital letter K to represent kilo instead of the lowercase letter k because it is easier to read. The correct symbol for the prefix *kilo-* is the lowercase k.

It is also important to understand that when you perform calculations with your calculator using micro, milli, kilo, and mega, your answer may contain ex-ponential values either positive or negative from 0 to 6. The problem with an answer that has an exponential value of 2 or 5 is that the VOM meter has only scales that recognize the exponents of 10^{-6}, 10^{-3}, 10^{+3}, and 10^{+6}. This means that you must convert all of the answers to your calculations to one of these exponents if you plan to make the measurement with the VOM.

4.6   Series-Parallel Circuits

As a technician you will work on circuits for residential and commercial heat-ing, air-conditioning, and refrigeration systems. These systems will have a num-ber of switches and loads that are connected in series and parallel. A typical cir-cuit is shown in Fig. 4–7. In this diagram you can see that a compressor motor and a condenser fan motor are connected in parallel. These two motors are con-nected in parallel so that they will each receive the same amount of voltage.

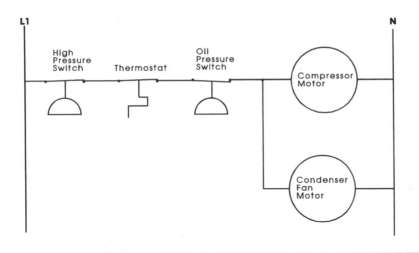

Figure 4–7   An electrical diagram of a compressor motor and condensor fan motor connected in parallel, and a high-pressure switch, thermostat, and oil pressure switch in series.

These loads are also connected in series with three control switches; a high-pressure switch, a thermostat, and an oil pressure switch. If any of these three switches open, the current to both loads will be stopped. The switches are connected in series for this reason. Since this circuit has some switches connected in series, and two loads connected in parallel, it is called a series-parallel circuit.

The parts of the circuit that are connected in series will follow all of the rules for series circuits, and the parts of the circuit that are connected in parallel will follow all of the rules for parallel circuits. This series-parallel circuit will provide the best of series circuits and the best of parallel circuits to supply voltage to the loads and control their operation. This type of series-parallel circuit is very common in field equipment that you will encounter.

It will be easier to see the operation of series-parallel circuits by following the voltage, current, and resistance of a circuit that has resistors connected in series-parallel. Fig. 4–8 shows an example of a typical series-parallel circuit that uses six resistors. You should notice that the overall shape of this circuit is similar to a large parallel circuit. You can better envision the outline of the parallel circuit if you combined R_1 and R_2 into one equivalent resistance, combine R_3 and R_4 into one equivalent resistance, and combine R_5 and R_6 into one equivalent resistance.

The first branch of this circuit consists of a 20 Ω resistor connected in series with a 30 Ω resistor. The second branch consists of a 10 Ω resistor with a 30 Ω resistor, and the third branch consists of a 40 Ω resistor with a 30 Ω resistor.

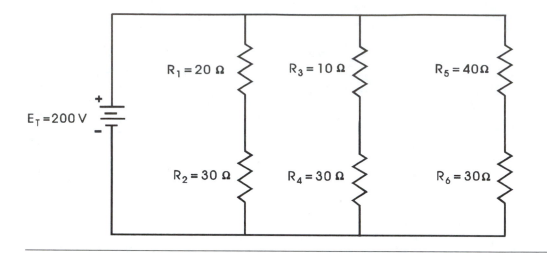

Figure 4–8   A series-parallel circuit consisting of six resistors.

The supply voltage for this circuit is 200 V. This means that each branch has 200 volts applied, but each individual resistor will have a different voltage applied because each is connected in series with one other resistor.

4.6.1   Calculating the Amount of Resistance for Each Branch

The amount of voltage that is measured across each resistor and the amount of current flowing through each resistor can be calculated by using the formulas that you have previously learned for series and parallel circuits. Fig. 4–9 shows the next step in calculating the voltage and current for each individual resistor. This step includes calculating the total resistance for each branch circuit and the total resistance for the circuit. In this diagram you can see that the equivalent resistance of each branch is calculated and shown at each branch. Equivalent resistance is very similar to making change for a dollar bill. For example, if you have two fifty-cent pieces, you could replace them with a dollar bill if you were solving a problem. The same is true if you had four quarters; you could show the equivalent amount of money by using two fifty-cent pieces, or you could use a dollar bill. When you are working with change and dollar bills, you basically use the form that is the simplest to handle for each different operation.

The same is true of combining resistor values to make a circuit easier to understand. For example, in Fig. 4–9 when you combine the 20 Ω and 30 Ω resistors, you can substitute their equivalent value of 50 Ω and the two resistors (R_1 and R_2) can be replaced with one resistance called **Req$_1$** (*Resistance equivalent*

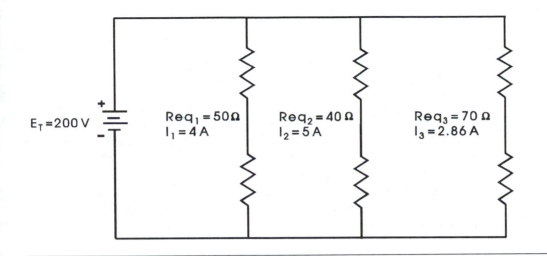

Figure 4–9   The series-parallel circuit that shows the series resistors in each branch combined. The value of the equivalent resistance is identified as Req_1, Req_2, and Req_3.

1) that has the same value as the two resistors in series (50 Ω). Resistors R_3 10 Ω and R_4 30 Ω are combined to form the equivalent resistance Req_2 that is equal to 40 Ω. Resistors R_5 (40 Ω) and R_6 (30 Ω) are combined to form the equivalent resistance Req_3 that is equal to 70 Ω.

4.6.2   Calculating the Current for Each Branch

The current in each branch circuit of the circuit in Fig. 4–9 can be calculated by Ohm's law ($I = E/R$). The following calculations show how the branch circuit current was determined. You should notice that since the overall design of the circuit is a parallel circuit, the supply voltage of 200 V is also supplied to each of the branch circuits.

$$I_1 = E_1/Req_1 \qquad I_1 = 200/50\ \Omega \qquad I_1 = 4\ A$$
$$I_2 = E_2/Req_2 \qquad I_2 = 200/40\ \Omega \qquad I_2 = 5\ A$$
$$I_3 = E_3/Req_3 \qquad I_3 = 200/70\ \Omega \qquad I_3 = 2.86\ A$$
$$I_T = I_1 + I_2 + I_3 \qquad I_T = 4\ A + 5\ A + 2.86\ A \qquad I_T = 11.86\ A$$

4.6.3   Calculating the Total Resistance for the Circuit

The total resistance for this circuit can be calculated by the formula

$$R_T = E_T/I_T \qquad R_T = 200/11.86\ A \qquad R_T = 16.86\ \Omega$$

or the formula:

$$R_T = \cfrac{1}{\cfrac{1}{Req_1} + \cfrac{1}{Req_2} + \cfrac{1}{Req_3}} \qquad R_T = \cfrac{1}{\cfrac{1}{50\ \Omega} + \cfrac{1}{40\ \Omega} + \cfrac{1}{70\ \Omega}} \qquad R_T = 16.87\ \Omega$$

You should notice a slight difference in the total resistance by using one calculation or the other, but this is due to rounding off the numbers and it really does not affect the circuit. Regardless of which method you use, the total resistance will determine the total current the circuit requires. The total resistance is determined by the connections of each load (resistor). Since the circuit looks overall like a parallel circuit, the more branch circuits that are added, the more current that is required for the loads. It should also be obvious that as more loads are added to this circuit in parallel, the total resistance will continue to drop.

4.6.4   Calculating the Voltage Drop Across Each Resistor

The voltage drop across each resistor can be calculated by determining the current through each branch circuit and multiplying it by the resistance in each branch circuit. Since the two resistors in each branch circuit are connected in series, the Ohm's law formulas will be valid. Fig. 4–10 shows the current and resistance for each resistor in each of the three branch circuits. The following formula will be used to calculate the voltage that is measured across each resistor. As you know, the voltage across each resistor is also called the *voltage drop*

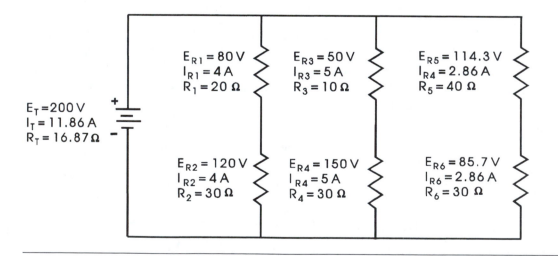

Figure 4–10   Series-parallel circuit with voltage, current, and resistance calculated for each resistor and for the total circuit.

for that resistor. The total current and total resistance is also shown in this circuit.

In the first branch circuit the current (I_1) is 4 A. When this current flows through the 20 Ω of resistor R_1, the voltage drop across R_1 is 80 V.

$$E_{R1} = I_1 \times R_1 \qquad E_{R1} = 4 \text{ A} \times 20 \text{ }\Omega \qquad E_{R1} = 80 \text{ V}$$

Since the second resistor (R_2) is connected in series with R_1, it will also see 4 A of current. The voltage drop across R_2 is 120 V. It is found by multiplying 4 A times 30 Ω.

$$E_{R2} = I_1 \times R_2 \qquad E_{R2} = 4 \text{ A} \times 30 \text{ }\Omega \qquad E_{R2} = 120 \text{ V}$$

In the second branch circuit the current is 5 A. When the 5 A current flows through the 10 Ω resistor R_3, the voltage drop across the resistor is 50 V.

$$E_{R3} = I_2 \times R_3 \qquad E_{R3} = 5 \text{ A} \times 10 \text{ }\Omega \qquad E_{R3} = 50 \text{ V}$$

Since the fourth resistor (R_4) (30 Ω) is connected in series with R_3, the current flowing through R_4 will also be 5 A. The voltage drop across R_4 is 150 V and it is found by multiplying 5 A $\times$ 30 Ω.

$$E_{R4} = I_2 \times R_4 \qquad E_{R4} = 5 \text{ A} \times 30 \text{ }\Omega \qquad E_{R4} = 150 \text{ V}$$

In the third branch circuit the current is 2.86 A. When this current flows through the 40 Ω fifth resistor (R5), the voltage drop across the resistor is 114.4 volts.

$$E_{R5} = I_3 \times R_5 \qquad E_{R5} = 2.86 \text{ A} \times 40 \text{ }\Omega \qquad E_{R5} = 114.4 \text{ V}$$

Since the sixth resistor (R_6) (30 Ω) is connected in series with R_5, the current flowing through R_6 will also be 2.86 A. The voltage drop across R_6 is 85.8 volts.

$$E_{R6} = I_3 \times R_6 \qquad E_{R6} = 2.86 \text{ A} \times 30 \text{ }\Omega \qquad E_{R6} = 85.8 \text{ V}$$

After all of the voltage current and resistance are calculated for each part of the circuit, you can begin to see patterns that will help you when you troubleshoot a series-parallel circuit that has motors and other types of loads instead of simple resistors. For example, in the previous circuit, a 30 Ω resistor was placed in each of the branches to show that in a series-parallel circuit the voltage drop across each of these resistors will not necessarily be the same. The only time the voltage drop across two resistors that have the same resistance value will be the same is when they have an equal amount of current flowing through them. This could only occur in a series-parallel circuit if the total resistance of two branch circuits is the same. The main point to remember is that it is generally possible to measure the voltage, current, and resistance of all components in a circuit. If you know some of these values, it is possible to use the Ohm's law formulas to calculate other values in the circuit to predict its behavior. These concepts will be built upon in later chapters when specific troubleshooting examples are provided.

4.7   Series-Parallel Circuits in Air-Conditioning and Refrigeration Systems

Series-parallel circuits exist in a variety of air-conditioning and refrigeration equipment. These type of circuits are also used in heating equipment. You can use the basic information that you learned about series-parallel circuits to make some simple observations about the size of wires and switches regarding the voltage and current for which they will need to be rated. Fig. 4–11 shows an example of a series-parallel circuit for an air-conditioning system. From this diagram you can see that the circuit has some switches in series with motors, and the motors are connected in parallel with each other. For example, the disconnect switch at the top of the circuit in both L1 and L2 is connected in series with the entire circuit. This means that if you open the disconnect switch, the entire circuit is disconnected.

The thermostat is connected in series with the evaporator motor and the parallel circuit that has the compressor motor connected to the condenser fan motor.

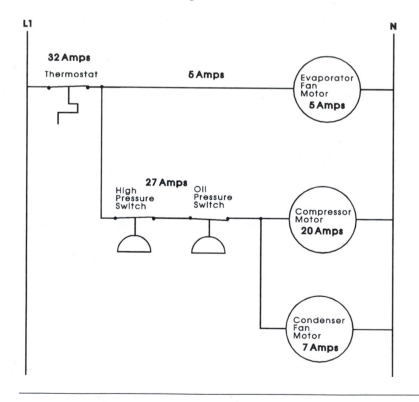

Figure 4–11   Electrical diagram of a typical series-parallel circuit for an air-conditioning system.

The evaporator motor draws 5 amps, the compressor motor draws 20 amps, and the condenser motor draws 7 amps; therefore, the thermostat must be rated for at least 32 amps because all of this current will flow through it. The high-pressure switch and oil pressure switch are connected in series with the compressor motor and the condenser motor, so they must be rated to carry the 27 amps used by these two motors. This also means the current to these motors will be interrupted anytime either of these switches is opened. All three of the motors are connected in parallel with each other so they will all receive the same amount of voltage, but the amount of current each will draw will be different unless their load characteristics are identical. The amount of total current that is needed to supply the three motors can be calculated by the formula $I_T = I_1 + I_2 + I_3$.

Questions for This Chapter

1. Explain why the voltage at each branch of a parallel circuit is the same as the supply voltage.
2. What is the advantage of using exponents when you solve Ohm's law calculations?
3. Explain what happens to the voltage at the condensor motor if the compressor motor has an open in the diagram in Fig. 4–11.
4. Determine the total current that flows through the thermostat in the diagram in Fig. 4–11.
5. Explain the prefixes M, k, m, and μ and provide an example of how each is used.

True or False

1. Current in each part of a parallel circuit will always be the same.
2. Voltage in the branches of parallel circuits will always be the same.
3. Total resistance in a parallel circuit becomes smaller as more resistors are added in parallel.
4. Total current in a parallel circuit becomes smaller as more resistors are added in parallel.
5. The prefix *milli-* (m) means that the value is one millionth.

Multiple Choice

1. Current in a parallel circuit _____
   a. increases as additional resistors are added in parallel.
   b. decreases as additional resistors are added in parallel.
   c. may increase or decrease when resistors are added in parallel depending on their size.

2. Voltage in a parallel circuit _____
   a. increases as additional resistors are added in parallel.
   b. decreases as additional resistors are added in parallel.
   c. stays the same across parallel branches as additional resistors are added in parallel.
3. Resistance in a parallel circuit _____
   a. increases as additional resistors are added in parallel.
   b. decreases as additional resistors are added in parallel.
   c. may increase or decrease when resistors are added in parallel depending on their size.
4. When one branch circuit of a multiple branch parallel circuit develops an open, voltage in other branch circuits _____
   a. decreases to zero.
   b. increases because fewer resistors are using up the voltage.
   c. stays the same as the supply voltage.
5. Electrical meters and circuits primarily use exponents 10^6, 10^3, 10^{-3}, and 10^{-6} because _____
   a. these numbers are easier to use than other exponents.
   b. these exponents are the values for the prefixes M, k, m, and μ.
   c. exponents must be multiples of 3 or 6.

Problems

Note: Please show the formula and all work for each problem.

1. Draw a circuit that has three resistors (20 Ω, 40 Ω, and 80 Ω) connected in parallel and calculate the total resistance for this circuit.
2. Calculate the current for each branch of the circuit that you made for Problem 1.
3. Calculate the total current for the circuit that you made for Problem 1.
4. Calculate the power consumed by each resistor in the circuit you made for Problem 1.
5. Determine the current capacity for each switch in the circuit for Fig. 4–11 if the evaporator motor uses 8 amps, the compressor motor uses 15 amps, and the condenser motor uses 5 amps.

5 Magnetic Theory

OBJECTIVES:

After reading this chapter, you will be able to:
1. Describe a magnet and explain how it works.
2. Explain the difference between a permanent magnet and an electromagnet.
3. Identify ways to increase the strength of an electromagnet.
4. Explain what flux lines are and where you would find them.
5. Explain electromagnetic induction.

5.0   Magnetic Theory

The theory of operation for all types of transformers, motors, and relay coils can be explained with several simple magnetic theories. As a technician who works on air-conditioning, heating, and refrigeration systems, you will need to comprehend fully all magnetic theories so that you will understand how these components operate. You must understand how a component is supposed to operate before you can troubleshoot it and perform tests to determine if it has failed. Understanding magnetic theory will make this job easier. It is also important to understand that some of the magnetic theories rely on AC voltage. These theories will be introduced in this chapter, and more detail about AC voltage and magnetic theories that use AC voltage will follow in the next chapter. In this chapter you will learn about permanent magnets first and then you will learn about electromagnets.

Magnet is the name given to material that has an attraction to iron or steel. This material was first found naturally about 4000 years ago as a rock in a city called Magnesia. The rock was called magnetite and was not usable at the time it was discovered. Later it was found that pieces of this material could be suspended from a wire and it would always orient itself so that the same ends always pointed in the same direction, which is toward the earth's North Pole and South Pole. Scientists soon learned from this phenomenon that the earth itself is magnetic. At first the only use for magnetic material was in compasses. It was many years later that the forces caused by two magnets attracting or repelling could be utilized as part of a control device or motor.

As scientists gained more knowledge and equipment became available to study magnets more closely, a set of principles and laws evolved. The first of these discoveries showed that every magnet has two poles that are called the north pole and south pole. When two magnets are placed end to end so that similar poles are near each other, the magnets repel each other. It does not matter if the poles are both north or both south, the result is the same. When the two magnets are placed end to end so that the south pole of one magnet is near the north pole of the other magnet, the two magnets attract each other. These concepts are called the first and second laws of magnets.

When sophisticated laboratory equipment became available, it was found that this phenomenon is due to the basic atomic structure and electron alignments. When the atomic structure of a magnet was studied, it was found that its atoms were grouped in regions called *domains* or *dipoles*. In material that is not magnetic or that cannot be magnetized, the alignment of the electrons in the dipoles is random and usually follows the crystalline structure of the material. In material that is magnetic, the alignments in each dipole are along the lines of the magnetic field. Since each dipole is aligned exactly like the ones next to it, the magnetic forces are additive and are much stronger. In material where the magnetic forces are weak, it was found that the alignment of the dipole was random and not along the magnetic field lines. The more closely this alignment is to the magnetic field lines, the stronger the magnet is. Today we refer to a piece of soft iron that has all of its dipoles aligned as a *permanent magnet*. The name *permanent magnet* is used because the dipoles remain aligned for very long periods of time, which means the magnet will retain its magnetic properties for long periods of time. Fig. 5–1a shows a diagram of nonmagnetic metal that has its dipoles randomly placed, and Fig. 5–1b shows a piece of metal that is magnetic and has its dipoles aligned to make a strong magnet.

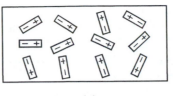

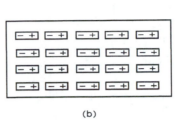

Figure 5–1a Example of metal that has its dipoles randomly placed, which makes it a very weak magnet or is not magnetic at all.
b Example of metal that has its dipoles aligned to make a very strong magnet.

(a)

(b)

5.1   A Typical Bar Magnet and Flux Lines

Fig. 5–2 shows a bar magnet that is made of soft iron that has been magnetized. The magnet is in the shape of a bar, and it has its north and south poles identified. Since the bar remains magnetized for a long period of time, it is called a permanent magnet. The magnet produces a strong magnetic field because all its dipoles are aligned. The magnetic field produces invisible *flux lines* that move from the north pole to the south pole along the outside of the bar magnet. The diagram in Fig. 5–2 shows these flux lines are lines of force that are in a slight arc as they move from pole to pole.

Since the flux lines are invisible, you will need to perform a simple experiment to allow you to see that the flux lines do exist and what they look like as

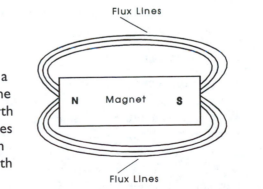

Figure 5–2 Example of a bar magnet. Notice that the poles are identified as north (N) and south (S). Flux lines are shown emanating from the south pole to the north pole.

they surround the bar magnet. For this experiment you will need a piece of clear plastic film such as the plastic sheets used for overhead transparencies, and some iron filings. Place the plastic sheet over a bar magnet. Make the plastic as flat as possible and sprinkle iron filings over it. The filings will be attracted by the invisible flux lines as they extend in an arc from the north pole to the south pole along the outside of the magnet. Since the flux lines begin at one pole and stretch to the other, the highest concentration of flux lines will be near the poles. The iron filings will also concentrate around the poles, but a definite pattern of flux lines can be seen along each side of the bar magnet. If an overhead projector is available, the image of the flux lines can be projected onto a projector screen or blackboard so that they can be seen more easily. The pattern of these filings will look similar to the diagram in Fig. 5–2. The number of flux lines around a magnet is directly related to the strength of the magnet. A stronger magnet will have more flux lines than a weaker magnet. The strength of a magnet's field can be measured by the number of flux lines per area. Since the strength of a magnet's field is based on the alignment of the magnetic dipoles, the number of flux lines will increase as the alignment of the magnetic dipoles increases.

Some materials, such as Alnico and Permalloy, make better permanent magnets than iron, since the alignment of their magnetic domains (dipoles) remains consistent even after repeated use. You may find these materials used in some expensive controls and motors, but normally the permanent magnet will be made of soft iron. The reason permanent magnets are useful in many types of controls, especially in motors and generators, is because the soft iron produces residual magnetism for long periods of time over many years. Permanent magnets have several drawbacks for use in some applications. One of these problems is that the magnetic force of a permanent magnet is constant and cannot be turned off if it is not needed. This means that if something is attracted to a magnet, it will remain attracted until it is physically removed from the force of the flux lines. Another problem with a permanent magnet's flux field being constant is that it cannot easily be made stronger or weaker if circumstances so require.

5.2   Electromagnets

An electromagnet is made by connecting a coil of wire to an electric cell (battery). The electromagnet has properties that are similar to a permanent magnet. When a wire conductor is connected to the terminals of the battery, current will begin to flow and magnetic flux lines will form around the wire like concentric circles. If the wire was placed near a pile of iron filings while current was flowing through it, the filings will be attracted to the wire just as if the coil were a permanent magnet. Fig. 5–3 shows several diagrams indicating the location of magnetic flux lines around conductors. Fig. 5–3a shows flux lines will occur

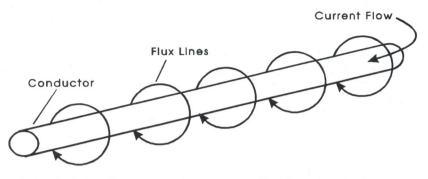

(a)   Few flux lines around conductor that is not coiled

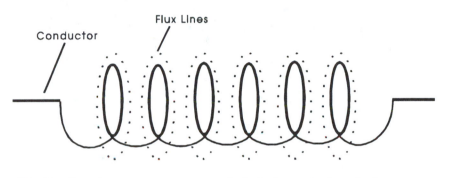

(b)   Flux lines become more concentrated when wire is coiled

---

Figure 5–3a   Example of flux lines around a straight wire that is carrying current. **b** Flux lines around a coil of wire that is carrying current. Notice that the number of flux lines increases when the wire is coiled.

around any wire when current is flowing through it. You can set up several simple experiments to demonstrate these principles. In one experiment you can insert a current-carrying conductor through a piece of cardboard and place iron filings around the conductor on the cardboard. When current is flowing in the wire, the filings will settle around the conductor in concentric circles showing where the flux lines are located. As the amount of current is increased, the number of flux lines will also increase. The flux lines will also concentrate closer and closer to the wire until the current reaches *saturation.* When the flux lines reach the saturation point, any additional increase of current in the wire will not produce any more flux lines.

When the straight wire is coiled up, the flux lines will concentrate and become stronger. Fig. 5–3b shows an example of flux lines around a coil of wire that has current flowing through it. Since the flux lines are much stronger in a coil of wire, most of the electromagnets that you will encounter will be in the form of coils. For example, coils are used in transformers, relays, solenoids, and motors.

One advantage an electromagnet has over a permanent magnet is that the magnetic field can be energized and deenergized by interrupting the current flow through the wire. The strength of the magnetic field can also be varied by varying the strength of the current flow through the conductor that is used for the electromagnet. This theory is perhaps the most important theory of magnetism since it will be used to change the strength of magnetic fields in motors, which causes the motor shaft to produce more torque so it can turn larger loads. Since the flux lines are invisible you may need to perform an experiment to prove that the magnetic field becomes stronger as current flow through the coil is increased. Fig. 5–4 shows how to set up this experiment so you can demonstrate it. Wrap a wire into a coil and connect the ends to a dry cell battery and place it near a pile of iron filings. Add a variable resistor to the circuit to increase or decrease the current. When the current is set to a minimum as in Fig. 5–4a, the magnetic field around the coil of wire will attract only a few filings. As the current is increased as in Fig. 5–4b, the number of filings the magnetic field will attract also increases until the current causes the magnetic field in the wire to reach *saturation*. When the saturation point is reached, any additional current flowing in the wire will not produce additional flux lines. When a switch is added to this circuit, the magnetic field can be turned on and off by turning the switch on and off to interrupt the flow of current in the coil. When the switch is opened, current is interrupted and no flux lines are produced so the magnetic field will not exist.

Components such as electromechanical relays and solenoids will use the principle of switching the magnetic field on and off. When current flows in the coil, the magnetic field will cause the contacts in the relay or the valve in the solenoid to close. When the current to the coil in these devices is interrupted and the magnetic field is turned off, springs will cause the relay contacts or valves to open. More information about relays and solenoids is provided in later chapters.

This principle is also used to turn motors on and off. When current flows through the coils of a motor, its shaft will turn. When current is interrupted, the magnetic field will diminish and the motor will stop rotating. This point is also important to remember if the wire that is used in the coil develops an open. When current stops flowing in the coil due to the open circuit, the shaft of the motor will stop turning.

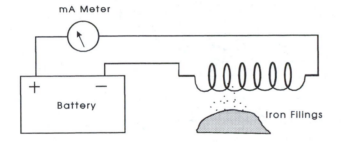

a) Few filings are attracted when small current passes
through the conductor.

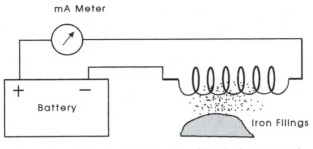

(b) Large number of filings are attracted as current
is increased.

Figure 5–4a Small amount of current flowing in a coil of wire creates a
small number of flux lines that attract iron filings. **b** Large amount of current
flowing in a coil of wire creates a large number of flux lines that attract iron
filings.

5.3   Adding Coils of Wire to Increase the Strength of an Electromagnet

Another advantage of the electromagnet is that its magnetic strength can be increased by adding coils of wire to the original single coil of wire. The increase of the magnetic field occurs because the additional coils of wire require a longer length of wire to be used, which provides additional flux lines. The magnetic field will be stronger when the coil is more tightly wound because the flux lines are more concentrated. This means that very fine wire is used in some electro-

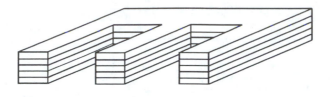

Figure 5–5   Example of pieces of steel pressed together to form a laminated steel core for use in electromagnets.

magnets to maximize the number of coils. You must be aware that as smaller wire is used, the amount of current flowing through it must be reduced so that the wire is not burnt open.

You will learn that some motors will use this principle to increase their horsepower and torque rating. These motors will have more than one coil that can be connected in various ways to affect the torque and speed of the motor's shaft. *Torque* is defined as the amount of rotating force available at the shaft of a motor. You will also learn that coils can be connected in series or in parallel to affect the torque and speed of a motor.

5.4   Using a Core to Increase the Strength of the Magnetic Field of a Coil

The strength of a magnetic field can also be increased by placing material inside the helix of the coil to act as a core. The farther the core is inserted into the coil, the stronger the flux field becomes. When the core is removed completely from the coil, it is considered to be an *air coil magnet* and the magnetic field is at its weakest point. If a soft iron is used as the core, it will strengthen the magnetic field, but it also creates a problem because it has excessive residual magnetism, which is unwanted. Residual magnetism means the core will retain magnetic properties when current is interrupted in the coil, which will make it like a permanent magnet. This problem can be corrected by using laminated steel for the core. The laminated steel core is made by pressing sheets of steel together to form a solid core. Fig. 5–5 shows an example of layers of laminated steel pressed together to form a core. When current flows through the conductors in the coil, the laminated steel core enhances the magnetic field in much the same way as the soft iron, and when current flow is interrupted, the magnetic field collapses rapidly because each piece of the laminated steel will not retain sufficient magnetic properties.

5.5   Reversing the Polarity of a Magnetic Field in an Electromagnet

When current flows through a coil of wire, the direction of the current flow through the coil will determine polarity of the magnetic field around the wire. The polarity of the magnetic field around the coil of wire is important because it determines the direction a motor shaft turns in an AC or DC motor. If the direction of current flow is reversed, the polarity of the magnetic field will be reversed, and the direction a motor shaft is turning will be reversed. In some motors used in heating, air-conditioning, and refrigeration systems, such as fan motors and pump motors, the direction of rotation is very important. In these applications you will be requested to make changes to the connections for the windings in the motor or to the supply voltage for a three-phase motor to make the motor rotate in the opposite direction. You should understand that the changes you are making are taking advantage of changing the direction of current flow through a coil or changing the polarity of the supply voltage with respect to the other phases so that the motor will reverse its rotation.

As you are learning advanced theories about motors and other electromagnetic components, diagrams will be presented to explain more complex concepts. When these diagrams are used to explain the direction of current flow through a wire, a method of identifying the direction of current flow has been developed and universally accepted. Fig. 5–6a shows a diagram that indicates the location of the flux lines in a coil of wire and shows the direction the flux lines travel around a straight current-carrying conductor. A dot or a cross (X) is used to mark the conductor to indicate the direction in which current is flowing. In Fig. 5–6a you should notice that a dot is used to indicate that current is flowing toward the observer. The flux lines in this diagram show their flow is in a clockwise motion. In Fig. 5–6b, a cross is used to indicate that current is flowing away from the observer and the flux lines are shown moving in the counterclockwise direction. If you know the direction of current flow, the advanced theories will allow you to determine the polarity of the magnetic field and predict the direction a motor's shaft will rotate.

Fig. 5–6c shows another way to determine the directions of the flux lines, the direction of current flow, and which pole is the north pole. This method is called the *left-hand rule*. From this diagram you can see that you need to know either the direction of the current flow or the direction of the flux lines to determine the polarity of the coil. Normally the direction of the current flow is easy to determine with a voltmeter by determining which end of the wire is positive and which is negative. Current (electron flow) is from negative to positive.

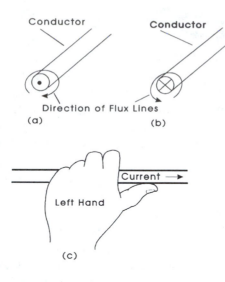

Figure 5–6a   The direction of flux lines around a wire when current is flowing in the wire toward you. The "dot" indicates the direction of current flow is toward the observer. **b** The direction of flux lines around a wire when current is flowing in the wire away from you. The "X" indicates current is flowing in the wire away from the observer. **c** An example of the left-hand rule. The thumb is pointing in the direction of the current flow in the wire, and the fingers are pointing in the direction of the flux lines as they move around the wire when current is flowing.

5.6   An Overview of AC Voltage

Before you can understand more about magnetic theories, you will need to know more about AC voltage and current. This section will provide an overview of AC voltage, and the next chapter will provide an in-depth explanation of AC voltage. *AC stands for alternating current.* This name is derived from the fact that AC voltage alternates positive for one half of a cycle and then it alternates negative for one half of a cycle. Fig. 5–7a shows the characteristic AC waveform. The AC waveform shown in this figure is called a *sine wave.*

The AC waveform is created by an alternator that looks similar to a motor. Fig. 5–7b shows a diagram of an alternator. The alternator has a rotating coil of wire that is mounted on a shaft and stationary coil that has current flowing through it so that it creates a magnetic field. The shaft of the alternator is rotated by an energy source such as a steam turbine. When the rotating coil of wire passes through the magnetic field, an electron flow is created in the coil of wire. Since

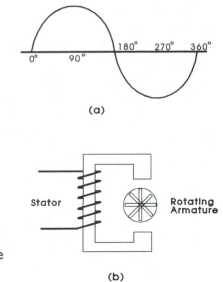

(a)

Figure 5–7a A sine wave for AC voltage. **b** A simple alternator that has a moving coil and a stationary coil. The alternator produces the AC sine wave.

Stator

Rotating Armature

(b)

the coil of wire rotates, it will pass the positive magnetic field and then the negative magnetic field during each complete rotation of 360°. This action causes the voltage sine wave to be produced and the sine wave is also identified as having 360°. In the United States, the speed of rotation of the alternator is maintained at a constant rate so that the sine wave will have a frequency of 60 cycles per second, which is also called 60 hertz (Hz).

5.7   Electromagnetic Induction

Electromagnetic induction is another magnetic principle that is used in transformer and motor theory. This theory states that if two individual coils of wire are placed in close proximity to each other, and a magnetic field is created in one of the coils so that flux lines are created around it, the flux lines will collapse across the second coil when the current is interrupted in the first coil. When these flux lines collapse across the coils of the second coil, they will cause a current to begin to flow in the second coil. This current flow will be 180° out of phase with the current flow that created the magnetic field in the first coil.

Since AC voltage periodically builds voltage to a positive peak and interrupts it when the sine wave reaches 180°, it is ideal to create the magnetic field that builds and collapses. When the current flows, it causes the flux lines to build in the coil and when the current flow is interrupted, it causes the flux lines to collapse. Since AC voltage in the United States is generated at 60 hertz, the magnetic field in the first coil will build and collapse 60 times a second. The flux lines will also build and collapse across the windings of the second coil 60 times a second.

One component that uses two coils is called a transformer. The first coil in the transformer is called the *primary winding,* and the second coil in the transformer is called the *secondary winding.* The ratio of the number of turns of wire in the primary winding and the secondary winding will determine the ratio of the voltage in the primary winding and secondary winding. Another important point to remember is that since the two coils are electrically isolated, the voltage produced in the second coil by induction is totally isolated from the voltage in the first coil. The isolation allows the number of turns of wire in the two coils to be different so that a different amount of voltage can be created in the second coil. If the number of turns of wire in the first coil and second coil are identical, the amount of voltage induced in the second coil will be approximately the same as the voltage in the first coil.

In motors, electromagnetic induction is used to create a second magnetic field in a part of the motor that rotates. Since the shaft of the motor must become magnetic and rotate freely from the stationary part of the motor, it would not be practical to make a permanent connection with a wire to the rotating member so that it could receive current to create its magnetic field. Electromagnetic induction creates the second magnetic field without any connections. In future chapters you will learn more about transformer turns, ratios, and magnetic theories for motors.

5.8   Review of Magnetic Principles

When you are learning about components that utilize electromagnetic principles, you can return to this section to review the basic concepts about them. The important electromagnetic principles follow:

1. A magnet has its dipoles aligned.
2. Every magnet has a north pole and a south pole.
3. Like magnetic poles repel, and unlike magnetic poles attract.
4. A permanent magnet is useful because its magnetic field is residual and can remain strong over a number of years.
5. When current passes through a conductor, magnetic flux lines will form around the conductor in concentric circles.
6. The strength of the magnetic field around a conductor will be proportional to the amount of current flowing through it until the amount of current reaches the saturation point of the electromagnet.

7. The magnetic field of an electromagnet can be turned on and off by interrupting the current flow through the coil of wire.
8. The polarity of an electromagnetic coil is determined by the direction of current flow through the conductor.
9. If the conductor is wrapped in coils, the strength of the magnetic field is increased.
10. When more coils of wire are added to an electromagnet, the magnetic field strength is increased.
11. A core will cause the field of an electromagnet to be strengthened.
12. The core must be made from laminated steel when AC voltage is used to power the coil so that the magnetic field can quickly build and collapse in the core as the AC voltage changes polarity every 1/60 of a second.

Questions for This Chapter

1. Explain the difference between a permanent magnet and an electromagnet.
2. Define the term *saturation* as it applies to an electromagnet.
3. Explain the principle of electromagnetic induction.
4. Identify two ways to increase the strength of an electromagnet.
5. Explain why it is important that the magnetic field of an electromagnet can be turned on and off.

True or False

1. When dipoles are aligned in a material, it is a good magnet.
2. Anytime current flows through a conductor, magnetic flux lines are created around the conductor.
3. The polarity of the magnetic field of an electromagnet is determined by the direction of current flow through the conductor.
4. Like poles of magnets attract.
5. The strength of the magnetic field for a permanent magnet can be changed easily.

Multiple Choice

1. _____ can have the strength of its field changed easily.
   a. A permanent magnet
   b. An electromagnet
   c. A dipole

2.   The polarity of an electromagnet can be changed by _____

   a.   changing the direction of current flow through the coil of wire.
   b.   changing the amount of current flow through the coil of wire.
   c.   changing the frequency of the current flow through a coil of wire.

3.   A laminated steel core is used in electromagnets that have AC voltage applied to them because _____

   a.   laminated steel is more economical to use than soft iron.
   b.   laminated steel is easily formed to any shape, which makes it more usable in complex components.
   c.   laminated steel allows the magnetic field to build and collapse quickly.

4.   A magnetic field is created in a coil of wire when _____

   a.   current flows through the wire.
   b.   current stops flowing through a wire.
   c.   the coil has a core.

5.   The left-hand rule is used to _____

   a.   determine the amount of magnetic flux in a coil.
   b.   determine the number of coils in an electromagnet.
   c.   determine the polarity of the magnetic field in an electromagnet.

Problems

1.   Draw a sketch of a permanent magnet and show where flux lines occur.
2.   Draw a sketch of an electromagnetic coil and show where flux lines occur.
3.   Draw a sketch of an electromagnetic coil with an open somewhere in the coil of wire. Explain what effect the open will have on the magnetic field.
4.   Draw two permanent magnets with their poles placed so the two magnets will attract each other.
5.   Perform the experiment presented in Section 5–1 and explain why the flux lines are concentrated around the ends of the permanent magnet.

6 Fundamentals of AC Electricity

OBJECTIVES:

After reading this chapter, you will be able to:

1. Explain the term *alternating current.*
2. Explain the terms *peak-to-peak* and *rms voltage.*
3. Calculate the frequency and period of an AC sine wave.
4. Explain the effects of a capacitor and inductor in an AC circuit.

6.0   What Is Alternating Current?

This section will review some of the principles about AC voltage that you learned in the previous chapter. *Alternating current* (AC) is identified by its characteristic waveform. The AC waveform is shown in Fig. 6–1. The sine wave has a positive half-cycle and a negative half-cycle. The device that creates AC electricity is called an alternator. The alternator has coils of wire mounted on its rotating part, which is called the rotor. The alternator also has coils of wire that are used to create strong magnetic fields. These coils of wire are mounted in the stationary part of the alternator, which is called the stator. The magnetic field in the stator, like all other magnets, has flux lines that move between its north and south poles. When the shaft of the alternator is rotated, the coils of wire in the rotor

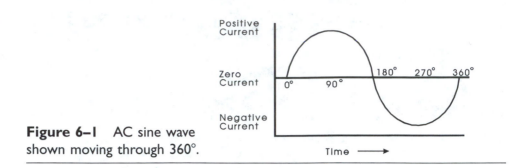

Figure 6–1 AC sine wave
shown moving through 360°.

move past the flux lines from the strong magnetic field in the stator. Since the coils pass the north pole of the magnetic field and then the south pole, they create a sine wave in the rotor which becomes the AC voltage. This voltage is used to power air conditioners and other electrical components. AC voltage is the primary voltage used in heating, air-conditioning, and refrigeration systems. AC electricity is generated by utility companies and transmitted to commercial and residential users.

6.1   Where Does AC Voltage Come From?

AC voltage is generated at a number of power stations around the country. The voltage that is used in your city is generated somewhere nearby within a range of 600 miles. The energy source that turns the rotor of the alternator to produce this voltage can come from burning coal, burning fuel oil, or burning other fossil fuels to make steam. In some areas of the country the steam is generated from nuclear power stations. The steam is used to turn a turbine wheel that turns an alternator shaft. In other parts of the country the turbine wheel is turned by water that is stored behind hydroelectric dams. When the alternator shaft turns, AC voltage is produced.

   The voltage that is produced at large utility power stations is generated from three equal fields in the alternator. This causes the generated voltage to be *three-phase voltage*. The voltage is sent to transformers at the generating station where it is stepped up to several hundred thousands volts so it can be transmitted over long distances. When the voltage arrives at a city, it is transformed down (stepped down) to a value of approximately 40,000 volts. When the voltage reaches an industrial or commercial area, the voltage is stepped down again to 480 or 230 volts. The voltage that is used in residential areas is stepped down to 230 volts.

6.2   Frequency of AC Voltage

The most important feature of AC voltage that is different from DC voltage is that AC voltage has a frequency. The typical frequency for AC voltage in the

United States and much of North America is 60 hertz (60 Hz). The frequency of 60 Hz is determined by the speed the alternator rotates when the voltage is generated. The frequency of AC voltage that is used to supply voltage for heating, air-conditioning, and refrigeration equipment is constant. *Frequency* is defined as the number of cycles (sine waves) that occur in 1 second. Fig. 6–2 shows a number of sine waves that occur in 1 second.

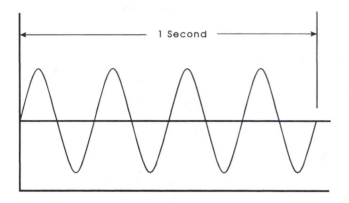

Figure 6–2   The frequency of AC voltage is calculated from the number of cycles that occur in 1 second.

The *period* of a sine wave is the time it takes one sine wave to start from zero point and pass through 360°, as seen in Fig. 6–3. This represents one complete cycle and it is one complete revolution of the alternator. Since the shaft of the alternator rotates one full circle to produce the sine wave, we will equate one complete cycle to 360°. This means that the sine wave can be described in terms of 360°. In Fig. 6–3 you can see the sine wave starts at 0°, reaches the positive

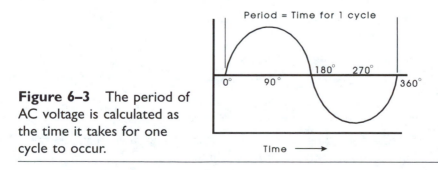

Figure 6–3   The period of AC voltage is calculated as the time it takes for one cycle to occur.

peak at 90°, returns to zero at 180°, then reaches the negative peak at 270°, and finally returns to zero volts at 360°. The number of degrees will be used to identify points of the sine wave in future discussions. The period of a sine wave can also be described as $P = \dfrac{1}{freq}$, and frequency can be defined as $F = \dfrac{1}{period}$.

EXAMPLE 6.1

Determine the period of a 60 Hz sine wave.

Solution:

Use the formula $P = \dfrac{1}{F}$ (where F = frequency).

$$P = \frac{1}{F} = \frac{1}{60} = 0.016 \text{ second}$$

♦

6.3   Peak Voltage and rms Voltage

Fig. 6–4 shows a sine wave with the point where peak voltage occurs. *Peak voltage* is the highest point of the sine wave. From this figure you can see that the peak voltage is 60 volts for this example. The negative peak voltage is also identified as minus 60 (−60) volts in this example. The total voltage peak to peak (PP) will be 120 volts. The peak voltage can only be measured with a peak-reading voltmeter or an oscilloscope.

The voltage that you will read with a VOM meter is called rms voltage. The term *rms* stands for root mean square, which is the mathematical function used to calculate the voltage that a typical rms meter will read. The major difference

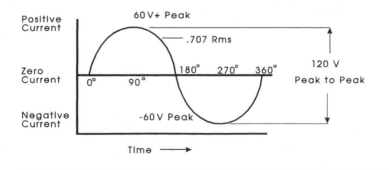

Figure 6–4   Sine wave with positive and negative peak identified. The peak-to-peak (P-P) voltage is also identified.

between root-mean-square voltage and peak voltage is that the meter compensates for the rms voltage being less than peak at various times in any given cycle. The formula for calculating rms voltage from peak voltage (V_P) is:

$$\text{rms} = 0.707 \times \text{peak voltage}$$

As you can see, the rms voltage is approximately 70% of the peak voltage. This means that if the peak voltage was 100 volts, the rms voltage would be 70 volts. In the example in Fig. 6–4, the peak voltage is 120, so the rms voltage is 84.84 V.

EXAMPLE 6–2

Calculate the rms voltage if the peak voltage is 156 volts.

Solution:

$$\text{rms} = 0.707 \times \text{peak}$$
$$\text{rms} = 0.707 \times 156$$
$$\text{rms} = 110 \text{ V}$$ ◆

EXAMPLE 6–3

Find the peak voltage if rms voltage is 120 volts.

Solution:

$$\text{Peak} = 1.414 \times \text{rms}$$
$$\text{Peak} = 1.414 \times 120$$
$$\text{Peak} = 169.68 \text{ volts}$$ ◆

6.4   The Source of AC Voltage for Heating, Ventilating, and Air-Conditioning Systems

When you are working on a job in a residence or in a commercial building, you will need to locate the source of incoming voltage. The source of incoming voltage at these locations is typically a *circuit breaker panel*. The circuit breaker panel is also called a *load center,* and it is shown in Fig. 6–5. The load center in a commercial building is typically three phase and the load center in a residential building is typically single phase.

Voltage from the load center is sent to a disconnect box that will be mounted near the air-conditioning or refrigeration system you are working on. Fig. 6–6 shows a picture of a disconnect box. If the disconnect box has fuses, it will be called a fused disconnect. The disconnect is used to turn power off to the unit when you are working on the equipment. The fuses in the disconnect also provide protection against overcurrent for the equipment that is connected to it.

Figure 6–5   A typical load center. *(Courtesy of Challenger Electrical Equipment Corp.)*

6.5   Measuring AC Voltage in a Disconnect

When you are installing or troubleshooting equipment in the field, you will need to measure the amount of AC voltage available at the load center or at the disconnect. You must be extremely careful when you take these readings because full voltage will be applied to the system at the time you are taking the measurement. Fig. 6–7 shows the points in the disconnect where you should place the

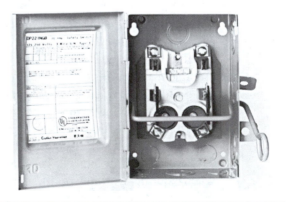

Figure 6–6   A fused disconnect that is mounted near the air-conditioning or refrigeration equipment. *(Courtesy of Challenger Electrical Corp.)*

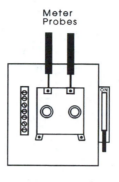

(a)

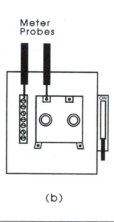

Figure 6–7a Location to place meter probes for reading 230 volts on the line side of a fusible disconnect. **b** Location to place meter probes for reading 115 volts on the line side of a fusible disconnect.

(b)

meter probes to measure high voltage and low voltage. Fig. 6–7a shows the points to place the meter leads when you are reading voltage line to line (L1 to L2). From this figure you can see that if the meter probes are placed on line 1 (L1) and line 2 (L2), incoming terminals and the applied voltage will be 230 volts if you are in a residence. If the fused disconnect is located in a commercial location, you may read 208 volts when you place the meter on L1 and L2 or you may measure 480 volts. These terminals are called the *line-side terminals*.

Fig. 6–7b shows the meter probes on line 1 (L1) and neutral. The meter should measure 115 volts if the disconnect is in a residence. This is the lower voltage that is used on smaller equipment. You may measure a higher voltage such as 208 V or 240 V in the disconnect in some cases.

The previous voltage measurements are made at the line-side terminals of the disconnect. You will need to make a second set of measurements at the *load-side terminals* that are located at the bottom of the disconnect. If you have voltage at the line-side, but not at the load-side terminals, the disconnect may have a blown fuse. In some cases you can determine that the fuse is blown if you can measure full voltage across the top and bottom of a fuse. The reason you can read full voltage across the fuse is that it has an open, and you are reading voltage that feeds back from the bottom of the fuse through a load such as a transformer winding to the other supply voltage line. If you suspect you have a faulty fuse, you can remove one or both of them for continuity.

6.6   Voltage and Current in AC Circuits

Voltage waveform and current waveform are in phase in an AC circuit that has only resistance in it. This means that both of the waveforms start at the same point in time. If the load in the circuit is a resistance heating element, the voltage waveform and current waveform will be in phase with each other. If a capacitor or inductor is used in an AC circuit such as a motor circuit, the voltage waveform and current waveform will not be in phase. Fig. 6–8 shows the voltage identified by the letter *v* and current waveform that is identified by the letter *i*. You can see that these waveforms are in phase because they start at the same point in time and end each of their cycles at the same time. The voltage in this circuit is larger than the current so the voltage waveform is shown larger than the current waveform in this example.

The important part about the voltage and current waveforms being in phase is that this means there are no losses in this circuit due to inductive reactance or capacitive reactance. If an AC circuit has an inductor in it, such as a motor coil, it will have *inductive reactance,* and if the circuit has a capacitor in it such as a start capacitor for a motor, it will have *capacitive reactance. Reactance* is an opposition to the AC circuit that is similar to the opposition in a DC circuit caused by resistance. *Inductive reactance* is the opposition caused by the inductor and *capacitive reactance* is the opposition caused by capacitors. If a circuit has inductors (transformer windings or motor windings) or capacitors, the voltage and current waveforms for that circuit will have a phase shift. The amount of phase

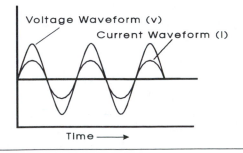

Figure 6–8 Voltage waveform (*v*) and current waveform (*i*) are shown in this example.

shift can be defined as a number of degrees or it can be defined as the total amount of reactance (opposition) in the circuit. As a technician you need to know that reactance exists in these types of circuits and that the effects of reactance can be calculated, predicted, and used to an advantage. For example, the more phase shift that can be created between the start winding and the run winding of an induction motor, the higher starting torque the motor will have. This is useful for helping larger compressor motors to start. The next sections will help you understand the theory about capacitive reactance and inductive reactance. Even though you may never complete reactance calculations when you are troubleshooting a compressor motor, you must comprehend the theory about capacitive reactance and inductive reactance to understand why additional capacitance may be required to make a compressor motor start more easily. You will also need a basic understanding of reactance to learn about electronic control circuits that are designed to make compressor motors and fan motors run more efficiently.

6.7   Resistance and Capacitance in an AC Circuit

In the previous section we learned that when capacitors or inductors are used in an AC circuit, they will create an opposition that is similar to resistance. The opposition caused by a capacitor is called *capacitive reactance,* and the opposition caused by an inductor is called *inductive reactance.* The combined opposition caused by capacitive reactance, inductive reactance, and resistance in an AC circuit is called *impedance.* The main difference between capacitive reactance, inductive reactance, and resistance is that a phase shift between the voltage and current waveform occurs. Fig. 6–9a shows a diagram of a capacitor in an AC circuit with a resistor. Fig. 6–9b shows the voltage and current waveforms for the this circuit, and Fig. 6–9c shows the vector diagram that is used to calculate the amount of phase shift for this diagram. A vector diagram is a graph that is derived from a trigonometric calculation that is used to determine phase angle.

When voltage encounters a capacitor in a circuit, it will take time to charge up the capacitor and then discharge it. This causes the reactance, which becomes an opposition to the voltage waveform. The capacitor does not bother the current waveform. In Fig. 6–9b you can see that the voltage waveform *lags* or starts later than the current waveform for this circuit.

6.8   Calculating Capacitive Reactance

The total amount of opposition caused by the capacitor is called capacitive reactance and it has the symbol X_C. Even though you may never need to calculate capacitive reactance, you should understand the effects of changes in capacitance

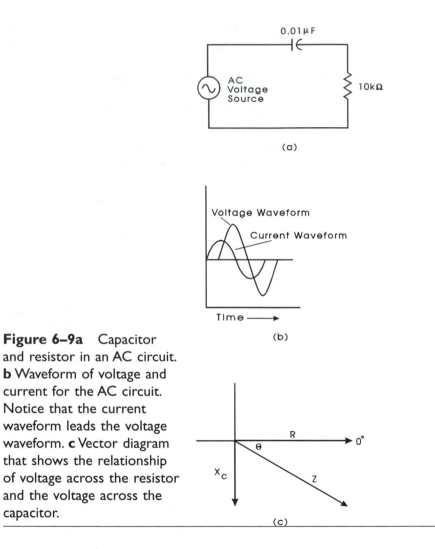

Figure 6–9a Capacitor and resistor in an AC circuit. **b** Waveform of voltage and current for the AC circuit. Notice that the current waveform leads the voltage waveform. **c** Vector diagram that shows the relationship of voltage across the resistor and the voltage across the capacitor.

and frequency on the amount of capacitive reactance. Capacitive reactance can be calculated by the following formula:

$$X_C = \frac{1}{2 \pi F C}$$

where $\pi = 3.14$

$F = $ frequency

$C = $ capacitance in microfarads

From this formula you can see that you must know the value of the capacitor and the frequency of the AC voltage. For example, if a 40 μF capacitor is used in a 60 Hz circuit, the amount of opposition (capacitive reactance) would be 66.35 ohms. Notice that since the capacitive reactance is an opposition, its units are ohms. The example of this problem follows:

$$X_C = \frac{1}{2\,\pi\,F\,C} = \frac{1}{2 \times 3.14 \times 60 \times 0.000040} = 66.35 \; \Omega$$

EXAMPLE 6.4

Calculate the capacitive reactance of a 60 Hz AC circuit that has a 90 μF capacitor.

Solution:

$$X_C = \frac{1}{2\,\pi\,F\,C} = \frac{1}{2 \times 3.14 \times 60 \times 0.000090} = 29.49 \; \Omega$$

◆

6.9   Calculating the Total Opposition for a Capacitive and Resistive Circuit

The amount of total opposition (impedance) caused by a capacitor and resistor in an AC circuit can be calculated. Fig. 6–10c shows the diagram that is used to determine the impedance for this type of circuit. For example, if a circuit has 70 ohms of resistance and 40 ohms due to capacitive reactance, the total impedance must be calculated with a trigonometric formula because the voltage and current in this circuit are out of phase. The formula for calculating impedance of this circuit is:

$$Z = \sqrt{R^2 + X_C^2} \qquad Z = \sqrt{70^2 + 40^2} \qquad Z = 80.62 \; \Omega$$

6.10   Resistance and Inductance in an AC Circuit

When inductors (coils of wire) are used in an AC circuit, they will create an opposition that is similar to resistance. The opposition caused by a inductor is called *inductive reactance*. The main difference between capacitive reactance, inductive reactance, and resistance is that a phase shift between the voltage and current waveforms occurs. Fig. 6–10a shows a diagram of an inductor in an AC circuit with a resistor. Fig. 6–10b shows the voltage and current waveforms for this circuit, and Fig. 6–10c shows the vector diagram that is used to calculate the amount of phase shift for this circuit.

When current encounters an inductor in a circuit, it will take time to charge up the inductor and then discharge it. This causes the reactance, which becomes an opposition to the current waveform. The inductor does not bother the voltage waveform. In Fig. 6–10b you can see that the current waveform *lags* or starts later than the voltage waveform for this circuit.

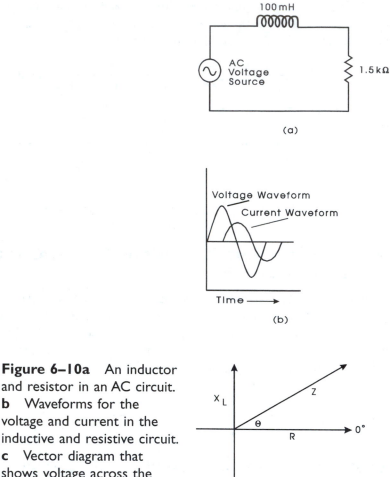

Figure 6–10a An inductor
and resistor in an AC circuit.
b   Waveforms for the
voltage and current in the
inductive and resistive circuit.
c   Vector diagram that
shows voltage across the
inductor leading the voltage
across the resistor.

6.11   Calculating Inductive Reactance

The total amount of opposition caused by the inductor is called inductive reactance and it has the symbol X_L. Like capacitive reactance, inductive reactance can be calculated, and you should remember that even though you will not calculate reactance in the field when you are troubleshooting, it is important that you understand the effects that changing the size of the inductor or changing frequency have on the total amount of inductive reactance. Inductive reactance can be calculated by the following formula:

$$X_L = 2 \pi F L$$

where
$$\pi = 3.14$$
$$F = \text{frequency}$$
$$L = \text{inductance in henries (H)}$$

From this formula you can see that you must know the value of the inductor and the frequency of the AC voltage. For example, if an 80 H inductor is used in a 60 Hz circuit, the amount of opposition (inductive reactance) would be 30,144 ohms (30.144 KΩ). Notice that since the inductive reactance is an opposition, its units are ohms. The example of this problem follows:

$$X_L = 2 \pi F L = 2 \times 3.14 \times 60 \text{ Hz} \times 80 = 30,144 \ \Omega$$

EXAMPLE 6–5

Calculate the inductive reactance of a 60 Hz AC circuit that has a 20 H inductor.

Solution:

$$X_L = 2 \pi F L = 2 \times 3.14 \times 60 \times 20 = 7536 \ \Omega \qquad \blacklozenge$$

6.12   Calculating the Total Opposition for an Inductive and Resistive Circuit

The amount of total opposition (impedance) caused by an inductor and resistor in an AC circuit can be calculated. Fig. 6–10c shows the diagram that is used to determine the impedance for this type of circuit. For example, if a circuit has 70 ohms of resistance and 50 ohms due to inductive reactance, the total impedance must be calculated with a vector diagram because the voltage and current in this circuit are out of phase. The formula for calculating impedance of this circuit is:

$$Z = \sqrt{R^2 + X_L^2} \qquad Z = \sqrt{70^2 + 50^2} \qquad Z = 86.02 \ \Omega$$

6.13   True Power and Apparent Power in an AC Circuit

In a DC circuit you could simply multiply voltage times the current and determine the total power of the circuit. In an AC circuit you must account for the current caused by any resistors and calculate it separately from the power caused by resistors and capacitors, or resistors and inductors. The reason for this is that when current is caused by a resistor, it is called *true power* (TP), and current caused by capacitive reactance and inductive reactance is called *apparent power* (AP). Apparent power is so called because it does not take into account the phase shift caused by the capacitor or inductor. The power that occurs when current flows through a resistor is called true power (TP) because the resistor does not cause a phase shift between the voltage waveform and the current waveform.

The main point to remember is that true power can be used to determine the heating potential in an AC circuit. This means that if you have 1000 watts due to true power, you can determine that you can get 1000 watts of heating power from this circuit.

6.14   Calculating Power Factor

The amount of true power and the amount of apparent power in an AC circuit combine to become a ratio that is called *power factor* (PF). The formula for power factor is:

$$PF = \frac{TP}{AP}$$

EXAMPLE 6–6

Calculate the power factor for a circuit that has 600 VA of apparent power and 500 watts of true power.

Solution:

$$PF = \frac{TP}{AP} = \frac{500 \text{ watts}}{600 \text{ VA}} = *.830 \text{ or } 83\%$$

The true power in a circuit will always be smaller than the apparent power, so the power factor value will always be less than 1. In a pure resistance circuit, the true power and the apparent power will be the same, so the power factor will be 1. When the power factor becomes too low, the power company accesses a penalty on the electric bill that will increase the bill substantially. When a large retail store, grocery store, or restaurant has an electric bill of over $20,000 per month, this penalty becomes important. In these types of applications, the power factor can be corrected by adding extra capacitance to an inductive circuit (having large or multiple motors), or by adding extra inductance to a capacitive circuit. Commercial power factor correction systems are available for these applications.

Questions for This Chapter

1. Draw a diagram of an AC sine wave and identify its peak voltage and peak-to-peak voltage levels.
2. Explain the term *frequency* as it refers to AC electricity.
3. Discuss the difference between a load center and a fused disconnect.

4. Explain how you would test for voltage in a fusible disconnect on the line-side and load-side terminals.
5. Trace the path of a generated AC voltage from the generating station where it is produced to the point where you would connect an air conditioner at a home in a residential area. Be sure to identify all of the important parts of the system between the generation station and the residence.

True or False

1. Impedance is the total opposition to an AC circuit caused by inductive reactance, capacitive reactance, and resistance.
2. The units of impedance are ohms.
3. A load center is the point where incoming voltage is connected in a residence and where circuit breakers are mounted.
4. Rms voltage is the voltage your VOM voltmeter measures.
5. If the rms voltage in an AC circuit is 110 volts, the peak voltage would also be 110 volts.

Multiple Choice

1. If the size of a capacitor in an AC circuit increases from 10 μF to 20 μF and the frequency stays the same, the capacitive reactance will

    _____

    a. increase.
    b. decrease.
    c. remain the same.
2. When a capacitor is added to an AC circuit, the voltage waveform will _____ the current waveform, which provides additional torque to a motor if it is connected to the circuit.

    a. lag
    b. lead
    c. remain in phase with
3. If a circuit has two large refrigeration or air-conditioning compressor motors and has a low power factor, the power factor can be raised by _____ to make the true power equal to the apparent power, which will avoid a penalty on the electric bill.

    a. adding capacitance
    b. removing capacitance
    c. adding inductance

4. The frequency of AC voltage is defined as _____
   a. the period of the voltage.
   b. the number of cycles per second.
   c. the peak voltage of the voltage.
5. The number of cycles per second in an AC voltage is called the _____ of the voltage.
   a. frequency
   b. period
   c. impedance

Problems

1. Calculate the rms voltage for an AC sine wave that has 80 volts peak.
2. Calculate the peak voltage for an AC sine wave that has 208 volts rms.
3. Calculate the period of a 60-cycle AC sine wave.
4. Calculate the frequency of an AC sine wave that has a period of 16 milliseconds.
5. Calculate the impedance of an AC circuit that has 40 ohms of resistance and 30 ohms of capacitive reactance.

7 Transformers: Three-Phase and Single-Phase Voltage

OBJECTIVES:

After reading this chapter, you will be able to:
1. Explain the operation of a transformer.
2. Discuss the difference between step-up and step-down transformers.
3. Explain how 440, 220, and 110 VAC are developed from a delta-connected transformer.
4. Explain how 440, 208, and 125 VAC are developed from a wye-connected transformer.

7.0   Overview of Transformers

Transformers are needed to step up or step down voltage levels. For example, when voltage is generated, it needs to be stepped up to several hundred thousand volts so its current will be lower when it is transmitted. Since the current is lower, the size of wire used to transmit the power can be smaller so the power can be transmitted over longer distances. When the voltage arrives at a city, it needs to be stepped down to approximately 40,000 volts so that it is less dangerous. This level of voltage is sufficient to transmit power throughout a city. When the voltage arrives at a residence, it must be further stepped down to 230 volts and 115 volts for all of the power circuits in the home. Transformers provide a means to step up or step down this voltage. Step-up transformers are used in oil furnaces to increase the line voltage (120 VAC) to several thousand volts to send to the igniter to cause the spark for ignition.

Another type of transformer is also used in heating and air conditioning equipment to step down 120 VAC or 230 VAC to 24 volts. The 24 VAC is used as control voltage for a residential heating or air-conditioning system so that higher voltages are not used in thermostats or other controls that people touch when they are setting them. The lower voltage provides a degree of safety against electrical shock and it also allows these controls to last longer. This type of transformer is called a control transformer and examples of a foot-mounted transformer and flange-mounted transformer are shown in Figs. 7–1a and 7– 1b.

As a technician you will encounter different voltages such as 440, 220, 208, 120, and 24 VAC while you are working on air-conditioning, heating, or refrigeration equipment. The equipment you are working with will require a specific amount of single-phase or three-phase voltage and you will be responsible to go to the disconnect or load center where supply voltage is provided and make the proper connections to provide the correct voltage. When the proper amount of voltage is not present, you will need to know the transformers that are required to step the voltage up or down to the proper level. This chapter will explain how transformers operate and how the various levels of voltage are derived from con-

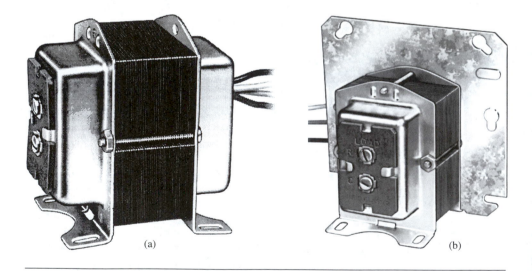

(a)　　(b)

Figure 7–1a　A foot-mounted control transformer that is used in residential heating and air-conditioning equipment. **b** A flange-mounted control transformer. *(Courtesy of Honeywell)*

necting transformers to provide the proper amount of three-phase and single-phase voltage. The chapter will also explain how to take voltage readings at the transformer and disconnect to troubleshoot the loss of a phase.

7.1 Operation of a Transformer and Basic Magnetic Theory

The transformer consists of two windings (coils of wire) that are wrapped around a laminated steel core. The winding where voltage is supplied to a winding transformer is called the primary winding. Line 1 (L1) and neutral are connected to the primary terminals. The winding where voltages come out of the transformer is called the secondary winding. The terminals for the secondary winding are identified by the letters R and C. The letter R stands for *red,* which is the color of wire that is connected to it and the letter C stands for *common.* Fig. 7–2 shows a diagram of a typical transformer with the primary coil and secondary coil identified. From this figure you can see that the windings are also identified.

A transformer works on a principle called *induction.* Induction occurs when current flows through the primary winding and creates a magnetic field. The magnetic field produces flux lines that emanate from the wire in the coil as current is flowing through it. When this current is interrupted or stopped, the flux lines will collapse and the action of the collapsing flux lines will cause them to pass through the winding of the secondary coil that is placed adjacent to the primary coil. You should remember that the AC sine wave continually starts at zero volts and increases to a peak value and then returns to zero volts and repeats the waveform in the negative direction. The process of increasing to peak and returning to zero provides a means to create flux lines in the wire as current passes through it, and then allowing the flux lines to collapse when the voltage returns to zero. Since the AC voltage follows this pattern naturally 60 times a second, it makes the perfect type of voltage to operate the transformer.

When the AC voltage returns to zero during each half-cycle, the flux lines

Figure 7–2 Diagram of a typical control transformer. The coil where incoming voltage is applied is called the primary, and the other coil is called the secondary.

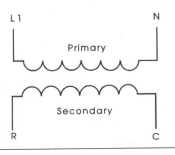

that were created in the primary winding begin to collapse and start to cross the wire that forms the secondary coil of the transformer. When the flux lines from the primary winding collapse and move across the wire in the coil of the secondary winding, the electrons in the secondary winding begin to move, which creates a current. Since the current in the secondary winding begins to flow without any physical connection to the primary winding, the current in the secondary winding is called an *induced current.*

The following provides a more technical explanation of the relationship between the AC voltage and the windings of the transformer. The magnetic field in the primary winding of the transformer builds up when AC voltage is applied during the first half-cycle of the sine wave (0° to 180°). When the sine wave voltage reaches its peak at 90°, the voltage has peaked in the positive direction and it begins to return to 0 volts by moving from the 90° point to the 180° point. When the voltage reaches the 180° point, the voltage is at 0 volts and it creates the interruption of current flow. When the sine wave voltage is 0 volts, the flux lines that have been built up in the primary winding collapse and cross the secondary coils, which creates a current flow in the secondary winding.

The sine wave continues from 180° to 360° and the transformer winding is energized with the negative half-cycle of the sine wave. When the sine wave is between 180° and 270°, the flux lines build again, and when the sine wave reaches the 360° point, the sine wave returns to 0 volts, which again interrupts current and causes flux lines to cross the secondary winding. This means the transformer primary energizes a magnetic field and collapses it once in the positive direction and again in the negative direction during each sine wave. Since the sine wave in the secondary is not created until the voltage in the primary moves from 0° to 180°, the sine wave in the secondary is *out of phase by 180°* to the sine wave in the primary that created it. The voltage in the secondary winding is called *induced* voltage because it is created by induction.

The induced voltage is developed even though the primary coil of the transformer does not make electrical contact with the secondary coil in any way. Fig. 7–3a–d shows the four stages of voltage building up in the transformer and the flux lines building and collapsing at each point as the sine wave flows through the primary coil.

7.2   Connecting a Transformer to a Disconnect for Testing

An easy way to test a control transformer is to connect it directly to a disconnect and apply power to the transformer primary circuit. When voltage is applied to the primary circuit, voltage should be available at the secondary. The amount of

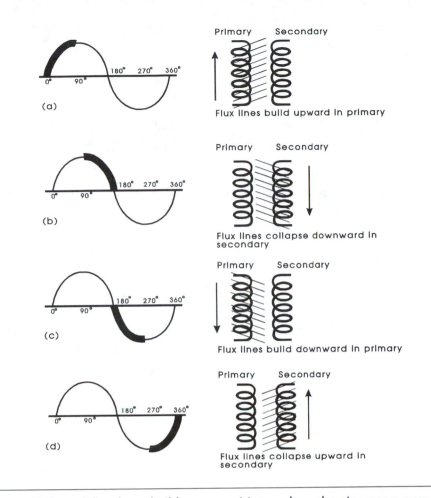

Figure 7–3a AC voltage builds to a positive peak as the sine wave moves from 0–90°. Flux lines are shown building in the primary coil during this time. **b** AC voltage drops off from peak at 90° to 180°. During this time, the flux lines in the primary coil begin to collapse and cut across the coils of the secondary. **c** AC voltage builds to a negative peak as the sine wave moves from 180° to 270° and the flux lines build in the primary coil again. **d** When voltage decreases from the negative peak back to zero as the sine wave moves from 270° to 360°, the flux lines collapse.

voltage at the secondary will depend on the rating of the transformer. Fig. 7–4 shows a transformer connected to T1 and N, which is the source for 115 volts at the bottom terminals of the disconnect. The secondary voltage available at this transformer at terminals R and C will be 24 volts. Fig. 7–5 shows a similar transformer connected to terminals T1 and T2 in a disconnect to provide 230 volts to the primary circuit. Since this transformer is rated as 230/24, the secondary voltage will also be 24 volts.

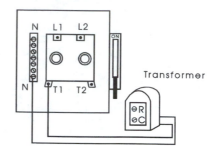

Fused Disconnect

Figure 7–4 A control transformer connected to T1 and N in a disconnect to test the transformer. The primary voltage will be 115 volts AC and the secondary voltage will be 24 volts AC.

7.3 Transformer Rated for 230 Volts and 208 Volts Primary

If the air-conditioning equipment is used in a residential application, the supply voltage is normally 230 volts and if the equipment is used in a commercial application, the supply voltage may be 230 or 208 volts depending on the transformer connections that are used by the utility company that supplies the power. Since the equipment manufacturer does not know where its equipment will be used, it may provide a control transformer that can be wired to either 230 volts or 208 volts. This is accomplished by providing a second connection that is *tapped* in the primary winding. Fig. 7–6 shows an example of this type of control transformer.

7.4 Transformer Voltage, Current, and Turns Ratios

The amount of voltage a transformer will produce at its secondary winding for a given amount of voltage supplied to its primary is determined by the ratio of the number of turns in the primary winding compared to the number of turns in the

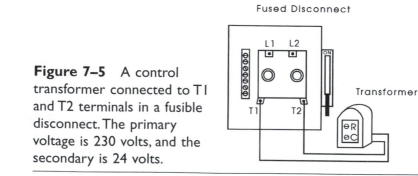

Fused Disconnect

Figure 7–5 A control transformer connected to T1 and T2 terminals in a fusible disconnect. The primary voltage is 230 volts, and the secondary is 24 volts.

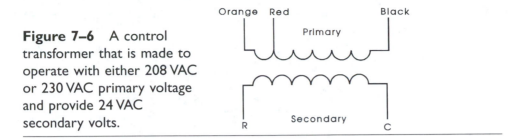

Figure 7–6   A control transformer that is made to operate with either 208 VAC or 230 VAC primary voltage and provide 24 VAC secondary volts.

secondary. This ratio is called the *turns ratio*. The amount of primary current and secondary current in a transformer is also dependent on the turns ratio.

Fig. 7–7 shows a diagram that indicates the primary voltage (E_p), the primary current (I_p), and the number of turns in the primary winding (T_p), the secondary voltage (E_s), the secondary current (I_s), and the number of turns in the secondary winding (T_s).

The primary and secondary voltage, primary and secondary current, and the turns ratio can all be calculated with the following formulas. The turns ratio for a transformer is calculated from the following formula:

$$\text{Turns ratio} = \frac{T_s}{T_p} \quad \text{or} \quad \frac{E_s}{E_p} \quad \text{or} \quad \frac{I_p}{I_s}$$

The ratio of primary voltage, secondary voltage, primary turns, and secondary turns is:

$$\frac{E_p}{E_s} = \frac{T_p}{T_s}$$

From this ratio you can calculate the secondary voltage with this formula:

$$E_s = \frac{E_p \times T_s}{T_p}$$

Figure 7–7   Diagram of a transformer that shows the primary voltage (E_p), primary current (I_p), and the number of turns in the primary winding (T_p). This diagram also shows the secondary voltage (E_s), secondary current (I_s), and the number of turns in the secondary (T_s).

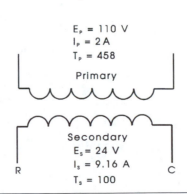

E_p = 110 V
I_p = 2 A
T_p = 458

Primary

Secondary
E_s = 24 V
I_s = 9.16 A
T_s = 100

The ratio of primary voltage, secondary voltage, primary current, and secondary current is:

$$\frac{E_p}{E_s} = \frac{I_s}{I_p}$$

(Notice that the ratio of voltage to current is an inverse ratio.)

From this ratio you can calculate the secondary current with this formula:

$$I_s = \frac{E_p \times I_p}{E_s}$$

Using these formula, you can see that if you were provided the information that the primary voltage is 110 V, the primary current is 2 A, the primary turns are 458, and the secondary turns are 100, you could easily calculate the secondary voltage, and the secondary current.

$$E_s = \frac{E_p \times T_s}{T_p} \qquad \frac{110 \text{ V} \times 110 \text{ turns}}{458 \text{ turns}} = 24 \text{ V}$$

$$I_s = E_p \times I_p E_s \qquad \frac{110 \text{ V} \times 2 \text{ A}}{24 \text{ V}} = 9.16 \text{ A}$$

EXAMPLE 7–1

Calculate the secondary current and secondary turns of the transformer shown in Fig. 7–8.

Solution:

Find the turns ratio:

$$\frac{E_p}{E_s} = \frac{240 \text{ V}}{24 \text{ V}} = 10 : 1$$

Since the number of turns in the primary is 458, the number of turns in the secondary will be 45.8.

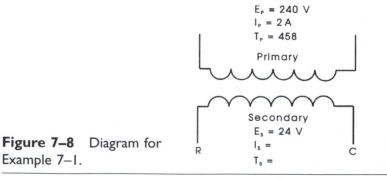

Figure 7–8 Diagram for Example 7–1.

The secondary current can be calculated from the formula:

$$I_s = \frac{E_p \times I_p}{E_s} \qquad \frac{240\text{ V} \times 2\text{ A}}{24\text{ V}} = 20\text{ A}$$

$$I_s = \frac{E_p \times I_p}{E_s}$$

◆

7.5   Step-up and Step-down Transformers

If the secondary voltage is larger than the primary voltage, the transformer is known as a *step-up transformer.* If the secondary voltage is smaller than the primary voltage, the transformer is known as a *step-down transformer.* The control transformer that was presented in the previous sections is an example of a step-down transformer because the secondary voltage is smaller than the primary voltage. Step-up transformers are used to boost the secondary voltage, which also has the effect of lowering the secondary current so that more power can be transferred on smaller-size wire. An application in heating and air conditioning where a step-up transformer is used is in the ignition system of an oil furnace. The oil furnace uses an ignition transformer that is a step-up transformer to increase the 120 VAC that is used to power the furnace to approximately 16,000 volts. The high voltage is supplied to the igniter points so that the high voltage can jump the gap between the ignition points to create a spark that will ignite the fuel oil as it comes out of the nozzle.

7.6   VA Ratings for Transformers

The VA (voltampere) rating for a transformer is calculated by multiplying the primary voltage and the primary current, or the secondary voltage and the secondary current. In the example in Fig. 7–8 you can see that the primary voltage is 240 volts and the primary current is 2 A. The VA rating for this transformer is 480 VA. The VA rating indicates how much power the transformer can provide. As a technician you can say that the primary VA is equal to the secondary VA. If you were a circuit designer, you would need to be more precise and you would find that the secondary VA is slightly less than the primary VA because of transformer losses. It is important that the VA rating of the transformer is large enough for the application. If the VA rating is too small, the transformer will be damaged and fail prematurely. Anytime you need to replace a transformer you must be sure the voltage ratings match and that the VA size of the replacement transformer is equal to or larger than the original transformer.

7.7   ## The 110 VAC Controls Transformer

Another type of control transformer provides 110 VAC in the secondary for use in commercial heating, air-conditioning, and refrigeration systems. You should remember that utility companies in different areas of the country provide voltage that is between 110 and 120 volts. This means that one part of the country may refer to their lower voltage as 110, while another part calls their voltage 115 or 120 volts. For this section the voltage will be referred to as 110 VAC. The 110 VAC secondary voltage is necessary because some of the control devices require 110 VAC and the larger loads in the system such as compressor and fan motors require 240 VAC. The control transformer allows the compressor and fan motors to be supplied with 240 volts while it provides the control circuit with 110 VAC.

The 240/110 VAC type of control transformer operates on a principle similar to the 110/24 VAC transformer. The diagram for this transformer is also provided in Fig. 7–9. From this diagram you can see that the primary windings of this transformer are identified with the letter H and the secondary windings are identified with the letter X. The primary winding for this transformer is constructed in two equal sections (windings). The first winding has its terminals identified as H1 and H2. The second winding has its terminals identified as H3 and H4. The primary winding is broken into two separate windings so that the transformer primary side can be powered with 480 volts or 240 volts. If these two sections are connected in parallel with each other, the primary side of the transformer will be powered with 480 volts, and if the two sections are connected in series with each other, the transformer will be powered with 240 volts.

7.8   ## Wiring the Control Transformer for 480 VAC Primary Volts

The control transformer can be powered with 480 VAC by connecting its two primary windings in series. Fig. 7–10 shows three diagrams of the control transformer. The first diagram in Fig. 7–10a shows the transformer as you would see it on a wiring diagram without any jumpers connected. You should notice that the H2 and H3 terminals are positioned so that they can be connected either in series or in parallel. In this application the jumper is connected between terminals H2 and H3 to connect them in series. Fig. 7–10b shows the physical location of the jumper when it is connected across these terminals. Fig. 7–10c shows the equivalent electrical diagram as you may see it in the electrical diagram of the air-conditioning unit. The reason the two windings are connected in series is because each winding is rated for 240 volts, and the primary voltage is 480 VAC. Since the windings are connected in series, each of them will receive 240 volts of the total 480 volts.

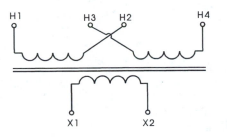

Figure 7–9   Diagram of a control transformer whose secondary voltage is 110 volts. The primary winding of this transformer is identified as H1, H2, H3, and H4. The secondary of this transformer is identified as X1 and X2.

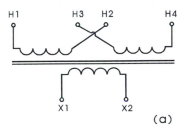

(a)

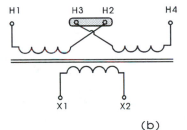

(b)

Figure 7–10a   Electrical diagram of the control transformer. Notice that the H2 and H3 terminals are located so that a jumper can be placed on them to connect them in series or in parallel. **b** Diagram showing the jumper in place connecting the two primary windings in series. **c** The equivalent electrical diagram that shows the two windings connected in series.

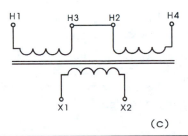

(c)

7.9 **Wiring the Control Transformer for 240 VAC Primary Volts**

The control transformer can also be connected to operate with 240 VAC primary volts and produce 110 VAC on the secondary terminals. Fig. 7–11 shows a set of three diagrams for the same control transformer that is connected for 240 VAC primary. Fig. 7–11a shows the electrical diagram of the control transformer as you would see it in a circuit diagram without any jumpers. The reason this diagram is shown is that some equipment manufacturers will show the control transformer in the electrical diagram for the air-conditioning system since they are not sure whether the system voltage will be 480 or 240. This diagram indicates that the control transformer is to be connected by the technician when the unit is

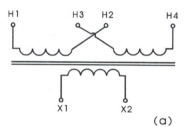

(a)

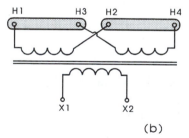

(b)

Figure 7–11a Electrical diagram of a control transformer. **b** Diagram that shows the jumpers in place for the transformer to have 240 VAC primary. **c** Equivalent diagram that shows the two primary windings connected in parallel.

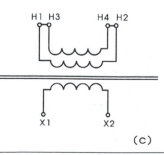

(c)

installed and the supply voltage can be verified. Fig. 7–11b shows the diagram with two jumpers in place to connect the two primary windings in parallel with each other. This diagram will help you to connect the jumpers to the correct terminals so that the windings are connected in parallel. Fig. 7– 11c shows the equivalent circuit of the two windings in parallel as you would see them in the system electrical diagram provided by the equipment manufacturer. Sometimes it is difficult to see that the connections of the two jumpers actually connect the two windings in parallel, so the diagram in Fig. 7–11c shows how the windings look when they are in parallel.

As a technician you will be responsible to make the proper jumper connections based on the primary voltage that the system will have. If the jumpers on the transformer are connected for 480 VAC primary when the unit is shipped, and the unit is connected to 240 volts, you will need to make the changes in the jumper so that the secondary winding of the transformer provides exactly 110 VAC.

7.10   Troubleshooting a Transformer

As a technician you will need to troubleshoot a variety of transformers. If the transformer you are testing is not connected to power, you can test each of the windings with an ohmmeter for continuity. Each of the coils should have some amount of resistance that indicates the amount of resistance in the wire that is in each coil. If you measure infinite resistance (∞), it indicates the winding has an open, and if you measure 0 Ω, it indicates one of the windings is shorted.

Another way that you can test a transformer is by applying power to the primary winding. It is important that you provide the continuity test first to detect any shorts before you apply power. If the transformer windings are not shorted, you can apply any amount of AC voltage that is equal to or less than the primary rating of the transformer. If the transformer is operational, a voltage will be present at the secondary terminals of the transformer. If no voltage is present at the secondary, check all of the connections to ensure they are correct. If no voltage is present at the secondary, the transformer is defective and should be replaced. It is important to remember that the transformer is a primary winding and a secondary winding that are placed in close proximity to each other and when AC voltage is applied to one winding, induction will cause voltage to be available at the other winding.

The transformer may have a problem where its secondary voltage is higher or lower than its rating. If this occurs, it is possible that the amount of primary voltage is incorrect or that the transformer jumpers are not connected properly. For example, if the primary voltage is 208 volts instead of 240 volts, the secondary voltage will be less than 110 volts, which will cause a problem. Be sure that the secondary is the proper amount before you use the transformer in a circuit.

7.11   Nature of Three-Phase Voltage

Air-conditioning and refrigeration equipment installed in commercial and industrial buildings may require three-phase AC voltage instead of single-phase voltage. Three-phase voltage is usually indicated on equipment data plates by the symbol 3φ voltage. Three-phase voltage is generated as three separate AC sine waves. Fig. 7–12 shows an example of the three sine waves. From this figure you can see that the sine waves are identified as Phase A, Phase B, and Phase C. Each of the phases is separated by 120° from each other. The 120° phase separation is caused by the generator windings being 120° out of phase with each other. Three-phase voltage is generated and transmitted because it is more efficient to produce than single phase, since the generator can have three separate windings.

7.12   Why Three-Phase Voltage Is Generated

Three-phase voltage is used more often than single-phase voltage for larger commercial air-conditioning and refrigeration systems because it provides more power in smaller components than single-phase voltage for an equal-size system. For example if a 10 hp (horsepower) motor is required as the air-conditioning compressor, a single-phase motor would need to be physically larger than the 10 hp three-phase motor because it has three sets of windings and it receives three equal sources of voltage (L1, L2, and L3) where a single-phase motor would receive only two sources of power. Since the power is shared by three circuits instead of two, the wire sizes, fuse sizes, and switch sizes are also smaller for a three-phase system.

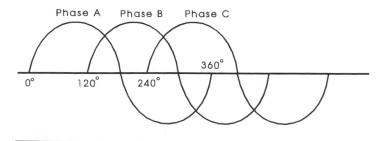

Figure 7–12   Sine waves for three-phase electricity. Notice that Phase A is 120° out of phase with Phase B, and Phase B is 120° out of phase with Phase C.

For example, if a 10 hp motor is connected to a single-phase 230 V power source, each line (L1 and L2) would need to be capable of providing 50 amps. A three-phase 10 hp motor connected to a three-phase power distribution would only need 28 amps from each wire. This means that the three-phase system could use much smaller wire, which would be lighter and less expensive.

7.13   Three-Phase Transformers

Fig. 7–13 shows a picture of several examples of three-phase transformers you will encounter on the job. The transformer may be mounted near the air-conditioning and refrigeration equipment, or it may be located in a transformer vault (a special room where transformers are mounted). In some cases the transformers are mounted on a utility pole just outside of the commercial site. Fig. 7–14 shows a three-phase transformer with its cover removed so you can see three separate transformer windings. The three-phase transformer operates exactly like a single-phase transformer in that AC voltage is applied to the primary side of the transformer and induction will cause voltage to be created in the sec-

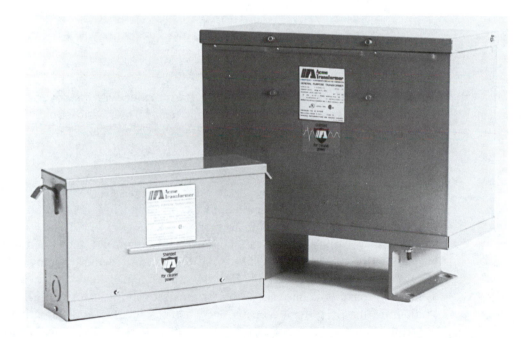

Figure 7–13   Typical three-phase transformers with their covers in place.
(Courtesy of Acme Electric, Acme Transformer Division)

Figure 7–14   A typical three-phase transformer with its cover removed so that you can see the three independent transformers and connection terminals. *(Courtesy of Acme Electric, Acme Transformer Division)*

ondary winding. In the case of the three-phase transformer, the 120° phase shift of the three-phase voltage applied to the primary windings will be maintained in the secondary side of the transformer.

When you are working on a commercial system, you may need to make connections on the secondary side if the transformer to the disconnect box, or between the secondary side of the transformer and a load center. You may also need to make voltage tests at the terminals of the transformer. You can use the diagram that is provided on the side of the transformer to identify the correct terminals to use. In most cases, you will not be requested to make any electrical connections at the primary side of the transformer, since the primary voltage may be rather high (1300–1700 VAC). The electrical technicians for the electric utility company or technicians from a high-voltage service company will make the primary voltage connections for the transformer. You will be expected to make the connections for the air-conditioning or refrigeration equipment at the disconnect or at the load center.

7.14   The Wye-Connected, Three-Phase Transformer

Fig. 7–15 shows a diagram of a wye-connected, three-phase transformer. You should notice the physical shape of the transformer windings looks like the letter Y, which gives this type of configuration its name. You should also notice that this diagram only shows the secondary coils of the transformer. It is traditional to show only the primary-side connections or only the secondary-side connections when discussing a transformer power distribution system, since showing both the primary and secondary connections in the same diagram tends to become confusing.

The amount of voltage measured at L1-L2, L2-L3, and the L3-L1 terminals on the secondary winding will be 208 volts for the wye-connected transformer if its turns ratio is set for low voltage. If the turns ratio of the transformers is set for high voltage, the amount of voltage between each terminal is 480 volts. The voltage indicated between each of the windings for the transformer shown in Fig. 7–15 is 208 volts, which indicates turns ratio for this transformer will provide the lower voltage. If 480 volts is needed, a transformer with a higher turns ratio would be used. The primary and secondary voltage for each transformer is provided on its data plate and can be specified when the transformer is purchased and installed.

7.15   The Delta-Connected, Three-Phase Transformer

Fig. 7–16 shows a diagram of a delta-connected, three-phase transformer. You should notice that the shape of the transformer windings in this diagram looks like a triangle (Δ). This shape is the Greek letter D which is named *delta*. This diagram also shows only the secondary side of the three-phase transformer.

Figure 7–15   A diagram that shows the secondary windings of a three-phase transformer connected in a wye configuration. The voltage available at L1-L2, L2-L3, and L3-L1 is 208 volts for this transformer.

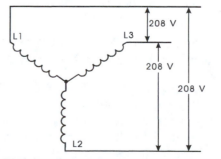

Figure 7–16 Electrical diagram of the secondary windings of a transformer connected in delta configuration. Notice that the voltage between L1-L2, L2-L3, and L3-L1 is 230 volts.

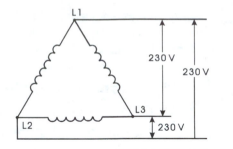

The amount of voltage measured at L1-L2, L2-L3, and L3-L1 is 230 volts for the delta-connected transformer if it is wired for its lower voltage. If the delta-connected transformer is wired for its higher voltage, the voltage between L1-L2, L2-L3, and L3-L1 is 480 volts. If the delta-connected transformer is wired for its lower voltage, it is very easy to differentiate it from a wye-connected transformer. If the delta-connected transformer and wye-connected transformer are connected for their higher voltage, you cannot tell them apart, since both will provide 480 volts between their terminals.

It is important to understand at this point that you will purchase the air-conditioning and refrigeration equipment so that its voltage requirements match the voltage supplied by the transformers at the commercial or industrial site. This means that you do not need to determine if the transformer is connected as in a delta configuration or a wye configuration. You will only need to measure the amount of voltage at the secondary of the transformer and if it is 480 volts, you will need to ensure that the air-conditioning system or refrigeration system is rated for 480 volts. If the secondary voltage is 230 volts, the air-conditioning system must be rated for 230, and if it is 208 the equipment must be rated for 208.

If you need to know if the transformer windings are connected as delta or wye, you can check the physical connections or you can make an additional voltage measurement between each line and the neutral terminal of the transformer if one is provided. The next section will explain these measurements.

7.16 Delta- and Wye-Connected Transformers with a Neutral Terminal

Fig. 7–17 shows the diagrams of a wye-connected transformer and delta-connected transformer, each with a neutral terminal. The neutral terminal on the wye-connected transformer is at the point where each of the three ends of each individual winding are connected together. This point is called the *wye point*.

You should notice that the neutral point for the delta-connected transformer is

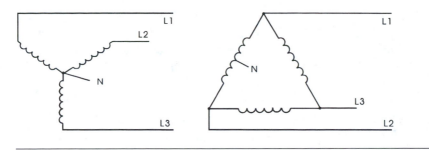

Figure 7–17   Electrical diagram of a wye-connected transformer with a neutral point and a delta-connected transformer with a neutral point.

actually the midpoint of the secondary winding that is connected between L1 and L2. This point is essentially the center tap of one of the transformer windings. Traditionally it will be the winding that is connected between L1 and L2 if a neutral connection is used with the three-phase transformer.

Fig. 7–18a shows the amount of voltage that you would measure between terminals L1-L2 of a wye-connected transformer and between terminals L1-N and L2-N. Notice that the voltage between L1-L2 is 208 volts, so this transformer is wired for the lower voltage. The voltage between L1-N and L2-N is shown as 120 volts. Fig. 7–18b shows the amount of voltage that you would measure between terminals L1-L2 and between L1-N and L2-N for a delta-connected transformer. The voltage between L1-L2 is 230 volts, so this transformer is

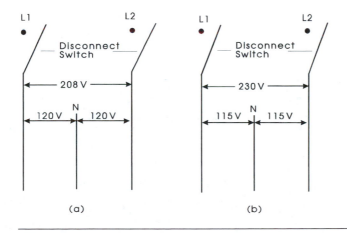

Figure 7–18a   Voltage from L1-L2 for a wye-connected transformer is 208 V, and from L1-N or L2-N is 120 V. **b** Voltage from L1-L2 for a delta-connected transformer is 230 V and from L1-N or L2-N is 115 V.

connected for its lower voltage. The voltage between L1-N and L2-N is shown as 115 volts. You should notice that since the neutral point for the delta-connected transformer is exactly halfway on the transformer winding, the voltage L1-N and L2-N will always be exactly half of the voltage between L1-L2.

This is the main difference between a wye-connected transformer and a delta-connected transformer. The voltage between L1-N and L2-N for any delta-connected transformer is always exactly half, and the L1-N or L2-N voltage for a wye-connected transformer will always be more than half the voltage L1-L2. The exact amount of voltage L1-N can be calculated by dividing the voltage between L1-L2 by 1.73, which is the square root of 3 ($\sqrt{3} = 1.73$). The square root of 3 is used because of the relationship between the phase shift of the three phases. You should notice that 208 divided by 1.73 is 120 V.

The same relationship of voltage between L1-L2 and L1-N exists when the transformers have turns ratios for their higher voltage. Fig. 7–19a shows a wye-connected voltage between L1-L2 is 480 volts, and between L1-N is 277 volts. The 277 volts can be calculated by dividing 480 by 1.73. The 277 volts that comes from L1-N or L2-N is generally used for fluorescent lighting systems in commercial buildings. Since the supply voltage originates from a three-phase transformer, the lighting system in a commercial or industrial building will also use L3-N, so that voltage is used from all three legs of the transformer. This voltage is referred to as single-phase voltage, since only one line of the three-

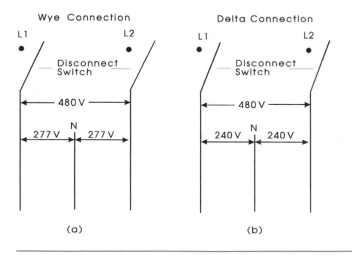

(a)

(b)

Figure 7–19a Voltage from L1-L2 for a wye-connected transformer is 480 volts and from L1-N or L2-N is 277 volts. **b** Voltage from L1-L2 for a delta-connected transformer is 480 volts and from L1-N or L2-N is 240 volts.

phase transformer is used at each circuit. For example, L1-N is a single-phase circuit.

It is also important to understand at this time that L1-L2, L2-L3, or L3-L1 could each be used to supply complete power for an air-conditioning system or refrigeration system. Even though this type of power supply would use two legs of the transformer, it is still called a single-phase power supply because only one phase of voltage is used at any instant in time. For example, if the system is powered with voltage from L1-L2, during any given half-cycle of AC voltage, the power source for the system would come from L1, and then during the next half-cycle it would come from L2. The source of power would continue to oscillate between L1 and L2 but only one phase is in use during any instant in time.

Fig. 7–19b shows the higher voltage for the delta system between L1-L2 is also 480 volts, but this time the voltage between L1-N or L2-N is 240 volts. Again the L1-N or L2-N voltage is exactly half of the supply voltage, since the neutral point on the transformer is the center tap of one of the transformers. It is important to understand that it is very easy to distinguish between a wye-connected power source and a delta-connected power source by measuring the voltage L1-L2 and L1-N. If the L1-N voltage is half of the L1-L2 voltage, the system is a delta-connected system. If the L1-N voltage is more than half, it is a wye-connected system. You should remember that the L1-N wye voltage can always be calculated by dividing the L1-L2 voltage by 1.73.

7.17   The High Leg Delta System

When a three-phase transformer system is used for the power source for an air-conditioning or refrigeration system, it may have the neutral tap. It is important to remember that a three-phase system does not need to have a neutral to operate correctly. The neutral is added only if the lower voltage (115 or 120 volts) is needed for some part of the system. Generally equipment manufacturers make all of the components in a three-phase system a higher voltage, or they may supply a small transformer inside the equipment power panel to drop the higher voltage to the necessary voltage level. For example, if the equipment needs 208 volts three phase for the compressor motor and fan motors, a control transformer can be provided in the power panel of the equipment to drop the 208 volts L1-L2 to 24 volts. The 24 volts is used to provide power for the control circuit and thermostat. Since the primary side of the control transformer can be powered by L1-L2 (208 volts), a neutral is not needed in this system. If the control transformer or fan motors required 120 volts for power, a neutral tap would be needed and these components would be connected between L1-N, L2-N, or L3-N.

If the transformer is connected as a delta transformer, it is important to understand that a different voltage becomes available between L3-N. Fig. 7–20 shows the diagram for this voltage. From this diagram you can see that the secondary winding of this three-phase transformer is connected as a delta transformer. The voltage from L1-N and L2-N is 115 volts, so the voltage between L1-L2, L2-L3, or L3-L1 is 230 volts.

The different voltage occurs between L3-N. Since the winding between L1-L2 has the center tap, it stands to reason that the voltage between L1-N or L2-N will be exactly half of the voltage L1-L2. Since the center tap is not between L3 and L1, the voltage for L3-N must come from one complete phase (230 volts) and half of the next phase (115 volts). Since this voltage uses two phases, the 230 volts and the 115 volts are out of phase, and the resultant voltage from them is 208. This voltage is not the same as the 208 voltage that occurs between L1-L2 or L2-L3 or L3-L1 of the wye-connected transformer because the neutral point is not at a midpoint.

Since the 208 volts between L3-N comes from two phases, it will cause the transformer to overheat if it is used to power any components that require 208 volts. For this reason the voltage is called the *high leg delta voltage* to indicate that it is derived from L3-N of a delta-connected transformer, and that it should not be used to power 208 V components. It is very important to understand that the L3 leg of the transformer is very usable when it is used with L1-L3, or L2-L3 as part of the three-phase system or 230 V single-phase system. The only problem occurs when the L3 terminal is used in conjunction with the neutral, which creates the L3-N voltage of 208 volts.

The L3 terminal in any power distribution box for a delta-wired system should always be marked with an orange wire to identify it as the high leg delta. In some areas of the country, the high leg delta is also called the *wild leg*. The high leg delta voltage of 208 volts may occur between L1-N if the neutral point is produced by a center tap of the L2-L3 winding, or it may occur between L2-N if the neutral point is produced by a center tap of the L1-L3 winding.

Figure 7–20 A high leg delta voltage occurs between L3-N because the center tap to produce the neutral is the center tap of the L1-L2 winding.

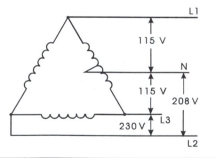

7.18   Three-Phase Voltage on Site

The on-site source for three-phase voltage will be a fusible disconnect or a load center. As you know, the voltage actually comes from the three-phase transformers, but as stated before, as an air-conditioning or refrigeration technician, you will not be expected to make connections right at the transformers. Instead technicians from the utility company will connect the wires between the transformers and the load centers or main disconnects. The load center may also be called a circuit breaker panel. Fig. 7–21 shows an example of a three-phase fusible disconnect, and Fig. 7–22 shows an example of a three-phase load center. In the picture of the three-phase disconnect, you can see the three fuses. Incoming voltage is connected to the line-side terminals at the top of the disconnect. The line-side terminals are identified as L1, L2, and L3. The disconnect has a handle on the right side that is used to turn its switch on and off. If the disconnect switch is depressed at the top, the switch is closed and the disconnect will have voltage at the load-side terminals at the bottom of the disconnect. The load-side terminals are identified as T1, T2, and T3. It is important to understand that the line-side terminals will always have power even when the disconnect switch is open (in the off position).

The three-phase load center is similar to the circuit breaker panel for a single-phase system, except it will have three individual circuits. The load center is specifically designed to mount three-phase circuit breakers or each individual circuit that will be used. This means that if you have four individual air-conditioning systems connected to the load center, each one will be connected to its own individual circuit breaker. When you are installing an air-conditioning system or refrigeration system, you will need to locate the load center and circuit breakers for the circuit you intend to use, or you must locate the fusible disconnect if one is installed near the unit location.

Figure 7–21 A three-phase fusible disconnect that is used to provide three-phase voltage to air-conditioning and refrigeration equipment in commercial and industrial applications. *(Courtesy of Challenger Electrical Equipment Corp.)*

Figure 7–22   A three-phase load center that contains the circuit breakers for the electrical system. *(Courtesy of Challenger Electrical Equipment Corp.)*

7.19   Installing Wiring in a Three-Phase Disconnect

When you are ready to connect the wires between the disconnect and the equipment you are installing, you will need to ensure that the source of three-phase voltage is turned off and a padlock is used to lock out the system. The padlock should also have a tag attached to it that has your picture and name on it to identify you as the technician who has locked out the system. The lockout and tagout procedure ensures that you can safely work on the system without anyone accidentally turning the power on when you are still working in the panel.

Fig. 7–23 shows a diagram of the locations in the disconnect where you will connect the equipment wires. The utility company will have the supply voltage wires previously installed at the three terminals at the top of the disconnect box. These terminals are called the *line-side terminals* and they are identified as L1, L2, and L3. You will make your connections at the bottom terminals that are identified as T1, T2, and T3. These terminals are called the *load-side terminals*. The other end of the wires will be connected in the power panel for the air-conditioning or refrigeration equipment. This process is called *connecting the field wiring*.

Sometimes you must get the supply voltage from a single-phase circuit breaker panel. The circuit breaker panel has single-phase voltage connected to its *main breaker* at the top of the panel as incoming voltage and it has individual circuit breakers that receive power anytime the main breaker is in the on position. You would need to add a circuit breaker for each piece of equipment for which you were providing power. Wires must be run from the circuit breaker to the incoming power terminals in the air-conditioning or refrigeration equipment.

7.20   Testing for a Bad Fuse in a Disconnect

At times you will be working on a system that receives its power from a three-phase disconnect. The system will not run and you may suspect one or more of the fuses are bad. You can check for a blown fuse by measuring the voltage or by removing the fuse and testing it for continuity with an ohmmeter. If you select to check the fuses with a voltage test, you will need to test the incoming voltage at L1, L2, and L3 to determine that the supply voltage is good. Check the voltage between terminals L1-L2, L2-L3, and L3-L1. The voltage at each of these sets of terminals should be the rated voltage for the system such as 480 volts. If you do not get the same voltage reading at each set of terminals, the incoming power has a problem and you will need to determine where the voltage has been interrupted.

The next test should be at the bottom terminals of the disconnect at T1-T2, T2-T3, and T3-T1. Each of these readings should be the same as the voltage

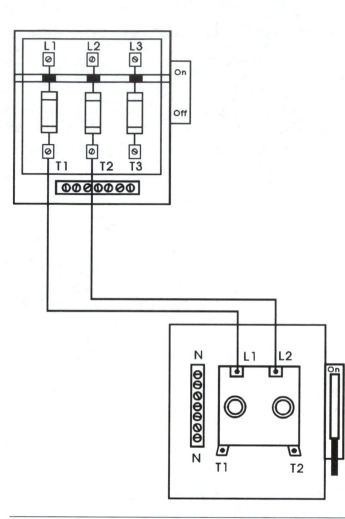

Figure 7–23 A single-phase disconnect is connected to the load-side terminals of a three-phase disconnect to provide a single-phase power source.

readings at the line side of the disconnect. If any of the voltage readings are lower than the line-side readings or 0 volts, you have an open fuse. At this point the simplest way to find the bad fuse is to turn the disconnect off and remove all of the fuses and test them for continuity. Be sure to use a fuse puller when you are removing or installing fuses in a disconnect. It is also important to remember that the line-side terminals of the disconnect are "hot" (powered) even though the disconnect switch is open.

The reason the load-side voltage may be lower than the line-side voltage rather

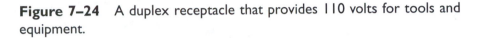

Figure 7–24　A duplex receptacle that provides 110 volts for tools and equipment.

than 0 volts is because voltage may feed back through two of the windings in any three-phase motor or three-phase transformer that is connected to the load side of the disconnect. It is difficult to determine the exact amount of voltage that is dropped when voltage feeds back through other windings, so you must test for an open fuse if the line-side voltage is not the same as the load-side voltage at each set of terminals where you take readings. If you find an open fuse, you will need to replace it with one that has the same current rating and same voltage rating. You should remember that you can also measure for voltage across each fuse and if you measure full voltage across any fuse, the fuse is open and the voltage is actually *back-feed voltage.*

7.21　Single-Phase Voltage from a Three-Phase Supply

At times you will need to provide a single-phase source of voltage for an air-conditioning or refrigeration system. The single-phase voltage can be either 480 volts, 208 volts, or 230 volts L1-L2. If the single-phase circuit requires a neutral, a neutral wire can be connected between the neutral terminal of the three-phase disconnect and single-phase disconnect, and between the three-phase disconnect and the transformer.

This type of circuit is used in many applications inside a commercial location where three-phase voltage is supplied to the system and you need to provide a single-phase voltage source to power a window-type air conditioner or a small refrigeration case. Once you understand that you can get single-phase voltage from the three-phase system, you will begin to see how the AC power distribution system works in industrial, commercial, and residential applications.

The remaining chapters of this book will continue to refer to the supply voltage as you would find it connected to the equipment at the line-side terminals. As a technician you will need to be aware of where this voltage comes from and how it is distributed so that you can trace the circuits if no voltage is present at the unit.

7.22   Wiring a Duplex Receptacle as a Utility Outlet

In some applications such as rooftop air-conditioning systems, it is necessary to provide one or more duplex receptacles so that you can plug in 110 V service-equipment like vacuum pumps, recovery pumps, and utility lights. It may also be necessary to provide power for electric tools such as drills or saws and the duplex receptacles provide a means to safely provide power for them. Fig. 7–24 shows a duplex receptacle. When you hold a duplex receptacle in your hand, you should notice that it has gold-colored screws on one side and silver-colored screws on the other side. The L1 wire should always be connected to a gold-colored screw and a neutral wire should be connected to the silver-colored screw. The ground wire should be connected to the green screw on the receptacle.

The openings in the face of the receptacle are different sizes. The larger opening on the left side of the receptacle face is connected to the neutral terminal, and the smaller opening on the right side of the terminal is connected to the L1 terminal. The round opening at the top is connected to the ground terminal. Since the openings in the face of the receptacle are different sizes, it will cause any plugs that are connected to it to be connected correctly. If a tool such as an electric drill is made of plastic, its plug will be manufactured so that it can be plugged into the receptacle either way. If the electric drill has a metal case, the plug will have a ground terminal and the plug will only connect to the receptacle one way.

When you are ready to connect L1 and N wires to the receptacle and disconnect, be sure the disconnect switch is open. It is also important to test the voltage from L1-N at the line side of the disconnect to ensure that you have 115 volts instead of the 208 volts of the high leg delta voltage. When all the connections have been made, you can close the main disconnect switch and test the two terminals of the receptacle with your voltmeter to ensure that 115 volts is present.

Questions for This Chapter

1. Explain how inductance is used in the operation of a transformer.
2. Provide an example of a step-up transformer in a heating system.
3. Discuss how it is possible to have a control transformer whose primary winding can be connected to 220 or 208 volts and provide 110 volts at its secondary.

4. Explain how you would test for 240 and 110 VAC in a fusible disconnect on the line-side and load-side terminals.
5. Explain why a control transformer is required in some air-conditioning systems.

True or False

1. If the voltage L1 to L2 is 240 volts and the voltage L1-N is 120 volts, the transformer is connected as a delta transformer.
2. The secondary voltage is larger than the primary voltage in a step-up transformer.
3. Voltage in a transformer moves from the primary winding to the secondary winding by conductance.
4. The VA rating for a transformer is determined by multiplying the voltage at the secondary winding by the current flow in the secondary winding.
5. The 208 volts from L3 to N for a high leg delta transformer is usable for compressor loads that require 208 volts.

Multiple Choice

1. The voltage at the secondary terminals of a _____ transformer will be larger than the applied voltage to the primary terminals.

   a. step-up
   b. step-down
   c. isolation
2. A transformer that is wired as a delta transformer has _____

   a. 440, 208, and 125 volts.
   b. 440, 220, and 110 volts.
   c. 440, 277, and 125 volts.
3. A transformer that is wired as a wye transformer has _____

   a. 440, 208 and 125 volts.
   b. 440, 220, and 110 volts.
   c. 480, 240, and 120 volts.
4. If a control transformer has a 2:1 turns ratio and it has 220 volts applied to its primary winding, the voltage at the secondary winding will be _____

   a. 440 volts.
   b. 220 volts because the secondary voltage is strictly controlled to equal the primary voltage.
   c. 120 volts.

5. The air-conditioning or refrigeration system is connected to the
    _____ of a fused disconnect.
    a. line-side terminals
    b. load-side terminals
    c. power company terminals

Problems

1. Calculate the secondary voltage of a transformer that has a 4:1 turns
    ratio and 110 VAC applied to the primary terminals.
2. Calculate the VA of a 24 V control transformer that requires 5 amps.
3. Determine the turns ratio of a transformer that has 4000 turns in its pri-
    mary windings and 800 turns in its secondary winding.
4. If the transformer in Problem 3 has 440 volts applied to its primary
    winding, how many volts will be measured at its secondary?
5. Calculate the amount of secondary amps for a control transformer that
    is rated for 800 VA and 110 volts at the secondary.

8 Symbols and Diagrams for HVAC and Refrigeration Systems

OBJECTIVES:

After reading this chapter, you will be able to:
1. Identify the electrical symbol for electrical components in HVAC systems.
2. Explain the difference between an electrical load and an electrical control.
3. Explain the difference between a ladder diagram and a wiring diagram.
4. Identify the components in the electrical system of an air-conditioning, heating, and refrigeration system.

8.0 Overview of Electrical Systems

All electrical systems for air-conditioning, heating, and refrigeration equipment have several things in common. For instance, you only need to learn about the way a motor works in an air-conditioning system, and the motor will operate the same way when used in a heating system or a refrigeration system. This makes it easier to understand all of the electrical components in these systems because the same components are used in all of the different equipment such as gas furnaces, oil furnaces, electric furnaces, refrigerators, ice makers, chillers, large commercial air conditioners, and even window-unit air conditioners. This chapter will help you begin to see how the same components and circuits are used in all of the equipment you will be expected to work on. This chapter will also introduce several types of electrical wiring diagrams and the symbols that are used in them. You will begin to see how the theory that was presented in previous chapters helps you to understand how the components operate individually and as a system.

8.1   Simple Air-Conditioning Electrical System

The air-conditioning system that you work on in the field consists of a compressor motor, a condenser fan motor, evaporator fan motor, and the controls that turn it on and off. The compressor motor, evaporator fan motor, and condenser fan motor are all examples of *loads* in the electrical system. The definition of an *electrical load* is a component that has electrical resistance and uses (consumes) electricity to convert it to another type of energy such as motion or heat. This electrical system also has components that are called *controls*. Controls for the air-conditioning system include the thermostat and the relays that energize and deenergize these motors. The definition of a control is a device or component that can switch electrical voltage and current on or off to energize or deenergize a load such as a motor.

Fig. 8–1 shows a diagram of a typical residential air-conditioning system. Fig. 8–2 shows a cut-away picture of the two parts of this system so that you can see inside their panels exactly where these components are located. In these pictures you can see that the compressor and condenser fans are located in the condenser unit that is found outside the house and the furnace fan, which is also called the evaporator fan in the air-conditioning system, is located in the furnace inside the house. Controls for this system include the thermostat that controls the temperature of the system. The thermostat is located inside the house and is usually mounted on a wall that is centrally located. The thermostat controls the compressor relay, which energizes the compressor and condenser motors, and the fan relay that is located inside the furnace and energizes the furnace fan. From this point on, the furnace fan will be called the *evaporator fan* for the air-conditioning system because that is the function it performs.

8.2   Electrical Symbols

Each electrical control and load has a unique symbol that represents that component in an electrical diagram. The electrical symbols are standardized and all manufacturers use similar symbols for each component. Some manufacturers alter the symbols slightly but overall each electrical component will have its own unique symbol when it is used in an electrical diagram. The symbols may not be used in wiring diagrams where the physical location and the outline of each part

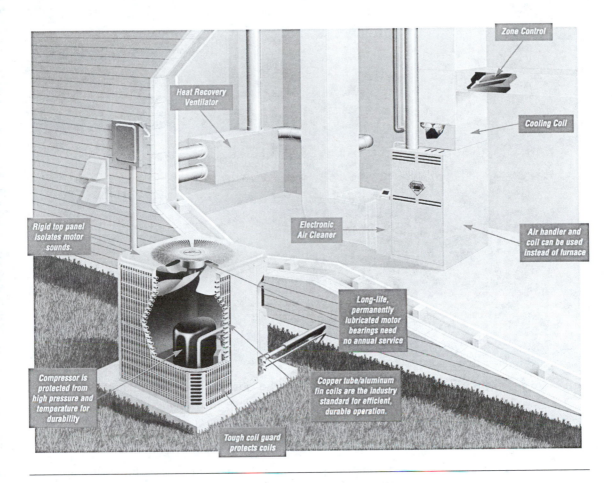

Figure 8–1   A typical residential air-conditioning system. The condenser section is located outside the house and the furnace part of the system is located inside the house. *(Courtesy of York International Corporation)*

is being presented, but in electrical ladder diagrams the symbols are generally used. Fig. 8–3 shows a complete list of the electrical symbols, and you should notice that all of the switches are shown with normally open or normally closed contacts. The symbol for each switch also shows the *operator* for the switch, which is the symbol for pressure, temperature, flow, and other actions and forces that cause the switch contacts to move open or closed. The symbol will also indicate whether the force or action will cause the contacts to open when the force or action increases or when it decreases. If the switch contact is located on the top of the output terminal of the symbol, and the contacts are shown as normally closed, the symbol is showing that an increase of action such as temperature will

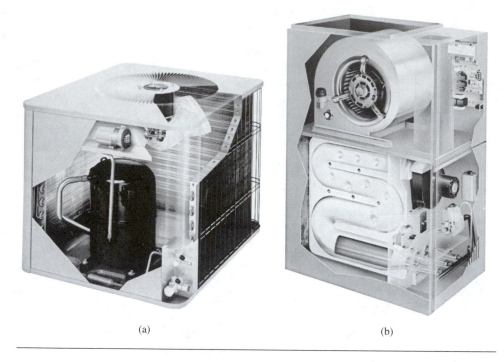

(a)   (b)

Figure 8–2a   A cut-away picture of the outdoor part of the air-conditioning system shown in Fig. 8–1. **b** A cut-away picture of the indoor part of the air-conditioning system. *(Courtesy of Armstrong Air Conditioning, Inc.)*

cause the switch contacts to open, and a decrease in temperature will cause the switch contacts to close. If the contact is located on the bottom of the output terminal, and the symbol is showing the contacts are normally closed, the symbol is showing that a decrease in temperature causes the contacts to open, and an increase in temperature causes the contacts to close.

8.3   Electrical Diagram of an Air-Conditioning System

As you know, the electrical components of the air-conditioning system consist of three main loads: the compressor motor, the condenser fan motor, and the evaporator fan motor. This system has a thermostat, a compressor relay, and a evaporator fan relay that acts as controls. The primary side of the control transformer will be treated as a load because it consumes power, and the secondary side will be treated as part of the control circuit because it provides voltage for the control circuit.

STANDARD ELEMENTARY DIAGRAM SYMBOLS

The diagram symbols shown below have been adopted by the Square D Company and conform where applicable to standards established by the National Electrical Manufacturers Association (NEMA).

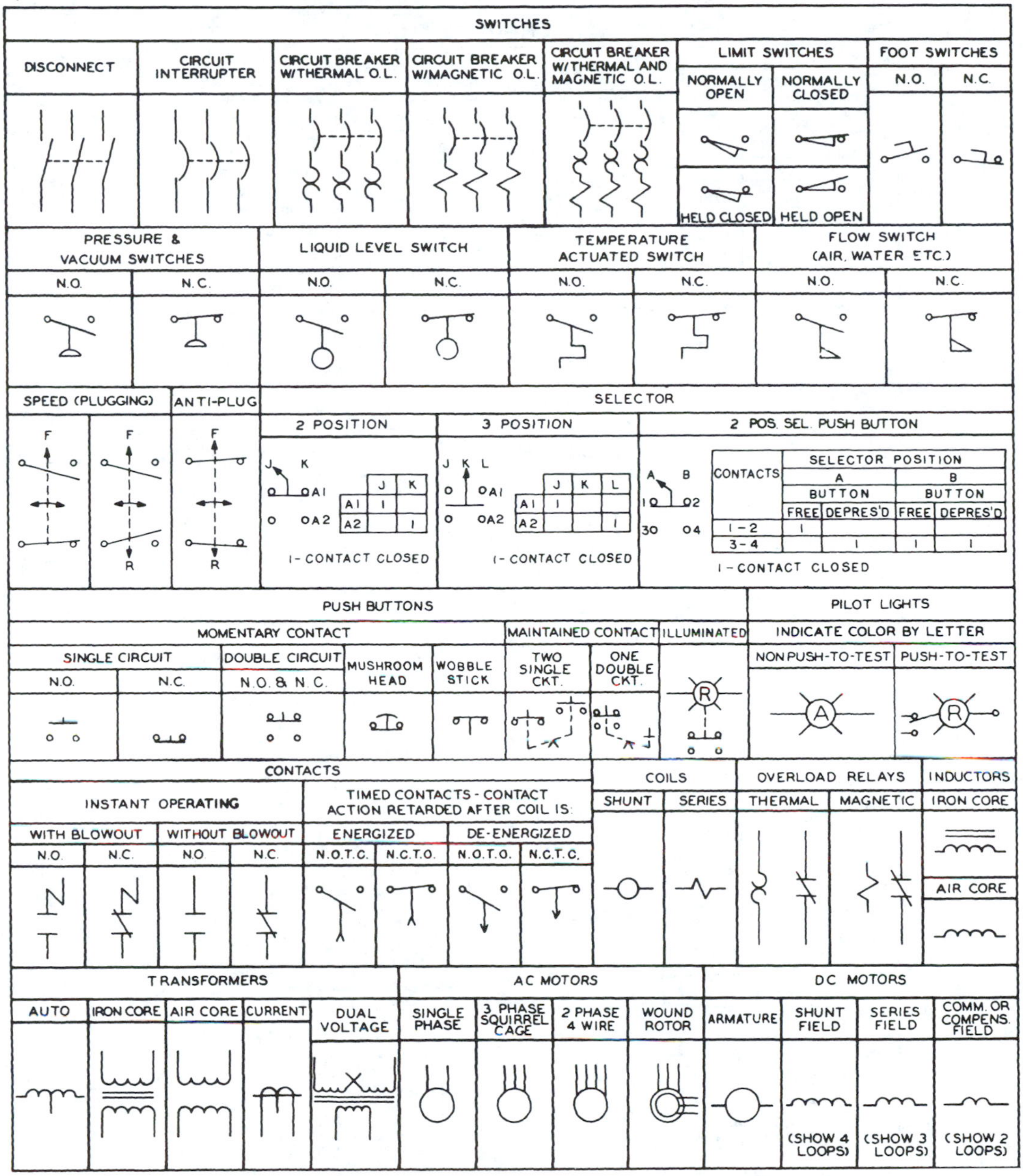

Figure 8–3 Electrical symbols for motors and controls used in air-conditioning, heating, and refrigeration systems. (*Courtesy of Square D/Groupe Schneider. Square D/Groupe Schneider assumes no liability for accuracy of information.*)

SUPPLEMENTARY CONTACT SYMBOLS

SPST, N.O.		SPST N.C.		SPDT		TERMS	
SINGLE BREAK	DOUBLE BREAK	SINGLE BREAK	DOUBLE BREAK	SINGLE BREAK	DOUBLE BREAK	SPST -	SINGLE POLE SINGLE THROW
						SPDT -	SINGLE POLE DOUBLE THROW
DPST, 2 N.O.		**DPST, 2 N.C.**		**DPDT**		DPST -	DOUBLE POLE SINGLE THROW
SINGLE BREAK	DOUBLE BREAK	SINGLE BREAK	DOUBLE BREAK	SINGLE BREAK	DOUBLE BREAK	DPDT -	DOUBLE POLE DOUBLE THROW
						N.O. -	NORMALLY OPEN
						N.C. -	NORMALLY CLOSED

Figure 8–3   *continued.*

Fig. 8–4 shows an electrical diagram of these components. This diagram shows the electrical symbols for each component and the physical location where those components exist. It also presents the physical outline of each component, and if electrical terminals or wires come into or out of the component, this diagram will indicate the exact location this occurs and the color of wire that is used. This diagram is called a *wiring diagram* and it is generally used to help locate components, and the electrical terminals on those components.

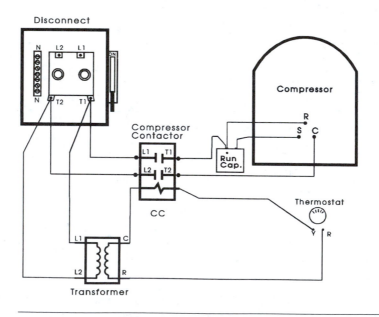

Figure 8–4   An electrical wiring diagram of the air-conditioning system shown in the previous figures.

Fig. 8–5 shows an example of a ladder diagram of the electrical system for the air conditioner. This diagram shows the same components as the wiring diagram, but it shows them in the sequence the events must occur for the system to operate correctly. You should notice that it is easier to understand the operation of the system when using the ladder diagram. The ladder diagram has derived its name because overall appearance is that of a wooden ladder. The ladder diagram is divided into two sections. The top section shows the high voltage that is used to energize the compressor fan, evaporator fan, and the condenser fan, which all use 230 volts. The evaporator fan is powered with 115 VAC and it is supplied voltage by L1 and neutral. The compressor and condenser fans can be operated on a different voltage from the evaporator fan because they are located in the outdoor part of the system called the *condenser,* which is powered by 230 volts,

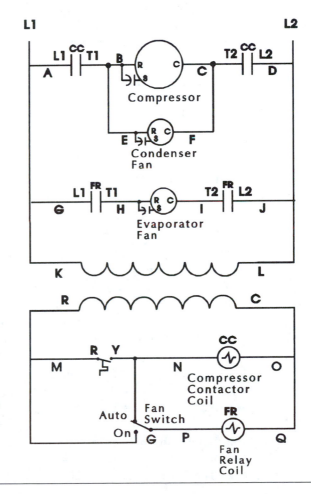

Figure 8–5 A ladder diagram of the air conditioner's electrical system. Notice the overall appearance of this diagram looks like a wooden ladder, which gives the diagram its name. This diagram shows the sequence of operation for the electrical system.

and the evaporator fan is located in the furnace, which is powered by 115 volts. The condenser unit is supplied higher voltage because this is the part of the system that will consume the most power when the system is operating.

The bottom part of the diagram shows the control circuit. The control circuit is powered by 24 VAC, coming from a step-down transformer called the *control transformer.* The transformer is physically located in the furnace and it steps 115 volts down to 24 volts. The thermostat is a switch that is located inside the house on a wall and it is activated by a change in room temperature. This switch energizes or deenergizes power to the relay coils for the compressor relay and evaporator fan relay. For this introduction of the electrical control system, you only need to understand that the relay coil becomes magnetized and the magnetic force causes its contacts to move from their normally open position to the closed position. The contacts of the compressor relay are wired in series with the compressor and condenser fan motor and when they close they provide power to these motors. The contacts of the evaporator fan relay are connected in series with the evaporator fan and when they close they provide power to the evaporator fan.

8.4   Loads for the Air-Conditioning System

The main load for the air-conditioning system is the compressor motor. In a residential system the compressor motor will be supplied with single-phase 230 volts. In a commercial system the compressor is supplied by either 230 volts three phase, or 480 volts three phase. The function of the compressor motor is to pump refrigerant and increase its pressure. The compressor is the only load required to cause the refrigerant to move through the system. Fig. 8–6a shows a picture of a compressor. The electrical diagram of the compressor is shown in Fig. 8–6b. The diagram in this figure shows the single-phase compressor that is powered by 230 volts.

The second major load in the air-conditioning system is the condenser fan. The condenser fan is generally powered by the same voltage as the compressor because they are mounted in the same cabinet. The condenser fan moves outside air over the fins of the condenser coil to provide sufficient cooling to cause the refrigerant to change from a vapor to a liquid. From the electrical diagram in Fig. 8–5, you should notice that the wires for the condenser fan are connected to the same terminals on the relay contacts as the compressor. This means that the condenser fan will run anytime the compressor is running. This means that the compressor relay also controls the condenser fan even though the name implies that this relay only controls the compressor motor. Fig. 8–7a shows a picture and Fig. 8–7b shows an electrical diagram of a typical condenser fan motor.

The third load for the air-conditioning system is the evaporator fan. The evapo-

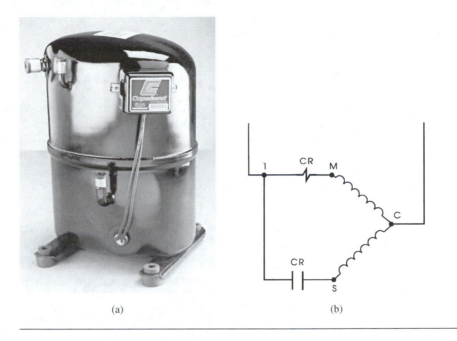

(a)

(b)

Figure 8–6a A typical compressor motor. **b** The electrical diagram of a compressor motor. *(Courtesy of Copeland Corporation)*

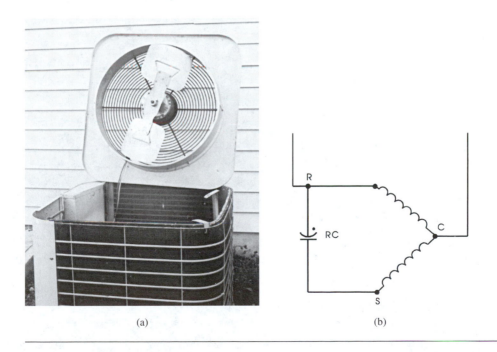

(a)

(b)

Figure 8–7a A typical condenser fan motor. **b** Electrical diagram of a typical condenser fan motor.

rator fan is mounted near the evaporator in the furnace. As you know, the evaporator fan is also used as the furnace fan, but it is referred to as the evaporator fan when you are working on the air-conditioning system. Fig. 8–8a shows a picture and Fig. 8–8b shows an electrical diagram of a typical evaporator fan.

From the diagram in Fig. 8–5, you should notice that the evaporator fan is energized by the contacts of the evaporator fan relay. The thermostat in this system supplies voltage to the coil of the evaporator fan relay anytime the temperature becomes warmer than the setpoint on the thermostat. The setpoint is the temperature setting on the thermostat. In the diagram in Fig. 8–5 you will notice that the terminals in the thermostat that provide this circuit are identified as terminal G. Since the "hot" side of a 24 V transformer is connected to terminal R of the thermostat, the circuit for the fan relay is between R and G. Anytime the thermostat switches power to terminal G, voltage will be supplied to the coil of the fan relay.

You should also notice from these diagrams that the evaporator fan relay is powered by 115 VAC (L1-N). The reason the fan is powered by 115 volts is that it is a part of the furnace. In some cases where a furnace is not part of a system, the evaporator fan will be housed in a unit called an air handler, which is basically an evaporator fan and evaporator coil. Since the furnace is located inside

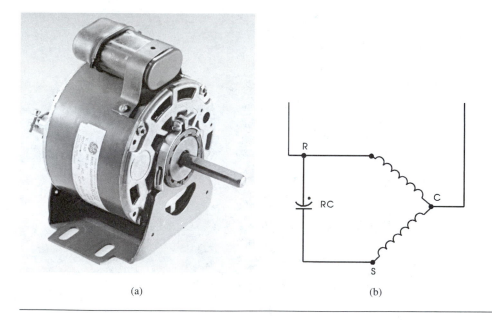

(a)   (b)

Figure 8–8a A typical evaporator fan. Since this fan is housed in the furnace, it is sometimes called the furnace fan. **b** An electrical diagram of an evaporator fan. *(Courtesy of GE Motors & Industrial Systems Fort Wayne, Indiana)*

the house and it does not consume a large amount of power, it generally is powered by 115 volts.

8.5   Controls for the Air-Conditioning System

The control circuit for the air-conditioning system provides the means to energize and deenergize the compressor, the condenser fan, and the evaporator fan. The control circuit is identified in the diagram in Fig. 8–5 as the 24 V circuit that comes from the secondary of the control transformer. The control circuit consists of the coil for the compressor relay, the coil for the evaporator fan relay, and the thermostat. You should notice in the diagram that the thermostat circuit that energizes the compressor relay coil is located between terminals R and Y in the thermostat. An in-depth presentation of information about the thermostat will be provided in Chapter 14.

Anytime the temperature inside the house becomes warmer than the setpoint on the thermostat, the circuit between R and Y is complete. The letters R and Y are a standard identification for thermostat terminals. These letters stand for the colors of the wires that are generally connected to these terminals: R is for red, and Y is for yellow.

The thermostat circuit that controls the evaporator fan is located between terminals R and G. G represents the color green. The coil of the evaporator fan relay can be energized in two different ways. First, it will be energized anytime the compressor is energized. This is accomplished through a switch in the thermostat called the *auto/on switch*. When the auto/on switch is in the *auto position,* it connects the G terminal of the thermostat directly to the Y terminal in the thermostat. This enables power to be sent to the G terminal and the Y terminal anytime the system calls for cooling, which will energize the compressor relay and the evaporator fan relay.

The second way the evaporator coil is energized is when the auto/on switch is moved to the *on position.* When the switch is moved to the on position, a circuit is provided between the R terminal in the thermostat and the G terminal. This provides power to the coil of the evaporator fan relay from terminal G. Anytime the fan switch is in the on position, the fan will be energized 100% of the time. Anytime the fan switch is in the auto position, it will be energized only when the system is calling for cooling, and the compressor relay is energized.

8.6   Electrical Diagrams of a Simple Heating System

Fig. 8–9 shows a picture of a typical residential gas furnace. This type of furnace is called an up-flow furnace because air is pulled in at the bottom of the furnace and is forced over the heat exchanger and out the top of the furnace into

Figure 8–9 Cut-away diagram of the furnace shown in the previous figure. Notice the location of the electrical loads and controls for this furnace. *(Courtesy of United Technologies Carrier Corporation)*

the duct work where it is sent to all parts of the house. The major electrical load for the gas furnace is the furnace fan. You should remember that this same fan is called the evaporator fan if the system is also used for air conditioning. This means that sometimes the fan will be referred to as the furnace fan, and sometimes it is referred to as the evaporator fan. In this section it will be called the furnace fan. Another load in this system is the gas valve. The gas valve is generally powered with 24 VAC, which comes from the transformer. The furnace fan is located near the bottom of this gas furnace, and the gas valve is located near the burners in the middle of the furnace.

The electrical wiring diagram for a simple gas furnace is shown in Fig. 8–10. Fig. 8–11 shows a ladder diagram of the same heating system. You should remember that the wiring diagram shows the component in the approximate position that you would physically find them when you inspect the furnace. You can refer to the cut-away picture of the furnace shown in Fig. 8–9 to help you determine the location of the major electrical components in the furnace. The lad-

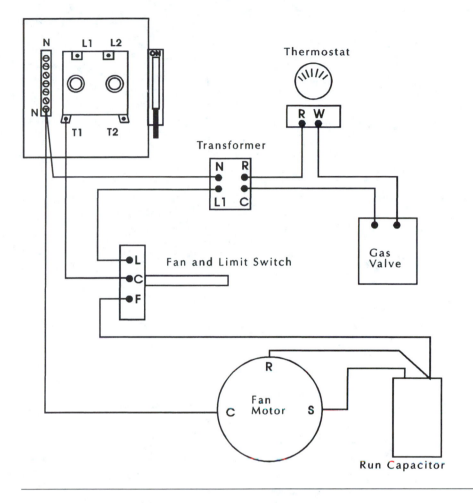

Figure 8–10   Wiring diagram of a typical gas furnace. Notice this diagram shows the basic location of each component in the electrical system.

der diagram presents the components in the sequence in which they operate. You should remember that the high-voltage circuit (115 VAC) is generally shown at the top part of the ladder diagram, and the control voltage circuit (24 VAC) is generally shown at the bottom part of the diagram.

8.7   Loads in the Heating System

The loads in the heating system consist of the furnace fan motor and the gas valve. From the diagram you can see that the furnace fan is energized and deen-

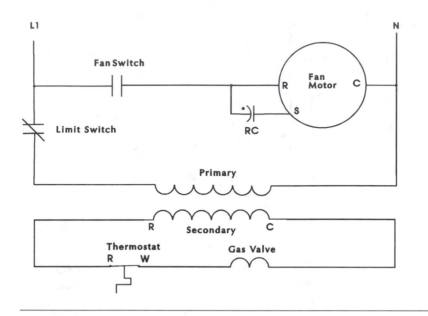

Figure 8–11 A ladder diagram of the gas furnace used in Fig. 8–10. Notice that this diagram shows the sequence of operation for the furnace.

ergized by the fan relay contacts. Anytime the contacts are closed the fan will be energized, and anytime the contacts are opened the fan will be deenergized. The furnace fan can be energized by the fan relay or by a fan switch. The fan switch is wired in series with line 1 and the fan motor. The switch is physically located in the furnace at a point where it will be activated when the temperature in the furnace increases to approximately 110° to 120°F. This means the fan can be energized to run continually by the fan relay or it can be run as needed when the temperature in the furnace is above 110°F. The temperature in the furnace should only be above 115°F if the burners are turned on.

The furnace fan is an open-type, fractional horsepower motor. It can be a direct-drive or belt-driven motor. Most newer furnaces use multispeed, direct-drive, permanent split-capacitor motors (PSC motors). Chapter 11 provides more in-depth information about the theory of operation and troubleshooting for these types of motors. The electrical diagrams in Fig. 8–10 and Fig. 8–11 show the identification of the motor terminals and their locations. Fig. 8–12a shows a picture and Fig. 8–12b shows an electrical diagram of the furnace fan motor. Fig. 8–13a shows a picture and Fig. 8–13b shows an electrical diagram of a typical gas valve.

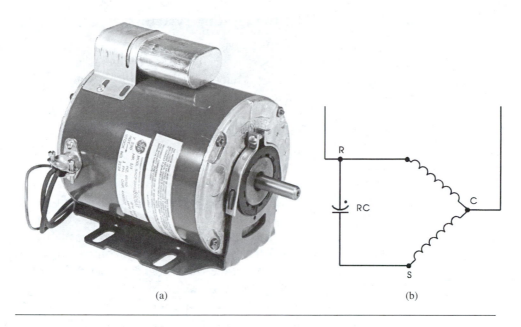

(a)   (b)

Figure 8–12a A furnace fan motor. **b** The electrical diagram of the PSC fan motor that is used as the furnace fan. *(Courtesy of GE Motors & Industrial Systems, Fort Wayne, Indiana)*

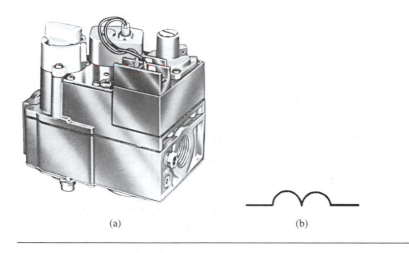

(a)   (b)

Figure 8–13a A typical gas valve. **b** Electrical diagram of the gas valve. *(Courtesy of Honeywell)*

8.7.1   Controls for the Heating System

The controls for the heating system consist of the thermostat, control transformer, and temperature switches. The gas valve is considered an electrical load for the furnace and in some cases it may be called a furnace control because it provides the gas that is burned to create the heat energy for the furnace. It is an electrical load by definition because it has resistance and consumes power. It can be classified as a control for the heating system because it energizes and deenergizes a solenoid valve, which allows gas to flow to the burners. The gas valve will be explained in more depth in Chapter 14.

Other controls in the heating system include the thermostat and the overtemperature safety switches. The thermostat provides a circuit to energize the gas valve when the temperature falls below the thermostat setpoint. This means that if the thermostat is set for 70°, and the temperature goes below 70°, the circuit between R and W in the thermostat will become completed, and voltage will be passed to the gas valve. When the room heats up, and the temperature increases above the setpoint, the circuit between R and W will become open in the thermostat, and the gas valve will become deenergized. Fig. 8–14a shows a picture and Fig. 8–14b shows the electrical symbol of a typical heating thermostat.

A typical furnace may include one or more overtemperature safety switches. In the electrical diagrams of the furnace in Fig. 8–10 and Fig. 8–11, you should notice that an overtemperature switch is wired in series with the gas valve. If the temperature inside the furnace exceeds 170°F the overtemperature switch will open and deenergize power to the gas valve. This provides a degree of safety for the furnace so it cannot cause a fire or other unsafe condition that could possibly

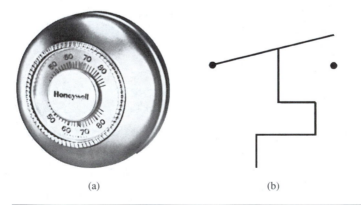

(a)   (b)

Figure 8–14a   A typical heating thermostat. **b** The electrical symbol of a heating thermostat. *(Courtesy of Honeywell)*

melt wires on the furnace. When the furnace experiences an overtemperature condition, the overtemperature switch will open and the gas valve will become deenergized so a hazard no longer exists. This overtemperature switch is sometimes called a *high limit switch*. The high-limit and other safety switches will be discussed in more depth in Chapter 14. Fig. 8–15a shows an example of a typical overtemperature switch and Fig. 8–15b shows its electrical diagram.

8.8   Electrical Diagram of a Simple Refrigeration System

Fig. 8–16 shows a picture of a typical refrigeration system. This picture shows a refrigeration display case often found in a grocery store. The electrical wiring diagram for a refrigeration system is shown in Fig. 8–18. From this diagram you can see that the main components of this system are very similar to the air-conditioning system. The compressor and condenser fan are powered by 230 VAC single phase. The evaporator fan is powered by 230 VAC or 115 VAC.

Fig. 8–17 shows the ladder diagram of the refrigeration system. You can see that it looks similar to the ladder diagram of the air-conditioning system. The main difference between the refrigeration system and the residential air-conditioning system is that the control voltage for the thermostat is 110 volts for the refrigeration system and 24 VAC for the residential air-conditioning system. The operation of the refrigeration system will be very similar to that of the air-conditioning system.

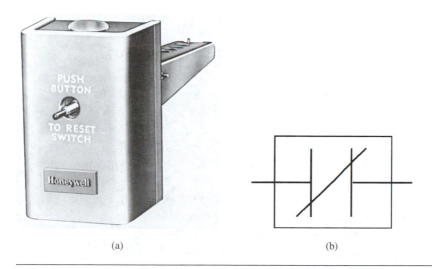

(a)   (b)

Figure 8–15a   A typical overtemperature switch. **b** Electrical diagram of a furnace overtemperature switch. *(Courtesy of Honeywell)*

Figure 8–16   A typical refrigeration system. This refrigeration system is part of a display case you would find in a grocery store. *(Courtesy of Tyler Refrigeration Corporation)*

8.9   Loads for the Refrigeration System

The electrical loads for the refrigeration system include the compressor, the condenser fan, and the evaporator fan. Some systems may also have a heating element to help defrost the system. The compressor will look very similar to the compressor in the air-conditioning system and electrically it will operate exactly like it. The compressor motor may be a single-phase or three-phase motor. The condenser fan will be slightly smaller for the refrigeration system than the condenser fan for the air conditioner. Fig. 8–19 shows the location of the condenser fan and evaporator fan for the refrigeration system and Fig. 8–20 shows the electrical diagram for these motors. The condenser fan can also be a three-phase motor or single-phase motor. It will normally be the same type of motor as the compressor because they operate from the same voltage.

The third load for the refrigeration system is the evaporator fan. The evaporator fan on smaller refrigeration systems is usually a blade-type fan that is powered by a small single-phase motor called a shaded-pole motor. In some systems the evaporator fans will use squirrel-cage blowers to move more air, and in other

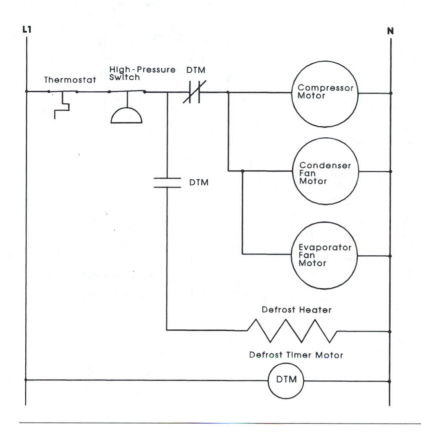

Figure 8–17   Electrical ladder diagram for the refrigeration case.

systems a fan is not needed because the refrigeration coils for the evaporator are built into a large plate and the products to be frozen are stacked on the plate. In these freezer applications the evaporator fan may not be required because the products will remain frozen without circulating evaporator air. Fig. 8–20a shows a shaded-pole motor that is used as a blade-type evaporator fan and Fig. 8–20b shows its electrical diagram.

8.10   Controls for the Refrigeration System

The controls for the refrigeration system consist of a temperature control (thermostat). The temperature control on some refrigeration systems may be a low-pressure switch that monitors the pressure of the refrigerant that is under pressure inside the tubing. It is important to understand that the amount of refrigerant pressure has a direct correlation to the temperature at the evaporator. When the temperature in the display case becomes warmer, the pressure will increase inside the tubing and the pressure switch will close and start the compressor.

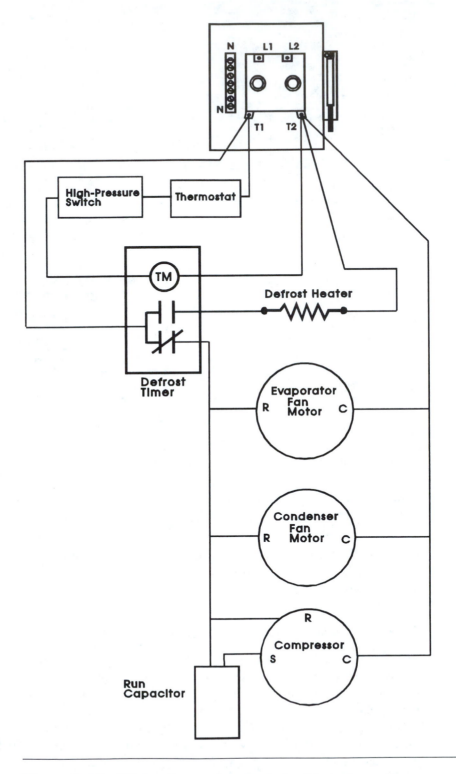

Figure 8–18 Wiring diagram for a typical refrigeration system.

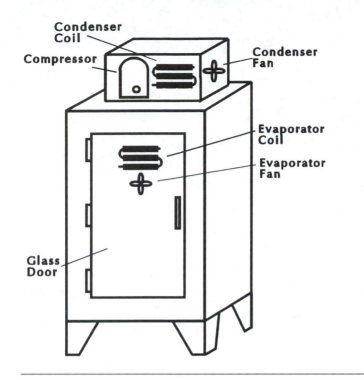

Condenser
Coil

Compressor

Condenser
Fan

Evaporator
Coil

Evaporator
Fan

Glass
Door

Figure 8–19 A refrigeration display case that shows the location of the compressor, condenser fan, and evaporator fan.

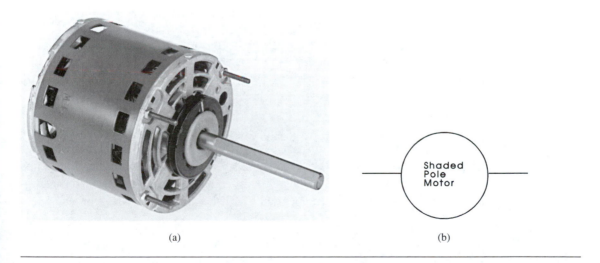

(a)

Shaded
Pole
Motor

(b)

Figure 8–20a Typical motor for blade-type evaporator fan for a refrigeration system. **b** The electrical diagram of the shaded-pole motor that is used as an evaporator fan.
(Courtesy GE Motors & Industrial Systems, Fort Wayne, Indiana)

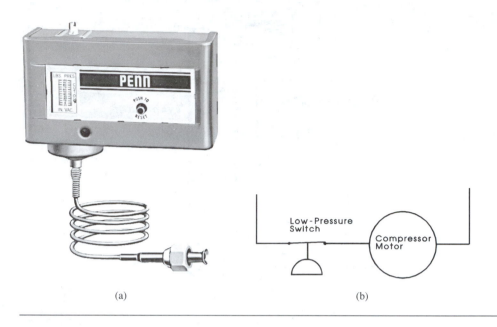

(a)   (b)

Figure 8–21a Low-pressure control used for temperature control for a refrigeration system. **b** Electrical diagram of a low-pressure switch controlling a compressor motor. *(Courtesy of Johnson Controls)*

When the compressor runs long enough, the temperature in the display case becomes cooler and causes the refrigerant pressure to drop, which causes the pressure switch to open and deenergize the compressor. Fig. 8–21a shows a picture of a low-pressure switch that is used as a temperature control and Fig. 8–21b shows a diagram of a low-pressure switch used as a temperature control for a refrigeration system.

The main difference between the temperature control in a refrigeration system and one used in a residential furnace is that the refrigeration temperature switch typically uses high voltage (115 VAC or 220 VAC) as the control voltage, and the residential furnace uses 24 volts. This means that the contacts in the refrigeration control and its overall appearance will be slightly larger than a residential thermostat.

Questions for This Chapter

1. Define the term *electrical load.*
2. Explain why the evaporator fan in an air-conditioning system may also be called the furnace fan.
3. Define the term *electrical control.*

4. Explain the difference between a wiring diagram and a ladder diagram.
5. Explain where you would typically find the evaporator and condenser on a refrigeration display case.

True or False

1. A wiring diagram shows the relative location of each component in a circuit.
2. An electrical load is a device that has resistance and uses current.
3. A ladder diagram shows the relative location of each component in a circuit.
4. A furnace fan can also be the condenser fan in some systems.
5. The thermostat for a furnace is an electrical load.

Multiple Choice

1. The control in the furnace that protects the furnace from overheating and causing a fire is the _____
   a. fan switch.
   b. high-limit switch.
   c. fan relay.
2. The "operator" of an electrical switch is the part of the switch that _____
   a. causes the action that makes the switch change from open to closed or closed to open.
   b. conducts current.
   c. is the same as the contacts.
3. The auto/on switch on the thermostat allows the fan to _____
   a. operate constantly or with the gas valve.
   b. operate constantly or with the transformer.
   c. operate constantly or with the compressor.
4. The thermostat terminals are identified with the letters R, Y, W, and G that stand for _____
   a. colors of the thermostat wires (red, yellow, white, and green).
   b. the operation of the thermostat (auto, manual, and semiautomatic).
   c. the primary and secondary windings on the control transformer.

5. The fan and limit switch in the gas furnace control _____

   a.   the temperature at which the fan turns on and the temperature at which the gas is ignited.

   b.   the temperature at which the fan turns on and the temperature at which the air conditioner turns off.

   c.   the temperature at which the fan turns on and the maximum temperature the furnace can reach before power to the gas valve is turned off.

Problems

1. Sketch a wiring diagram or a ladder diagram of a simple air-conditioning system and identify the loads and controls.

2. Sketch a wiring diagram or a ladder diagram of a simple heating system and identify the loads and controls.

3. Sketch a wiring diagram or a ladder diagram of a simple refrigeration system and identify the loads and controls.

4. Sketch the wiring diagram symbol and the ladder diagram symbol of any load or control and explain why they are different.

5. Sketch a wiring diagram or ladder diagram of the fan auto/on switch on a thermostat and explain what causes the fan to turn on and off in each mode.

9 Relays, Contractors, and Solenoids

OBJECTIVES:

After reading this chapter, you will be able to:
1. Identify the contacts and coil in a relay and contactor.
2. Explain the operation of a relay and contactor.
3. Identify the components in a control circuit.
4. Select the proper size of contactor from a NEMA table.
5. Identify the basic parts of a solenoid and explain their functions.

9.0 Overview of Relays and Contactors in the Control Circuit

The *control circuit* in air conditioning, heating, and refrigeration is the heart of the system. It consists of the control devices such a thermostat or pressure switch to determine when to energize the system or deenergize each system. Each control circuit uses relays or contactors to control the compressor motor, evaporator fan motor, condenser fan motor, and furnace fan motor. This chapter will introduce the main components that are used in the control circuits of these systems and their theory of operation. The components in the control circuit are energized and in turn they energize components such as compressors and motors in the *load circuit*. At the bottom of Fig. 9–1 shows an example of a typical circuit

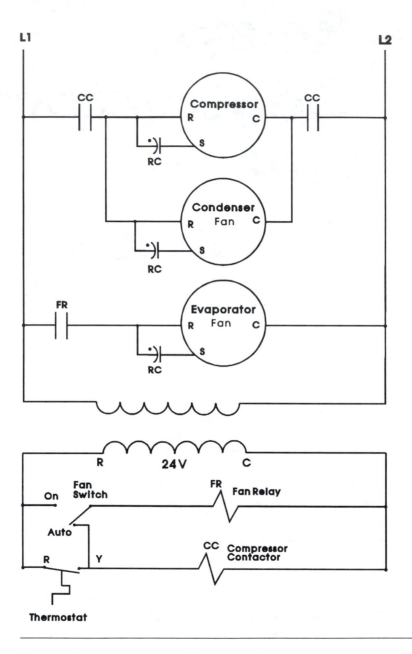

Figure 9–1   A typical control circuit for an air-conditioning system is shown at the bottom of this diagram.

for an air-conditioning system. Notice that it is the bottom part of the circuit and the load circuit is the top part of the electrical diagram.

9.1   The Control Transformer

The control transformer was introduced in Chapter 7. In that chapter you learned that the transformer converts line voltage to low-voltage power for the control circuit. The load-side terminals of the control transformer are labeled R and C. In commercial air-conditioning systems and in refrigeration systems, the control transformer steps down 240 VAC to 110 VAC, and in residential air-conditioning and heating control systems it steps down 120 VAC to 24 VAC. The lower voltage is used in each case to provide power for the control circuit.

In the diagram in Fig. 9–1 the control circuit is shown as the bottom part of the diagram and it starts at the secondary side of the transformer where it is identified by the letters R and C. The terminal that is identified as the R terminal is considered the source of power and it will be called the *hot* side of the circuit, and the right side of the diagram that is identified by the transformer terminal C is considered to be the *common* side of the circuit. It is important to understand that since the voltage in the transformer is AC voltage, both sides of the circuit have the same potential during every other half-cycle. The concept of hot side and common side is defined to make it easier to understand the operation of the control circuit.

9.2   The Theory and Operation of a Relay

The *relay* is a magnetically controlled switch that is the main control component in air-conditioning, heating, and refrigeration systems. Fig. 9–2 shows a picture of a typical relay and Fig. 9–3 shows a cut-away diagram of a typical relay, which consists of a *coil* and a *number of sets of contacts*. The coil becomes an electromagnet when it is energized and its magnetic field causes each set of normally open contacts to close and each set of normally closed contacts to open. The contacts are basically a switch that is operated by magnetic force. The part of the relay that moves and causes the contacts to move is called the *armature*. Power is applied to the coil of the relay first and the magnetic flux causes the armature to move and cause the contacts to change position. The coil is part of the control circuit, and the contacts are part of the load circuit.

Fig. 9–4 shows the electrical symbol for the coil and contacts of a relay. The armature is not shown in the electrical symbols, since it is considered part of the contacts. When you encounter a relay in the control panel of a system, you must relate the physical parts with the electrical symbols that are used to identify the components. You must also learn to envision its operation as two separate pieces,

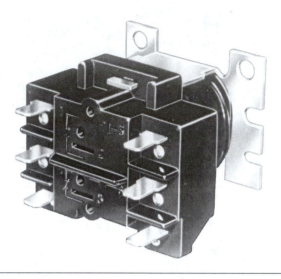

Figure 9–2   A typical relay that is used to control an air-conditioning, heating, and refrigeration system. *(Courtesy of Honeywell)*

the coil and contacts, even though they are mounted near each other and operate almost simultaneously. It is important to understand that the coil must be energized first, and a split second later the magnetic field built up in the coil will cause the contacts to move.

9.3   Types of Armature Assemblies for Relays

You will encounter many different types of relays and different name brands of relays that look like they all use different theories of operation. In reality the operation of all relays can be broken into four basic methods of operation because one of four basic armature assemblies is used in all relays. Fig. 9–5 shows an example of these four basic armature assemblies. A few relays will have hybrid armatures that use some features of more than one type of armature assembly.

Fig. 9–5a shows an example of a *horizontal-action armature* for a relay. In this example a set of stationary contacts are mounted in the horizontal position. A set of movable contacts is mounted on the armature assembly directly across from the stationary set. The armature is mounted inside the coil in such a way that when the coil is energized, its magnetic field will cause the mass of the armature to center in the middle of the coil. Since the armature is slightly offset when it is placed in the coil, it will move to the left to center itself directly in the middle of the coil when the magnetic field is developed in the coil. The movement of the armature to the left will shift the movable contacts to the left until they come into contact with the stationary set of contacts. When the movable contacts and stationary contacts touch each other, they will complete a circuit that

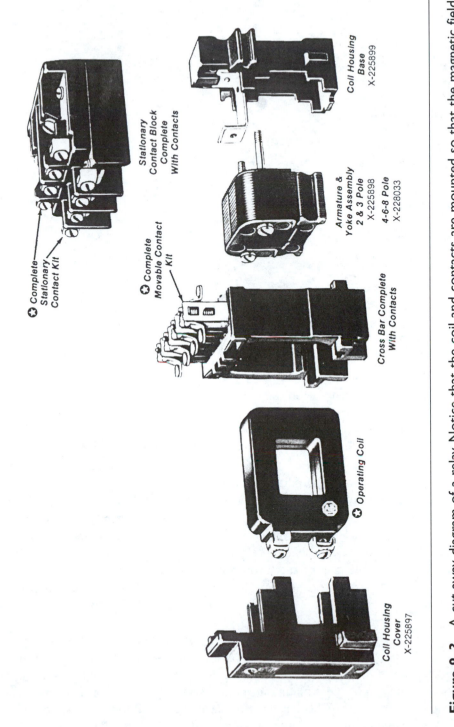

Complete Stationary Contact Kit

Stationary Contact Block Complete With Contacts

Coil Housing Base X-225899

Complete Movable Contact Kit

Armature & Yoke Assembly 2 & 3 Pole X-225898
4-6-8 Pole X-228033

Cross Bar Complete With Contacts

Operating Coil

Coil Housing Cover X-225897

Figure 9–3 A cut-away diagram of a relay. Notice that the coil and contacts are mounted so that the magnetic field of the coil will make the contacts move. (*Courtesy of Rockwell Automation's Allen Bradley Business*)

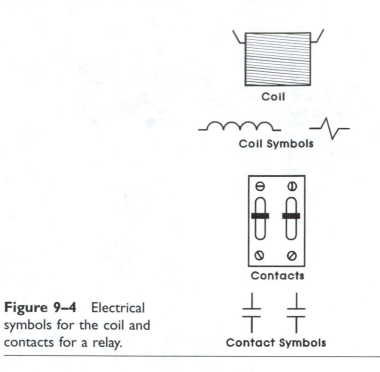

Figure 9–4   Electrical symbols for the coil and contacts for a relay.

allows large amounts of current and voltage to pass through them like a normal switch when it is in the closed position. When the current to the coil is deenergized, a small spring will cause the armature to shift back to the left to its original position.

The diagram in Fig. 9–5b shows an example of a *bell-crank armature assembly*. In this example the coil is mounted above the armature, and when it is energized it produces a magnetic field that pulls the armature upward. The moveable contacts are connected to an arm that is bent at a right angle. When the end of the arm is pulled upward by the magnet, the other end of the arm is moved to the left. This action shifts the movable contacts against the stationary contacts to complete the electrical circuit. When the coil is deenergized, gravity will cause the armature to drop downward away from the coil. This movement causes one end of the arm to move downward also, causing the left end of the arm to shift back to the left, which causes the contacts to move to the open position again.

The diagram in Fig. 9–5c shows an example of a *clapper armature assembly*. The armature is a large arm on the right side of the coil. The armature (arm) has a pin through the bottom to act as an axis. The movable contacts are mounted to the top of the armature at the left side. When the coil is energized, it will create a magnetic field that pulls the armature to the left toward the coil. This action causes a small amount of travel at the bottom of the arm and a large amount of

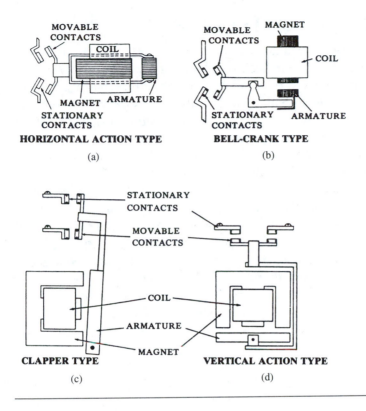

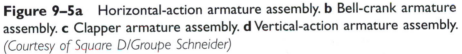

Figure 9–5a Horizontal-action armature assembly. **b** Bell-crank armature assembly. **c** Clapper armature assembly. **d** Vertical-action armature assembly. *(Courtesy of Square D/Groupe Schneider)*

travel at the top of the arm, since the bottom of the arm is held in place at the axis. Since the movable contacts are mounted near the top of the arm, they will move to the left a significant distance until they come into contact with stationary contacts. When the coil is deenergized, a small spring will pull the armature to the left, which will cause the contacts to return to their open position.

Fig. 9–5d shows an example of the *vertical-action armature assembly.* The armature is mounted to a bracket that is shaped like the letter C. The bottom part of the C-shaped bracket is mounted directly to the armature and a set of movable contacts is mounted directly to the top of the C-shaped bracket. When the coil is energized, it creates a magnetic field that pulls the entire bracket upward until the armature is pulled tight against the coil. This movement causes the contacts that are mounted to the top of the bracket to shift upward until they touch the stationary contacts. When the coil is deenergized, gravity will cause the complete bracket to drop downward and move away from the coil, which causes the movable contacts to move away from the stationary contacts.

9.4   **Pull-In and Hold-In Current**

When voltage is first applied to a coil of a relay, it will draw excessive current. This occurs because the coil of wire presents only resistance to the circuit when current first starts to flow. As the flow of current increases in the coil, inductive reactance begins to build, which will cause current to become lower. When the current is at its maximum, it will create a strong magnetic field around the coil, which will cause the armature to move. When the armature has moved, it will cause the induction in the magnetic coil to change so that less current is required to maintain the position of the armature.

Fig. 9–6 shows a diagram of the pull-in current and the hold-in current. The pull-in current is also called the *inrush current* and the hold-in current is also called the *seal-in current*. The pull-in current is typically three to five times larger than the hold-in current. At times the supply voltage will be slightly lower in the summer. This condition is called a brownout and it may cause a problem with activating relays. If the voltage is too low, it may not be able to supply sufficient current to pull the armature into place when the relay is first energized because the pull-in current is too small. If this occurs, the air-conditioning or refrigeration system will not start because the starting relay cannot close.

If the system is running when the brownout condition occurs, the relay coil will probably remain energized because the amount of hold-in current is sufficient to keep the armature in place even though the voltage is low. Since the brownout condition occurs when it is very hot outside, most air conditioners will remain running through this condition. If the air conditioner ever gets the room cool enough so it can cycle off, it will probably not energize again because of the low voltage and the large amount of pull-in current required.

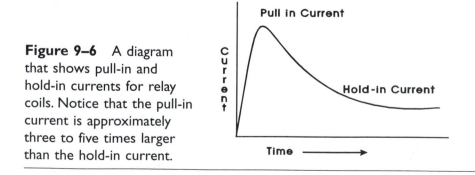

Figure 9–6   A diagram that shows pull-in and hold-in currents for relay coils. Notice that the pull-in current is approximately three to five times larger than the hold-in current.

9.5   **Normally Open and Normally Closed Contacts**

A relay can have normally open or normally closed contacts. It is important to understand that the word *normal* for contacts indicates the position the contacts are in when no voltage is applied to the coil. The contacts can be held in their normal position by a spring or by gravity. The contacts of a relay will move from their normal position to their energized position when power is applied to its coil.

Some types of contacts can be changed or converted from normally open to normally closed in the field. Other types are manufactured in such a way that they cannot be changed. Fig. 9–7 shows examples of converting normally open contacts to normally closed contacts while the relay is installed. The contacts can be converted in the field by the technician by simply removing them from the relay and turning them upside down. When a set of normally open contacts is inverted, the contacts become normally closed, and when normally closed contacts are inverted, they will become normally open contacts. This means that a technician in the field can change the contacts in a relay to get the exact number of normally open or normally closed contacts needed for the application.

9.6   **Ratings for Relay Contacts and Relay Coils**

When you change a relay that is worn or broken, you must ensure that the coil of the new relay matches the voltage of the control circuit exactly. This means that if the voltage for the control circuit is 24 VAC, the coil must be rated for 24 volts. If the control voltage is 120 VAC, the coil must be rated for 120 volts. The voltage rating for a relay coil is stamped directly on the coil. If the coil is rated for 24 VAC, it will be color coded black. If the coil is rated for 110 VAC, it should also be color coded red or have a red-colored stamp on the coil. If the relay coil is rated for 208 or 230 VAC, it will be color coded green or identified with a green stamp or green printing on the coil. DC coils are color coded blue. It is important to understand that the current rating for a relay coil is seldom listed on the component. If it is important to know the current rating for the coil, you can look for it in the catalog or on the specification sheet that is shipped with the new relay. If you change a relay, you must also make sure that the rating for the contacts meets or exceeds the current rating and the voltage rating of the load to which it will be connected. For example, if the contacts of the relay are used to energize a 240 VAC fan motor that draws 3 amps, the contacts must be rated for at least 240 volts and 3 amps. It is permissible to have the contacts in this example rated for more voltage and current, such as 600 volts and 10 amps.

ADDING or CONVERTING CONTACT CARTRIDGES HAVING "SWINGAROUND" TERMINALS

General Instructions (Specific cases below.)

1.1 **Adding a contact cartridge:**

As received, accessory cartridges are in the normally open mode with terminal screws adjacent to N.O. symbols. If normally closed mode is desired, convert contact as indicated in Step 1.2 below. When cartridges are inserted, the terminal screws must face the front. The clear cover may face either side. **Do not install more than 8 N.C. contacts per relay.** When installing one cartridge, locate it at an inner pole position. When installing 2 cartridges, locate both in inner or outer (balanced) positions.

1.2 **Converting a contact to its alternate mode (N.O. ⇌ N.C.):**

Withdraw an assembled cartridge for replacement or conversion by inserting the blade of a suitably-sized screwdriver under a terminal screw pressure plate. Slide cartridge out. See Figure 2. Back the terminal screws out of the cylindrical nuts a sufficient amount (approximately 2 turns for a fully-tightened screw) to permit rotation of each screw and nut assembly to its alternate position. See Figure 3.

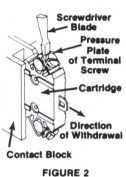

FIGURE 2

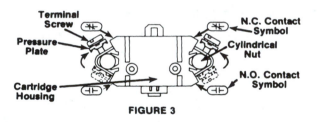

FIGURE 3

Figure 9–7 Example of changing normally open contacts to normally closed contacts in the field by turning them over. Normally closed contacts can also be converted to normally open contacts. *(Courtesy of Rockwell Automation's Allen-Bradley Business)*

Contact ratings are grouped by voltage and by current. The voltage ratings are generally broken into two groups, of 300 volts and 600 volts. This means that if you are using the contacts to control 240 or 208 VAC, you would use contacts that have a 300 V rating. If you are using the contacts to control a 480 VAC motor, you would need to use 600 V-rated contacts.

The current rating of contacts is listed in amperes or horsepower. The current rating or horsepower rating must exceed the amount of current the relay is controlling. This means that if the relay is controlling 12 amps, the contacts would need to be rated for over 12 amps of current. The current rating of the contacts and the voltage rating for the contacts will be printed directly on the contacts or on the side of the relay.

9.7   Identifying Relays by the Arrangement of Their Contacts

Some types of contact arrangements for relays have become standardized so that they are easier to recognize when they are ordered for replacement or when you are trying to troubleshoot them. The diagrams in Fig. 9–8 show examples of some of the standard types of relay arrangements. Fig. 9–8a shows a relay with a set of normally open contacts. This type of relay could also have a single set of normally closed contacts instead of normally open contacts. Since this relay has only one contact and it can only close or open this type of relay, it is called a *single-pole, single-throw (SPST) relay.* The word *pole* in this identification refers to the number of contacts, and the word *throw* refers to the number of terminals to which the input contacts can be switched. Since the contact in this relay has one input and it can only be switched to a single output terminal, it is said to have a single throw.

Fig. 9–8b shows a relay with two sets of normally open contacts. Since this relay has two sets of single contacts, it is called a *double-pole, single-throw (DPST) relay.* The *double-pole* part of the name comes from the fact that the relay has two individual sets of normally open contacts, and the *single-throw* part of the name comes from the fact that each contact has only one output terminal. When the coil is energized, both sets of contacts will move from their normally open position to the normally closed position.

Fig. 9–8c shows a relay with a set of normally open and a set of normally closed contacts that are connected on the left side. The point where this connection is made is called the common point and it is identified with the letter C. When the relay coil is energized, the normally open part of the contacts will close, and the normally closed part of the contacts will open. Since these contacts basically have a common point as the input terminal, and two output terminals [one normally open (NO) and one normally closed (NC)], it is called a *single-pole, double-throw relay (SPDT).* The most important part of this relay is that the contacts have two terminals on the output side so it is called a double-throw relay. The single-pole, double-throw relay is used where two exclusive conditions exist and you do not want them both to ever occur at the same time. For example, if this relay is controlling the cooling (air conditioning) and the heating (furnace) for a residential system, you would never want them both to be on at the same time. By connecting the air-conditioning system to the normally open terminal on the right side of the relay, and the furnace to the normally closed

Figure 9–8a A relay with a single set of normally open contacts. This type of relay is called a single-pole, single-throw (SPST) relay. **b** A relay with two individual set of contacts. This relay is called a double-pole, single-throw (DPST) relay. **c** A relay with two sets of contacts that are connected at one side at a point called the common (C). The output terminals are identified as normally open (NO) and normally closed (NC). This type of relay is called single-pole, double-throw (SPDT). **d** A relay with two sets of SPDT contacts. This relay is called a double-pole, double-throw (DPDT) relay. **e** A relay with multiple sets of normally open contacts. **f** A relay with a combination of normally open and normally closed contacts.

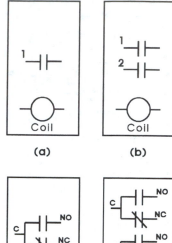

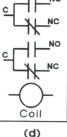

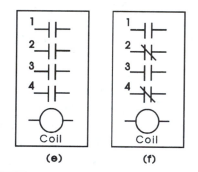

terminal, you have created the conditions so that the furnace and air-conditioning system cannot be on at the same time.

Fig. 9–8d has two sets of single-pole, double-throw contacts so it is called a *double-pole, double-throw relay (DPDT)*. In this case the term *double throw* is used because two sets of normally open/normally closed contacts are provided. Each set has a common point on their left side (input side) and a terminal that is connected to the normally open (NO) set and a terminal that is connected to the normally closed (NC) set on their right side. This type of relay would be used where the exclusion is needed and the loads are 208 VAC or 230 VAC where they need power from both L1 and L2. In this type of application, L1 would be connected to the common terminal (C) of one set of contacts, and L2 would be connected to the common terminal (C) of the other set. This would cause L1 and L2 to be switched the same way in both conditions.

Fig. 9–8e shows multiple sets of normally open contacts. This type of relay can have any number of sets of normally open contacts. The additional sets of contacts can be added to original contacts in some types of relays. If the original relay is manufactured with this provision, you can purchase the additional contacts and add them to the original relay by placing them on top of the original relay and tightening the mounting screws to make the additional contacts operate with the relay armature. The contacts for this type of relay could all be normally closed if the application required it. The main feature of this type of relay is that it can have any number of contacts.

Fig. 9–8f shows a relay with multiple sets of individual normally open and normally closed contacts. The combination of normally open and normally closed contacts can be any mixture of the two. This type of relay is similar to the one shown in Fig. 9–8e, except in this type of relay the contacts can be any combination of normally open or normally closed sets. In most cases, the contacts in this type of relay are convertible in the field and as a technician, you could add sets of contacts and change them from normally open to normally closed or vice versa as needed. In most installations, the relays are provided in the original equipment and you will only need to identify them for installation and troubleshooting purposes. Later if the equipment is modified or if additional equipment such as electronic air cleaners, humidifiers, or other similar equipment is added, you may need to locate additional contacts on a relay to connect the add-on equipment so that it will operate correctly with the original system.

9.8   Examples of Relays Used in Air-Conditioning, Heating, and Refrigeration Systems

Relays are used in a variety of applications in air-conditioning, heating, and refrigeration systems. For example, one of the most common relays is the fan relay that is used in heating systems. Another common relay is the sequencing relay used in some electric furnaces. Fig. 9–9a–d shows examples of several types of relays that you will find in these systems. In some newer systems, the relays use plug-in terminals so that the relay can be removed and replaced without changing any wiring. The relay is simply plugged into a socket-type base. An example of a plug-in relay with its socket is also shown in this figure. You will begin to become familiar with relays and recognize them in the control panels of the equipment you work on.

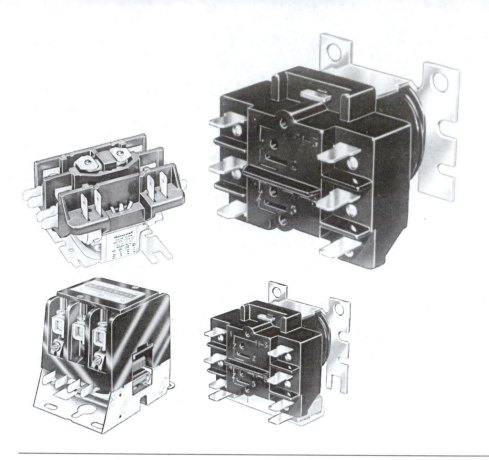

Figure 9–9a–d Variety of relays used in air-conditioning, heating, and refrigeration applications. *(Courtesy of Honeywell)*

9.9   Current Relays and Potential Relays for Starting Single-Phase Compressors

Current relays and potential relays are special relays that are designed to energize the start winding of single-phase compressor motors during starting, and then immediately de-energize as soon as the motor is running. The current relay has a low-impedance coil that takes advantage of pull-in and hold-in current to energize a set of normally open contacts. The coil for the current relay is connected in series with the run winding of a single-phase motor, and the contacts of the relay are connected in series with the start winding of the motor. In some cases a capacitor is also connected in series with the contacts and start winding. When voltage is applied to the motor, current can only flow through the run winding, which causes very large current to be drawn. This current is large enough to exceed the pull-in current level of the contacts, which causes them to close. When the current relay contacts close, voltage is applied to the start winding of the motor, which will provide additional power to allow the motor to start and to continue running. After the motor has started and begins to run, the large initial current that is flowing through the run winding and current relay coil diminishes to a minimal level, which is below the hold-in current level. When this occurs, the current relay coil is not able to keep its contacts closed, so they open and deenergize the start winding of the motor. This action is exactly what the single-phase motor needs, since its start winding can only be energized a few seconds while the motor is starting. Fig. 9–10a shows a drawing and Fig. 9–10b shows an electrical diagram of the current relay. The current relay is also available as an electronic relay. You will learn more about current relays in Chapter 13, which covers single-phase motors.

The *potential relay* uses a high-impedance coil with a set of normally closed contacts to energize a start capacitor for a capacitor-start, capacitor-run (CSCR) single-phase compressor motor. Fig. 9–11a shows a picture and Fig. 9–11b shows an electrical diagram of a typical potential relay. When power is applied to the compressor motor, current will flow through the run winding and through the normally closed contacts of the potential relay to the start winding. When the motor begins to run, its windings create a back voltage called back EMF. The back EMF is large enough to cause the coil of the potential relay to energize and pull its contacts open, which deenergizes the start winding. When main power is turned off to the motor, it will stop running and the normally closed contacts of the potential relay will return to their normally closed position, so the relay is ready to start the motor again. More information about the potential relay will be presented in Chapter 13 with other information about single-phase compressor motors.

9.10   The Difference Between a Relay and a Contactor

A contactor is similar to a relay in that it has a coil and a number of contacts. The main difference is that the contactor is larger and its contacts can carry more current. A *relay* is generally defined as magnetically controlled contacts that carry

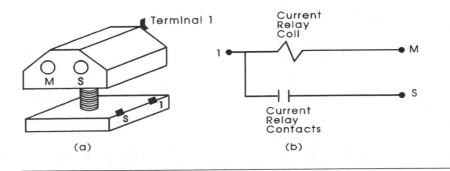

Figure 9–10a   A typical current relay. **b** An electrical diagram of a current relay connected to a single-phase compressor motor.

(a)

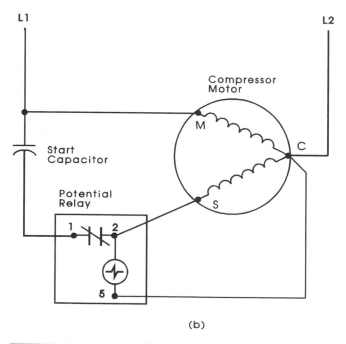

(b)

Figure 9–11a   A potential relay used to start single-phase compressors. **b** A potential relay connected to a CSCR compressor motor. *(Courtesy of Supco Sealed Unit Parts Co. Inc.)*

current less than 15 amps. A *contactor* is defined as having contacts that are rated for 15 amps or more. Some manufacturers do not follow the 15 A rating so sometimes you will find a relay that has the current rating for its contacts in excess of 15 amps, and you may also find a contactor with contact ratings less than 15 amps. In general the main difference is that a contactor is specifically designed so its contacts can carry a larger amount of current, up to 2250 amps. Contactors are rated by the National Electrical Manufacturers Association (NEMA) and their sizes range from a size 00 to a size 9.

In most air-conditioning and refrigeration systems the compressor motor is controlled by a contactor instead of a relay because the amount of current the compressor motor draws is usually in excess of 15 amps. When you look at the compressor contactor, you will see that it looks like a very large relay. Since its contacts can carry current larger than 15 amps, it will technically be called a contactor instead of a relay.

9.11   NEMA Ratings for Contactors

Fig. 9–12 presents a table of NEMA ratings for contactors, showing that a size 00 contactor is rated to safely carry up to 9 amps for a continuous load. You should also notice that the contacts are rated for up to 575 volts. You should remember that the current rating for contacts depends on the size of the contacts, and the voltage rating of the contacts depends on the way the relay is manufactured so that arcs do not jump between different terminals. This means that contacts that are rated for a higher voltage have more plastic or insulating material between the sets of contacts so arcs do not jump from the contacts to other parts of the relay or to other sets of contacts. This table also identifies the load as a maximum horsepower rating. This means that you may identify the load from its current rating or its horsepower rating and select the proper contactor size to safely control the load.

You should notice from this table that the next larger size of contactor is a size 0. This contactor is rated for up to 18 amps on a continuous basis. The size 9 contactor is the largest and its current rating is 2250 amps.

Air-conditioning, heating, and refrigeration systems usually use contactors up to a size 4. One application for contactors is to control electric heating elements for electric furnaces. These contactors are called heating contactors. Since heating elements may draw up to 100 amps, the size 4 contactor can easily carry this current. If larger heating elements are used, a larger contactor is used. Fig. 9–13a shows an example of a size 3 contactor and Fig. 9–13b shows a size 5 contactor. The physical size of the size 3 contactor is approximately 5 inches tall and the size 5 contactor is approximately 7 inches tall.

NEMA Size	Continuous Ampere Rating	Maximum Horsepower Rating Full Load Current Must Not Exceed "Continuous Ampere Rating"			
		Motor Voltage			
				50 Hz	
		200V	230V	380V-415V	460V-575V
		3 ∅ ● 4 Power Poles ● 600V AC Maxi			
00	9	1-1/2	1-1/2	2	2
0	18	3	3	5	5
1	27	7–1/2	7-1/2	10	10
2	45	10	15	25	25
3	90	25	30	50	50
4	135	40	50	75	100
5	270	75	100	150	200
6	540	150	200	300	400
7 ❷	810	–	300	600	600
8 ❷	1215	–	450	900	900
9	2250	–	800	1600	1600

Figure 9–12   NEMA ratings for contactors. *(Courtesy of Rockwell Automation's Allen-Bradley Business)*

9.12   Using a Relay to Control the Furnace Fan

One application of using a relay in a heating system is to control the furnace fan. Since the furnace fan is called the evaporator fan if the system is also used for air conditioning, this same relay may be called the evaporator fan relay. Fig. 9–14 shows an example wiring diagram of the electrical system with the fan relay. The fan relay coil is in the control system, and its contacts are in the load circuit connected in series with the load. The fan relay is identified by the letters FR. Notice the coil is identified as FR and the contacts are identified as FR. You should also notice that there are other sets of contacts identified as CC, which means those contacts are controlled by the compressor contactor. The coil of the compressor contactor is identified as CC also.

When you are troubleshooting an electrical system, it is important to locate each coil in the system and see how it is identified. Then you should look around the diagram to locate all sets of contacts that have the same letters identifying them as the coil. Anytime you locate contacts and a coil in a diagram that have the same identification such as FR, you will know they are the coils and contacts

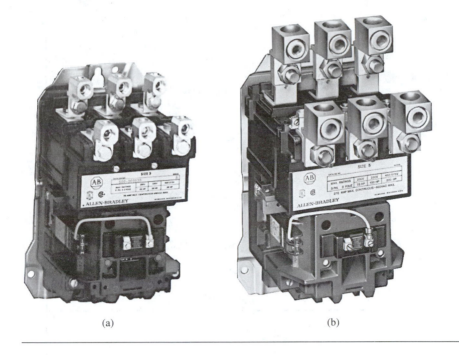

(a)   (b)

Figure 9–13a Example of NEMA size 3 contactors. **b** Example of NEMA size 5 contactors. *(Courtesy of Rockwell Automation's Allen-Bradley Business)*

of the same relay. If you find a coil symbol that is identified as CC, and contacts that are identified as FR, you will know that the coil is part of the compressor contactor, and the contacts are part of the fan relay.

It will be difficult when you first use a ladder diagram to troubleshoot an electrical system because the coils and contacts of a relay are not shown near each other. Instead the contacts are shown in the circuit they are controlling, and the coil is shown in the circuit that controls it. Fig. 9–15 shows a wiring diagram of the same circuit. In this diagram you should notice that the physical outline of each relay is shown, and the contacts for each relay are shown as they would physically be found near the coil. It is important to remember that the wiring diagram is intended to show physical locations and relationships between all of the controls, and the ladder diagram is intended to show the sequence of operation.

9.13   Using a Contactor to Control a Compressor Motor

The electrical ladder diagram presented in Fig. 9–14 and the wiring diagram presented in Fig. 9–15 also show the compressor contactor. The coil of the com-

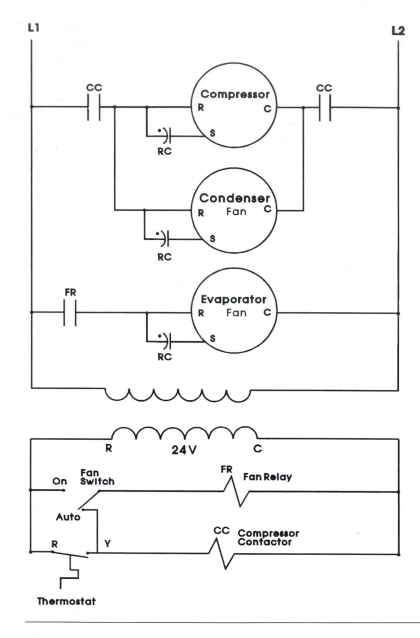

Figure 9–14 Ladder diagram of an air-conditioning system that shows the relay coils in the control circuit, and the contacts of the relays in the load circuit.

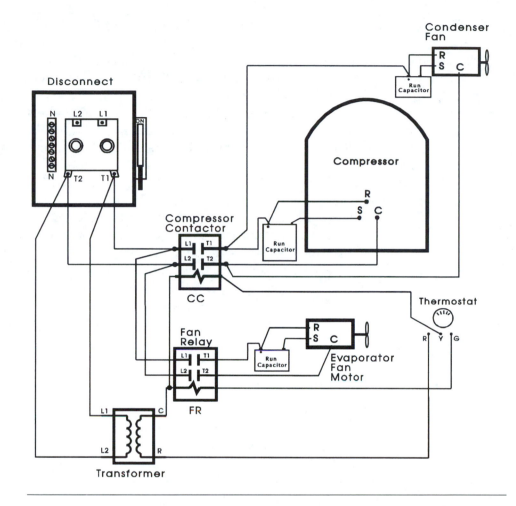

Figure 9–15 A wiring diagram of the air-conditioning system shown in the previous figure. Notice in the wiring diagram that the outline of each relay shows the physical relationship between each of the relays. This diagram also shows all contacts for a relay closed by the relay coil.

pressor contactor is shown in the control circuit of the ladder diagram and it is identified by the letters CC (compressor contactor). You should notice the contacts that are identified as CC are shown connected in series with the compressor motor and condenser fan motor shown in the top line of the diagram.

The compressor contactor is larger than the fan relay and it also has two sets of contacts instead of one, like the fan relay. The wiring diagram is useful to identify the location of the compressor contactor in the control panel. According to the wiring diagram, it is located above the fan relay and to the right of the disconnect. You should also notice that it is rather difficult to determine the circuits and the sequence of operation in a wiring diagram because of the number of wires that are shown and the way that they cross over each other several times. You should also notice that the control circuit and load circuit are not separated in the wiring diagram, which makes it more difficult to understand the operation. The wiring diagram is necessary and it does serve a very important purpose in that it shows the physical relationship of all components in the circuit.

On the other hand, you should notice how easy it is to determine the sequence of operation in the ladder diagram. For example, if you are asked to determine what conditions must exist for the compressor contactor coil to be energized, you could simply look at the ladder diagram and see that terminals R-Y on the thermostat must close to energize the coil. The main drawback of the ladder diagram is that it does not give a quick indication of the location of components, and it does not easily indicate how many contacts a relay has. You can now begin to see why you will need to use both a wiring diagram and a ladder diagram if they are provided. An exercise is provided in the Appendix that shows how to convert a wiring diagram to a ladder diagram.

9.14   The Function of the Control Circuit

As you have seen in this section, the function of the control circuit in air-conditioning, heating, and refrigeration systems is to control the conditions that cause the equipment to become energized or deenergized. The components in the control circuit of residential furnaces and air conditioners are generally low-voltage devices, since the control circuit operates on low voltage. The relay coil is located in the control section of the electrical system and the relay contacts are located in the load section; they are typically connected in series with the load they are controlling. It is important to remember that the function of the control circuit is to energize the relay coil so power can be sent through the relay contacts to the load. The action in the control circuit must always occur before the action in the load circuit.

If you must troubleshoot an electrical system for an air-conditioning, heating, and refrigeration system, you should make two checks. First, you should see if the loads are energized and operating correctly. If the loads are not energized, you should suspect the control circuit is not energized, and move on to the second test and make checks in the control circuit. As you become more familiar with the control circuit and load circuit, you will find it is almost automatic to check the control circuit to ensure that the relay coil is energized to make sure its contacts are energized. The next sections of this book will present in-depth material concerning the operation of typical loads, such as the compressor motor and the fan motors.

9.15   Solenoids Used in HVAC and Refrigeration Systems

A solenoid is a *magnetically controlled valve*. You learned in earlier chapters about the power of current flowing through a coil of wire and causing a strong magnetic field to develop. The strong magnetic field in the solenoid valve is used to make the valve plunger move from the open to the closed position or from the closed to the open position. The valve is part of the refrigerant system plumbing or the plumbing for the water system. Solenoid valves allow the plumbing system to be interfaced with the electrical system in order for the plumbing system to be automated. For example, in residential heat pumps, a solenoid valve is used as a reversing valve. The reversing valve allows the system to change from a heating system to a cooling system and it also allows the refrigerant flow to be reversed automatically during the defrost cycle to provide hot gas to the outside coil to defrost it.

9.16   Basic Parts of a Solenoid Valve

Fig. 9–16 shows typical solenoid valves and Fig. 19–17 shows a cut-away diagram of one of the solenoid valves. From these figures you can see that the so-

Figure 9–16   A typical solenoid valve. *(Courtesy of Danfoss Automatic Controls)*

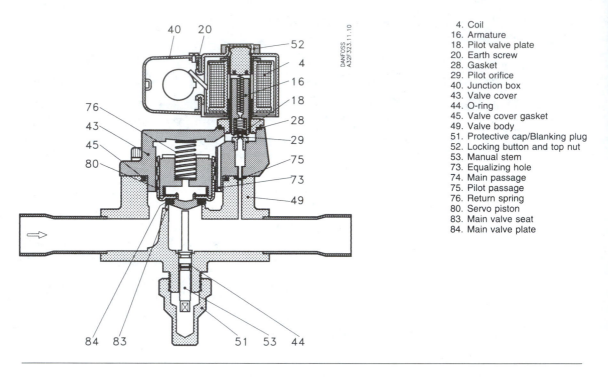

4. Coil
16. Armature
18. Pilot valve plate
20. Earth screw
28. Gasket
29. Pilot orifice
40. Junction box
43. Valve cover
44. O-ring
45. Valve cover gasket
49. Valve body
51. Protective cap/Blanking plug
52. Locking button and top nut
53. Manual stem
73. Equalizing hole
74. Main passage
75. Pilot passage
76. Return spring
80. Servo piston
83. Main valve seat
84. Main valve plate

DANFOSS
A32F323.11.10

Figure 9–17 Diagram of a typical solenoid valve. *(Courtesy of Danfoss Automatic Controls)*

lenoid valve is a plumbing-type valve that has a movable plunger used to close off the valve. The valve can be designed so that the plunger is held in either the open position by spring pressure, or it can be designed to be held in the closed position by spring pressure. When electric current is applied to the coil in the valve, the strong magnetic field will cause the plunger to move upward. If the valve is the type where spring pressure is causing the plunger to stay in the closed position, the magnetic field will cause the plunger to open the valve passage. If the valve is the type where spring pressure is causing the plunger to stay in the open position, the magnetic field will cause the plunger to close the valve passage.

The coil for the solenoid is exactly like the coils that you learned about previously for relays and contactors. Since the coil is basically a long wire that is coiled on a plastic spool, it will have two ends. You can test the coil for continuity and you should notice that each type of solenoid coil has some amount of resistance that will vary from approximately 50 Ω to 1500 Ω. Remember the exact amount of resistance does not matter because the coil is made from one long piece of wire. It is doubtful that the amount of resistance will change. The main problem a coil will have is an open somewhere along its wires.

9.17   Troubleshooting a Solenoid Valve

At times you will be working on a system that has one or more solenoid valves and you will suspect the solenoid valve is not operating correctly. The first test you should make when troubleshooting a solenoid valve is to read its data plate and determine the amount of voltage the valve should have and then measure the amount of voltage that is available across the two wires that are connected to the coil. It is important to remember that the coil of the reversing valve must have the correct amount and type of voltage applied to its two wires if it is to conduct sufficient current to cause the plunger to move. This means that if the valve is rated for 120 VAC, you should measure 120 VAC across the two coil wires. If the voltage is missing or is the incorrect amount, the coil will not operate correctly. If the proper amount of voltage is not present, you will need to use troubleshooting procedures to locate the open in the circuit that is causing the loss of voltage before it reaches the coil.

If the proper amount of voltage is present and the coil will not activate, you must make several additional tests. You can test to see if the coil has developed a magnetic field by placing a metal screwdriver near the coil to see if it is attracted by the magnetic field. If the screwdriver is attracted to the coil by the magnetic field but the valve is not opening or closing, the valve seat may be frozen in place and the valve will need to be replaced.

The other problem that the coil may have if the correct amount of voltage is present at the coil wires is that the coil may have an open in its wire. You can test the coil wire with a continuity test. Be sure to turn off all voltage to the solenoid and disconnect the coil wires so that the coil is isolated from back feeding to the remainder of the circuit. If the continuity test indicates infinite (∞) resistance at the highest resistance range, you can suspect the coil has an open and it should be replaced. You should notice that the coil can be replaced without unsoldering the valve.

9.18   Reversing Valve for Heat Pumps

The reversing valve is a specialized solenoid valve. Fig. 9–18 shows a picture of a reversing valve and Fig. 9–19a and Fig. 9–19b show diagrams of a reversing valve in a circuit. The heat pump is basically an air-conditioning system that has the flow of its refrigerant reversed in cold weather to route the hot, high-pressure refrigerant gas from the compressor through the indoor coil where its heat is used to heat the indoor space. In this mode the outdoor coil is used as the evaporator.

In warm weather, the reversing valve switches so that the flow is returned to a normal air-conditioning system where the compressor pumps the hot, high-pressure gas to the outdoor coil to be condensed. The indoor coil in the normal

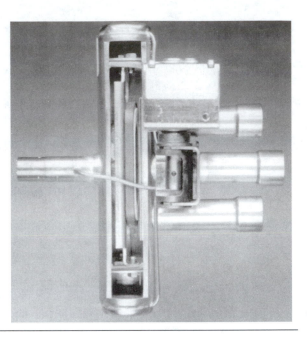

Figure 9–18 A reversing valve is a specialized solenoid valve that is used in heat pumps. *(Courtesy of Alco Controls Division, Emerson Electric Company)*

air-conditioning mode is used to evaporate the liquid refrigerant, which takes heat from the indoor space and causes it to cool.

In the diagram you can see that the indoor coil is on the right side and the outdoor coil is on the left side. The top diagram represents a typical cooling cycle. The reversing valve switches to route the refrigerant as a hot, high-pressure gas to the outdoor coil where it condenses and turns to a liquid. The liquid refrigerant in this diagram is represented by the solid line. The liquid refrigerant in the cooling cycle moves through the indoor coil where the heat in the indoor space is used to cause the liquid to evaporate.

The bottom diagram shows the route of the refrigerant when the reversing valve is switched to the heating cycle. In this diagram you can see that the refrigerant is a hot, high-pressure gas as it comes from the compressor and it is directed to the indoor coil. The heat is removed from the hot, high-pressure gas and it is used to heat the indoor space. The refrigerant condenses while it is moving through the indoor coil and it moves to the outdoor coil where it evaporates. Again the path of the liquid refrigerant is shown as a dark line.

9.19   The Internal Operation of the Reversing Valve

The reversing valve is a specialized solenoid valve that is called a pilot-operated valve, which consists of a small solenoid-operated valve that moves a small amount of vapor and liquid to make the larger main valve switch position. This

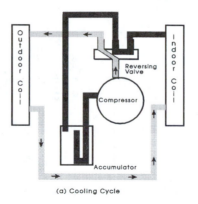

(a) Cooling Cycle

Figure 9–19a A reversing valve is switched so that the heat pump is in its cooling cycle. The indoor space is cooled when the heat pump is in this mode. **b** The reversing valve is switched so that the heat pump is heating the indoor space.

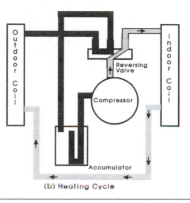

(b) Heating Cycle

means that the small solenoid valve switches a small amount of liquid refrigerant and vapor refrigerant directly to the main valve. The main valve is the part of the reversing valve that actually switches to change the direction of the refrigerant flow.

Fig. 9–20a and Fig. 9–20b show diagrams of the internal operation of the reversing valve. In the top diagram the reversing valve has switched to the cooling cycle, and it is switched to the heating cycle in the bottom diagram. The diagrams for this figure show that the main valve has one port on one side of the valve and three ports on the other side. The single port shown on the top part of the valve diagram is soldered (welded) to the compressor discharge line. The port on the left side of the diagram is soldered to the condenser coil, the middle port is soldered to the compressor suction line, and the port on the right side is connected to the evaporator coil.

The solenoid valve is shown on the right side of the diagrams and it has three lines connected to it. The solenoid in the reversing valve is the pilot valve. This

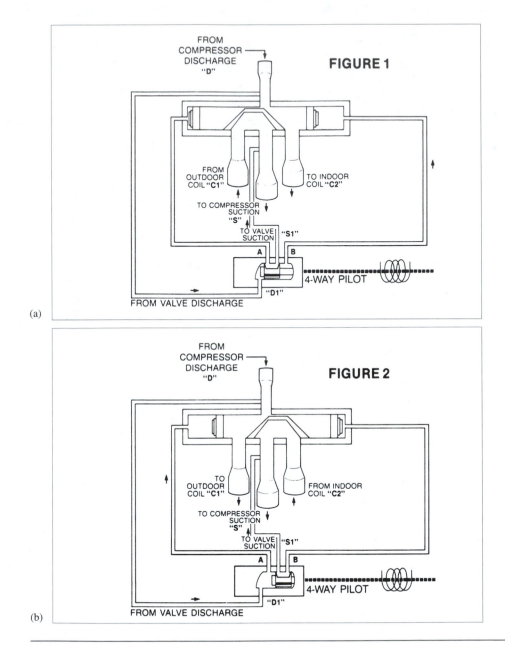

Figure 9–20a   The internal diagram of a reversing valve in the cooling cycle.
b The internal diagram of a reversing valve in the heating cycle. *(Courtesy of Alco Controls Division, Emerson Electric Company)*

means that it takes a small amount of refrigerant and directs it to the main valve to cause the main valve to switch positions. The middle line in the solenoid valve is connected to the compressor suction line where it takes low-pressure refrigerant that is a vapor to use as the pilot fluid. In the top diagram you can see the reversing valve is switched for cooling, and the solenoid valve moves the pilot fluid to the left side of the main valve, causing the valve to shift to the right. You should remember that even though the pilot fluid comes from the compressor suction line, it will have 40–70 psi, which is sufficient pressure to cause the main valve to change positions.

When the main valve is switched to the right, it makes a plumbing connection between the compressor suction line and the indoor coil. A plumbing connection is also made between the compressor discharge line and the outdoor coil. In the bottom diagram the solenoid valve has switched the pilot fluid so that it activates the right side of the main valve, which causes the main reversing valve to switch to the left side. When the main valve is switched to the left side, it makes a plumbing connection between the compressor suction line and the outdoor coil, and another plumbing connection between the compressor discharge line and the indoor coil.

The reason the reversing valve is a pilot-operated valve is because a smaller coil can be used to ensure the small pilot part of the valve changes position. When the pilot part of the valve switches, it can accomplish much more work from the advantage of the pressure of the pilot fluid. This ensures the valve is as small as possible, yet it will switch positions when the electrical coil of the solenoid is energized and deenergized.

Questions for This Chapter

1. Identify the main parts of a relay and explain their operation.
2. Explain the operation of the four types of armature assemblies presented in this chapter.
3. Explain the operation of a current relay and a potential relay.
4. Discuss the differences between a contactor and a relay.
5. Identify the main parts of a solenoid and explain their operation.

True or False

1. Plug-in relays are used in air-conditioning and refrigeration systems because they are easy to change when they become faulty.
2. The main difference between a relay and a contactor is the contactor usually has more contacts.

3. A solenoid is similar to a relay in that it has a coil that becomes magnetic and moves the plunger of a valve.
4. It is possible to change normally open contacts to normally closed contacts in the field for some relays.
5. For a relay to operate correctly, its contacts must close first to provide current to its coil.

Multiple Choice

1. The normally closed contacts of a relay _____

    a. pass current when the coil is not energized.
    b. pass current when the coil is energized.
    c. pass current at all times whether or not the coil is energized.

2. A current relay is _____

    a. a special relay designed to control only current.
    b. a relay that is designed to be used as a relay or a contactor.
    c. a relay that is designed to start single-phase compressors.

3. A solenoid is _____

    a. a valve that is controlled (opened and closed) by gravity and springs.
    b. a reversing relay that is used to start heat pumps.
    c. a valve that is controlled (open and closed) by springs and a magnetic field produced by a coil.

4. A reversing valve is _____

    a. a special type of solenoid used in heat pumps.
    b. a special relay used to reverse the direction in which the compressor motor runs.
    c. a special valve that can be used as either a fill valve or a drain valve for a heat pump.

5. An SPDT relay has _____

    a. one set of contacts that connects to two output terminal points.
    b. two sets of contacts that connect to one terminal point.
    c. one set of contacts and one output terminal point.

Problems

1. Draw a waveform that shows pull-in and hold-in current for a relay coil.
2. Use the NEMA table provided in Fig. 9–12 to select a NEMA starter size that can control 45 continuous amps.

3. Draw an example of single-pole, double-throw (SPDT) contacts and explain their operation.
4. Draw an example of double-pole, single-throw (DPST) contacts and explain their operation.
5. Draw an example of double-pole, double-throw (DPDT) contacts.

10 Single-Phase Open Motors

OBJECTIVES:

After reading this chapter, you will be able to:
1. Explain the theory of operation of a simple induction AC motor.
2. Identify the main parts of an AC motor and explain their functions.
3. Identify the parts of the centrifugal switch and explain their operations and functions.
4. Explain the data found on a typical motor data plate.

10.0 Overview of AC Single-Phase Induction Motors

The main loads that you will work on in most air-conditioning, heating, and refrigeration systems are motors of all types. Motors are used to move air over evaporator coils, condenser coils, and over heat exchangers. They are also used as compressors to pump refrigerant. The motors that are used in compressors are sealed inside a housing so that the refrigerant does not leak out to the atmosphere and they are called *hermetic motors* or *hermetic compressors*. The motors that are used in fan applications are called *open-type motors*. Fig. 10–1 shows a picture of an open-type fan motor. The theory of operation, installation, and troubleshooting for open-type motors that are used to power fans, pumps, and

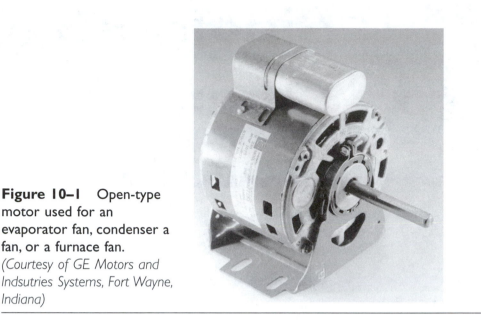

Figure 10–1 Open-type motor used for an evaporator fan, condenser a fan, or a furnace fan.
(Courtesy of GE Motors and Indsutries Systems, Fort Wayne, Indiana)

belt-driven loads will also be explained in this chapter. The material in these sections will show that some motors require special switches and components to start and operate correctly. These devices will be explained so that when you must troubleshoot a motor you will fully understand their function. A diagram and picture will be provided for each type of motor, so that you will learn to identify each type of motor from its physical appearance and from its electrical diagram. As a technician you will be expected to identify, install, troubleshoot, and repair or replace any motors you encounter. This chapter will provide a comprehensive overview for all single-phase, open-type motors you would find in HVAC and refrigeration systems.

10.1   AC Split-Phase Motor Theory

The AC split-phase motor is widely used in HVAC applications and it has many of the basic parts of the other types of open motors. This motor has three basic parts: the stationary coils that are called the *stator,* the rotating shaft called the *rotor,* and the *end plates* that house the bearings that allow the rotor shaft to turn easily. Fig. 10–2 shows an exploded view diagram of a capacitor-start, induction-run open-type motor so that you can see where these parts are located inside the motor. The capacitor-start, induction-run motor is a variation of the split-phase motor that has a start capacitor mounted on it to provide more starting torque. Fig. 10–3 shows an exploded view diagram of a larger three-phase motor so that you

Figure 10–2 Exploded view picture of a capacitor-start, induction-run motor that is a typical single-phase motor. *(Courtesy of Leeson Electric Corporation)*

can see more easily each of the parts. From these diagrams you can see that the rotor is supported by the bearings in the end plates of the motor. The rotor is also mounted directly inside the stationary windings so that when voltage is applied to them, their magnetic field can be induced into the rotor. The parts of a

Figure 10–3 Exploded view picture of a three-phase motor. This picture shows the relative location of all of the parts of the motor. *(Courtesy of Reliance Electric)*

hermetic motor are similar but they are not visible because they are housed inside the hermetic dome.

If the AC split-phase motor requires a *centrifugal switch* to help it get started, it will be mounted inside the end plate. The actuator for the centrifugal switch is mounted on the end of the rotor's shaft, and it has flyweights that swing out to open the centrifugal switch when the motor reaches 75–85% full rpm. In the next sections of this chapter you will see how these parts operate with each other to provide a rotating force to the motor's shaft to turn a fan or pump.

10.2   The Rotor

The rotating part of the motor is called the *rotor.* When voltage is applied to the coils of wire in the stator, they will produce a very strong magnetic field. This magnetic field is passed to the rotor by induction in much the same way as voltage is passed from the primary winding to the secondary winding of a transformer. After the rotor becomes magnetized, it will begin to rotate. The speed of the rotor is determined by the number of poles and the frequency of the AC voltage that is applied to the motor. The rotor has a shaft that rotates to do the work of the motor, such as to turn a fan blade or to move the compressor piston to pump refrigerant. Fig. 10–4a shows a diagram and Fig. 10–4b shows a picture of a rotor that is used in most AC single-phase motors. As a rule, the rotor of an AC motor does not have any wire in it (only repulsion start or some synchro-

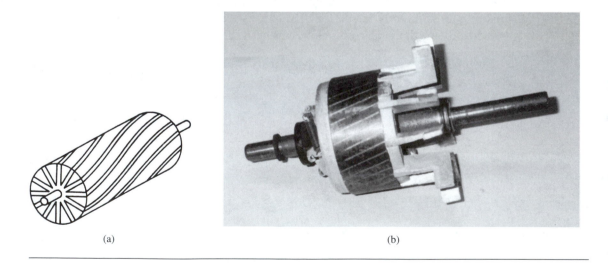

(a)   (b)

Figure 10–4a   Diagram of a squirrel-cage rotor for an AC motor. The rotor bars will become multiple bar magnets when the rotor is magnetized. **b** Picture of a squirrel-cage rotor for an AC motor.

nous AC motors have wire in their rotors, but these motors are not used too often in HVAC applications). The rotor is made from pressing laminated steel plates on the frame. The frame of the rotor looks like the wire-frame exercise wheel that is used by hamsters. For this reason it has been named a *squirrel cage,* and when it is used in the rotor it is called the *squirrel-cage rotor.* The squirrel cage is actually the frame for the rotor, and the part that looks like the squirrel cage is actually made up of rotor bars that act exactly like multiple bar magnets. The laminated steel plates that are pressed onto the frame of the squirrel-cage rotor allow the rotor bars to be magnetized by induction when current is passed through the stator. This means that the permanent field in the stator acts like the primary of a transformer, and the rotor acts like the secondary windings of a transformer. The result of this is that the induction motor does not need brushes to pass current to the rotor to get its magnetic field to build. This is also why these types of motors are called *induction motors.* It is also important to understand that the laminated steel sections in the rotor allow it to become magnetized very easily by the AC voltage, and then quickly change the polarity of its magnetic field as the AC voltage oscillates from positive to negative as the AC sine wave changes. The polarity of the magnetic field in the rotor of the AC motor gets changed because the sine wave of the AC voltage changes from positive to negative naturally during each cycle of the sine wave. This causes the magnetic fields in the rotor continually to change polarity and spin as long as AC voltage is applied to the motor.

10.3   Locked-Rotor Amperage and Full-Load Amperage

When voltage is first applied to the motor, the run and start windings will draw a large amount of current called *locked-rotor amperage* (LRA). This current is called locked-rotor amps because the rotor has not started to turn at this time. The amount of current is large because the amount of resistance in the windings is very small, and because the rotor has not yet started to rotate and create *counterelectromotive force* (CEMF voltage). When the rotor starts to spin and comes up to full rpm, the current will return to normal and it will be called *full-load amperage* (FLA). Fig. 10–5 shows a graph of the large LRA current that occurs when the motor is first started and it also shows the current returning to lower levels (FLA) when the motor is running at full-load speed.

The reason the current returns to its normal value after the rotor is spinning at full speed is due to the rotor producing CEMF. When the rotor starts to spin, its magnetic field will begin to pass the coils of wire in the run winding and a voltage will be generated when the magnetic field passes the coils. This generated voltage will be out of phase with the applied voltage, so it will be called counter EMF (CEMF). Counter EMF is also sometimes called *back EMF.*

The amount of CEMF will be equivalent to the speed the rotor is turning. If

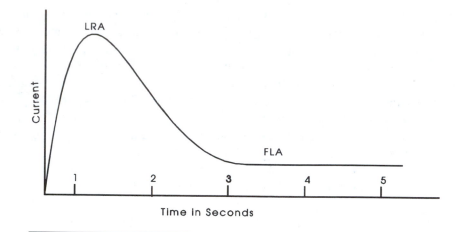

Figure 10–5 Graph of the locked-rotor amperage (LRA) that occurs when a motor is first started, and the full-load amperage (FLA) that occurs when the motor is running at full speed.

the rotor is turning at 90% full speed, the CEMF will be approximately 90% of the applied EMF. For example, if the applied voltage is 230 VAC, and the rotor is turning at approximately 90% rpm, the CEMF may be 200 VAC. If the resistance of the run winding is 10 ohms, the full-load current would be determined by dividing it by the difference in voltage (230 – 200), which is 30 VAC. This means the FLA current would be approximately 3 amps.

10.4   Rotor Slip and Torque in an AC Induction Motor

The amount of LRA and FLA that an AC induction motor draws will also determine the amount of torque the motor's shaft will have. *Torque* is defined as the amount of *rotating force* that the rotor shaft has. You should remember that this force is needed to pump the piston of the compressor and turn the fan blades for a condenser fan. If the motor has a large amount of LRA during starting, it will also have a lot of starting torque. If the motor draws a large FLA when the motor is running, it will have a lot of running torque.

The amount of FLA will be determined by the amount of slip a motor has. *Slip* is defined as the *difference in the rated speed of a motor and its actual running speed*. For example, a four-pole motor is rated for 1800 rpm, but its actual running speed will be approximately 1725 rpm. The difference of 75 rpm is called slip. At first glance you would think that slip is not good for a motor because the motor is not running as fast as it should be, but you will see that the induction motor relies on slip to create the amount of current necessary to turn the load at the end of the motor shaft. If the amount of slip was minimal and the rotor is spinning at close to the rated speed, the amount of CEMF would be large, and the difference between the CEMF and the applied EMF may be as small as 5 VAC. When this occurs, the amount of current in the motor is reduced to less than 0.5 amp, and consequently the amount of torque the rotor has is also reduced. It is also important to remember that the rotor gets the current required to build its magnetic fields through induction from the stator windings. The induction can only occur if the rotor is not turning at the same speed as the magnetic field is rotating in the stator.

When the rotor's torque is minimal, the shaft will not be able to turn its load, and it will begin to slow down, which decreases the amount of CEMF. This makes the difference between the CEMF and the applied voltage larger, which will make the motor draw more current and produce more torque. All induction motors rely on slip to operate correctly and produce sufficient torque. If it is important that the motor run at its rated speed, a special type of motor called *synchronous motor* could be used because it runs at its rated speed with no slip. The synchronous motor is not used very often in HVAC or refrigeration applications because it requires DC voltage to be supplied to its field.

It is also possible to use a variable-frequency drive to adjust the frequency of the voltage above 60 hertz to get the motor to run at its rated speed even if it has slip. Variable-frequency drives have been introduced into HVAC and refrigeration applications since the 1980s. You will learn more about variable-frequency drives in the chapter that covers electronic devices. As a technician you will encounter a variety of variable-frequency drives in newer residential air-conditioning and heating systems where they are used to control the speed of compressors and fan motors.

10.5   End Plates

The end plates play a very important part in the operation of the motor. They house the bearings that support the rotor to allow it to spin with a minimum amount of friction, and one end plate houses the end switch that is used to help the single-phase motor to get started. The bearings in a single-phase motor may be the ball-bearing type or a bushing type. The ball bearing is more expensive and is generally used in larger single-phase motors. The bushing is used in smaller and less expensive single-phase motors. Fig. 10–6 shows an armature with one end plate and Fig. 10–7 shows an example of a ball bearing and several types of bushing. The ball bearing shown in Fig. 10–7a uses a number of small steel balls that create a rolling surface between the inner hub and outer hub. The shaft of the motor is pressed securely into the inner hub of the bearing so that the shaft and inner hub of the bearing move together as though they were welded. As the

Figure 10–6 An end plate shown with a rotor. The bearings are mounted in the end plates and they support the rotor shaft so that it can spin freely.

rotor shaft spins, the inner hub rolls on the balls and the balls make contact with the outer hub. This means that only a small portion of each ball is touching the inner and outer hubs at any time, which keeps friction to a minimum. The ball bearing for smaller motors is lubricated with grease when it is manufactured and usually it is sealed for the life of the bearing, which means that motors with these types of ball bearings will not need to be lubricated. Some larger motors use bearings that must be greased periodically. A grease fitting is provided so that the ball bearing can be greased once or twice a year, depending on the number of hours the motor runs.

The bushing shown in Fig. 10–7b is made of a brass or bronze alloy sleeve with a lubrication hole in it. An *oiler pad,* which is also called the *oil wick,* is mounted on the outer side of the bushing, so that oil from the pad can find its way through the hole to the inside of the bushing. The oil creates a "bearing"

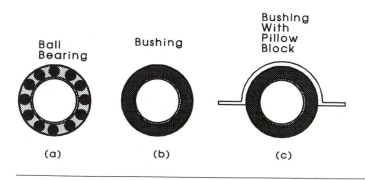

Ball Bearing   Bushing   Bushing With Pillow Block

(a)   (b)   (c)

Figure 10–7a An oil-type bearing (bushing) with insulator. **b** Sealed-type bronze bushing and sealed ball bearing. **c** Pillow block-type bearing with bronze bushing.

surface between the shaft of the rotor that spins, while the bushing remains stationary. As long as a sufficient amount of oil remains on the inside of the bushing, between it and the rotor shaft, the amount of friction will be held to a minimum. If the bushing ever runs dry, the rotor shaft will be allowed to touch the bushing surface and both surfaces will begin to wear severely. It is vitally important to ensure that the lubrication ports for the bushing-type motor are always located on the top side of the motor so that gravity can draw the lubrication through the lubrication pad into the bushing. It is also important to remember that the motor that uses a bushing must be lubricated at least once a year. Fig. 10–7c shows a bushing with a pillow block to hold it in place when it is used on the shaft of a squirrel-cage fan.

10.6   The Stator

The stator is actually made of two separate windings called the *run winding* and the *start winding*. Each of these windings is subdivided into poles which represent the poles of a magnet. We will use a four-pole motor for our examples in this chapter. This means that the run winding will have four poles and the start winding will have four poles. Fig. 10–8 shows a picture of a stator so that you can more easily see the coils of wire that are mounted in it. From this picture you can see the four poles of the run winding and the four poles of the start winding. The windings are pressed over pole pieces that also become magnetized when the coils have current flowing in them. The pole pieces are made

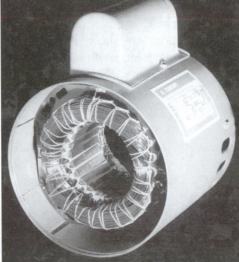

Figure 10–8   Close-up picture of start and run windings in the stator. The run winding has larger wire than the start winding.
(Courtesy of Century Electric, Inc.)

from laminated steel like the rotor so the poles can quickly change polarity when AC voltage is applied.

The AC sine wave starts at 0 volts and peaks positive, returns to 0 volts and peaks negative, and returns to 0 volts again. It continues to oscillate positive and negative as long as voltage is applied. This action causes the magnetic field in each winding in the stator to be positive and then negative and continually change. The changing magnetic field actually results in the magnetic field rotating around the poles inside the stator. Since the rotor also has a magnetic field, it will begin to spin and follow the rotating magnetic field in the stator. It will be easier to understand the function of the stator if you study the operation of the start winding and run winding separately.

10.7   The Start Winding

If the AC induction motor only had one winding, the rotor would tend not to rotate when single-phase AC voltage is first applied to the stator windings in the AC induction motor. The rotor needs a small amount of phase shift between the applied voltage and current in the magnetic field created in the stator to begin to spin. There are several ways to provide the phase shift required to get the rotor to start to spin. One way to provide the phase shift is to physically offset two windings in the stator. A second way to cause a phase shift in the stator is to put more wire in one set of windings than the other. For this reason, the start winding is made from smaller wire than the run winding, so that it has more wire in each coil than the run winding.

Fig. 10–8 shows a close-up picture of a stator that clearly shows the run and start windings. From this picture you can see that the run winding is made of larger wire and it is physically offset from the start winding. Fig. 10–9a shows a sketch of how the run winding and start winding are placed in the stator, offset from each other. Fig. 10–9b also shows an electrical diagram of the start winding and run winding for an eight-lead, AC single-phase motor and Fig. 10–9c shows a four-lead, single-phase motor.

The electrical diagram in Fig. 10–9a shows that the start winding is made of smaller wire and has more turns than the run winding. The start winding coil has more turns of wire than the run winding coil so that more feet of wire will be used in the start winding. Since the start windings has more feet of wire, it takes the current longer to pass through it than the run winding. It is important to understand that when voltage is applied to the AC induction motor, voltage starts into the run winding and the start winding at the same time. Since the start winding has more wire and it is offset physically from the run winding, a phase shift is created. (See Fig. 10–10.)

This phase shift gives the rotor a sufficient amount of starting torque (rota-

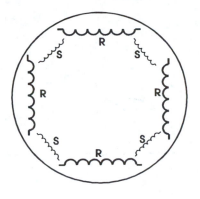

(a)

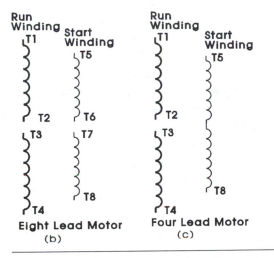

Eight Lead Motor
(b)

Four Lead Motor
(c)

Figure 10–9a A diagram that shows the four-pole run winding and four-pole start winding in the stator of an AC induction motor. You should notice that the start winding is physically offset in the stator from the run winding to help provide more of a phase shift. The run winding is made of larger-gauge wire and fewer turns than the start winding. **b** A diagram of the run winding and start winding of an eight-lead induction motor. Notice that the two sections of the run winding are identified as T1-T2 and T3-T4, and the two sections of the start winding are identified as T5-T6 and T7-T8. **c** A diagram that shows a four-lead induction motor. The run winding is identified as T1-T4, and the start winding is identified as T5-T8.

tional force) to get the rotor to start turning. Once the rotor begins to turn, it will cause the magnetic field to continue to build as the motor comes up to full speed. The phase shift occurs between the run winding and the start winding, but it also can be described as the phase shift that occurs between the voltage waveform and the current waveform because inductors (motor windings) cause the current waveform to lag the waveform of voltage. In later sections you will see how

Figure 10–10 A diagram of the phase shift that occurs when voltage is applied to the start and run windings of an AC motor. Since the start winding has more wire than the run winding, a phase shift will be created between the magnetic fields in the two windings.

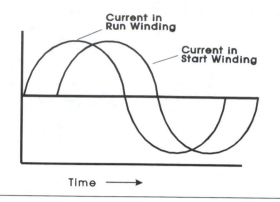

capacitors can be added to the start windings to enhance the amount of phase shift so the motor will have more starting torque.

Since the start winding is made from very small wire so that its coils can have many more turns of wire than the run winding, the start winding will draw a large amount of current when voltage is applied to it. If the start winding was allowed to remain in the circuit continually, this large current would damage it and eventually burn a hole in the windings. For this reason a centrifugal switch is used to disconnect the start winding after the motor's rotor is spinning. The centrifugal switch can determine when the speed of the rotor has reached 75–85% full rpm and disconnect the start winding from the circuit until it is needed the next time the motor is started.

10.8   The Operation of the Centrifugal Switch

The centrifugal switch is also called the *end switch* in a motor because it is mounted in the end plate. Even though the centrifugal switch is physically mounted in the end plates, it is important to understand that it is electrically connected to the start winding that is located in the stator. The centrifugal switch is actually part of a larger assembly that consists of an activation mechanism that has flyweights that begin to swing out when the rotor reaches 75% full rpm. The activation mechanism is mounted on the end of the rotor shaft so it can turn when the rotor shaft is turning. The activation mechanism is mounted on the shaft in such a way that it makes contact with the centrifugal switch and holds the switch contacts in the closed position when the shaft is not rotating.

Fig. 10–11 shows a series of pictures of a centrifugal switch and the activation mechanism. Fig. 10–11a shows a centrifugal switch that is removed from an end plate, and an activation mechanism removed from the armature shaft. Fig. 10–11b shows a centrifugal switch as you would find it mounted in the end plate.

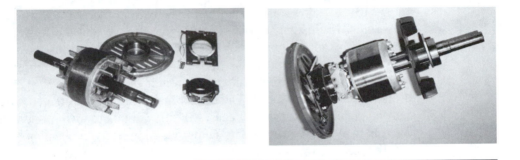

Figure 10–11a   Centrifugal switch removed from an end plate and activation mechanism removed from the armature shaft. **b** Centrifugal switch mounted in the end plate and the end plate is mounted on the end of the motor shaft so that the activation mechansm is touching the centrifugal switch.

This figure also shows the location of the activation mechanism as it touches the centrifugal switch. You can see its flyweights at rest as you would find them when the rotor is not turning.

The centrifugal switch is a set of electrical contacts mounted on spring steel. Spring steel is used because it can be flexed so that the spring tension in the steel tends to keep the contacts in the open position. The activation mechanism that is mounted on the rotor shaft will come into contact with the spring steel of the switch when the end plate is attached to the stator. The rotor shaft goes through the bearing in the end plate and when the end plate is drawn tight against the stator with screws, the activation mechanism will touch the spring steel and cause the centrifugal switch contacts to press against each other to complete the electrical circuit through the switch.

When voltage is applied to the windings in the motor, current will flow through the run winding and through the start winding, since the activation mechanism is holding the contacts of the centrifugal switch closed. As the rotor begins to spin, centrifugal force begins to pull the flyweights outward. Since the flyweights are held together with a spring on each side, they will not swing out all the way until the rotor is spinning at approximately 75–85% full rpm. The spring tension that holds the flyweights inward will determine the speed at which the flyweights will swing out completely.

When the flyweights swing out completely, they allow the spring tension on the activation mechanism to move it to a retracted position, which allows the spring tension in the contacts of the centrifugal switch to move them to the open position. When the contacts of the centrifugal switch open, current to the start winding is interrupted and the motor continues to run with current flowing only through the run winding of the motor.

Fig. 10–12a shows a diagram of the activation mechanism when the rotor is not turning. In this part of the diagram you can see that the flyweights are held in position by spring tension. When the flyweights are in this position, the activation mechanism is pressed against the switch contacts holding them in the closed position so current can flow through them. In Fig. 10–12b the rotor is turning at nearly full speed. The centrifugal force that is created when the rotor shaft is spinning at full speed causes the flyweights to swing out away from the rotor shaft. When the flyweights swing out completely, the activation mechanism is pulled away from the contacts, which allows the tension of the spring steel to open the contacts.

10.9   The Run Winding

The run winding is the part of the stator that remains in the motor circuit at all times. Fig. 10–13 shows an electrical diagram for a single-phase motor and you

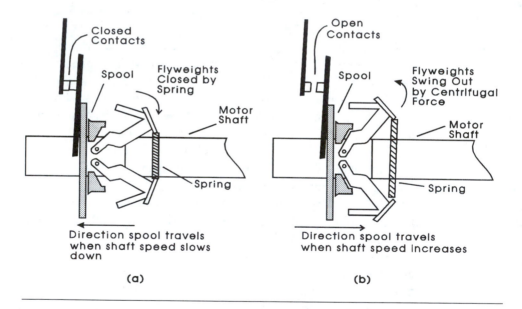

(a)   (b)

Figure 10–12a   Flyweights and activation device for centrifugal switch. The activation device is mounted on the end of the rotor shaft so that it can come into contact with the centrifugal switch. A spring holds the flyweights against the shaft when the rotor is not turning. **b** When the rotor is spinning at 75–85% full rpm, the flyweights move outward from the shaft because of centrifugal force. The action of the flyweights causes the activation device to drop away from the switch contacts so that they can move to their open position.

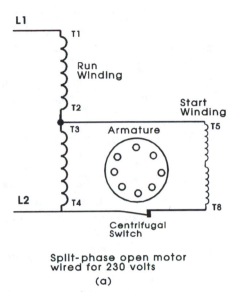

Split-phase open motor
wired for 230 volts

(a)

Figure 10–13a Electrical diagram of an AC induction motor wired for high voltage (230 or 208 VAC). The two sections of the run winding numbered T1 and T2, and T3 and T4 are connected in series. **b** Electrical diagram for an AC induction motor wired for low voltage (115 VAC). The run winding in this diagram is shown in two sections that are connected in parallel.

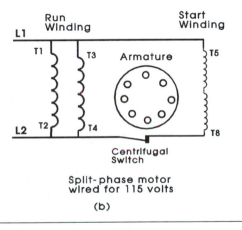

Split-phase motor
wired for 115 volts

(b)

can see that the run winding is easily identified since it uses larger wire and fewer coils in its windings. In most single-phase motors, the run winding is numbered as shown in the diagram. If the motor is a dual-voltage motor, its run winding is made in two sections, one with terminal ends T1 and T2, and the other with terminal ends T3 and T4. If the motor is wired for high voltage (230 or 208 VAC), the windings will be wired in series with each other. (See Fig. 10–13a.)

Fig. 10–13b shows the two sections of the run winding connected for low voltage (115 VAC). If the motor is wired for low voltage, the two sections of the run winding will be connected in parallel with each other. These two diagrams will be used in all electrical system diagrams that you will use to install and troubleshoot motors in heating, ventilating, air-conditioning, and refrigeration systems.

You can test the run winding and start winding with an ohmmeter to determine which one is which. For example, in the diagram each part of the run winding has 4 ohms of resistance and the start winding has 12 ohms.

10.10   Multiple-Speed Motors

The coils of wire in the run winding can be separated into sections called poles. The motor will have an even number of poles. The speed of an induction motor is determined by the number of poles and the frequency. The speed can be calculated from the following formula:

$$\text{rpm} = \frac{\text{freq} \times 120}{\text{poles}}$$

A typical motor will have two, four, six, or eight poles. The speed of each type of motor is listed in the following table.

Number of poles	rpm
2	3600
4	1800
6	1200
8	900

Motors for HVAC and refrigeration applications need to be able to operate at a variety of speeds. This means that the motor may be manufactured with a set number of windings to provide a set speed, or the motor may be reconnected in the field to give the motor a different speed.

Another way to change the speed of a motor is to tap the run winding so that more or less inductive reactance is used. This means that if you have a very long run winding, you can place taps on it just like the multiple-tapped transformer. Each tap uses less of the total run winding and, consequently, the change in inductive reactance will allow the speed of the motor to change. The amount of change in rpm for each tap is approximately 15–20%. Fig. 10–14 shows an example of a motor with a multitapped run winding. From the diagram you can see the taps are in the run winding, and you can see the color code for the wires. If you want the motor to run at high speed, you would use the black wire for the run winding. If you wanted the motor to run at medium speed, you would use the blue wire, and the red wire would allow the motor to run at low speed. Generally, the multiple speeds are used for fan motors, where the high speed is used for cooling and the low speed is used for heating. The medium speed is used

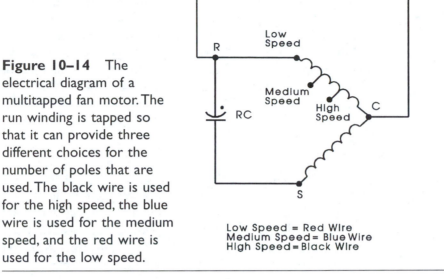

Figure 10–14   The electrical diagram of a multitapped fan motor. The run winding is tapped so that it can provide three different choices for the number of poles that are used. The black wire is used for the high speed, the blue wire is used for the medium speed, and the red wire is used for the low speed.

Low Speed = Red Wire
Medium Speed = Blue Wire
High Speed = Black Wire

when the customer wants a slightly lower speed for air conditioning, or a higher speed than the lowest speed for heating. It is important to remember that as the tap is changed for this motor, it will use less of the winding to run faster, but it will also lose power.

10.11   Motor Data Plates

The motor's *data plate* lists all the pertinent data concerning the motor's operational characteristics. It is sometimes called the *name plate*. Fig. 10–15 shows an example of a data plate for a typical AC motor. The data plate contains information about the ID (identification number), FR (frame and motor design), motor type, phase, horsepower rating, rpm, volts, amps, frequency, service factor (SF), duty cycle (time), insulation class, ambient temperature rise, and NEMA design and code. Each of these features will be discussed in detail in the next sections.

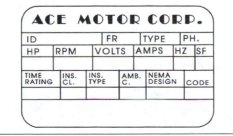

Figure 10–15   Data plate for typical AC motor.

10.11.1   Identification Number (ID)

The identification number for a motor will basically be a model and serial number. The model number will include information about the type of motor, and the serial number will be a unique number that indicates where and when the motor was manufactured. These numbers will be important when a motor is returned for warranty repairs, or if an exact replacement is specified as a model for model exchange.

10.11.2   Frame Type

Every motor has been manufactured to specifications that are identified as a *frame size*. This information includes the distance between mounting holes in the base of the motor, the height of the shaft, and other critical data about physical dimensions. These data are then given a number such as 56. The frame number indicates that any motor that has the same frame number will have the same dimensions even though it may be made by a different company. This allows users to stock motors from more than one manufacturer or replace a motor with any other with the same frame information and be sure that it will be an exact replacement.

10.11.3   Motor Type

The motor-type category on the data plate refers to the type of ventilation the motor uses. These types include the open type, which provides flow-through ventilation from the fan mounted on the end of the rotor. In some motors that are rated for variable-speed duty (used with variable-frequency drives), the fan will be a separate motor that is built into the end of the rotor. The fan motor will be connected directly across the supply voltage, so it will maintain a constant speed to provide constant cooling regardless of the motor speed.

Another type of motor is the *enclosed type,* which is not air cooled with a fan. Instead it is manufactured to allow heat to dissipate quickly to and from the inside of the motor outward to the frame. In most cases the frame has fins built into it on the outside to provide more area for cooling air to reach.

10.11.4   Phase

The phase of the motor will be indicated as single phase or three phase. The number may be indicated as 1 or 3, or the number may be listed by itself.

10.11.5 Horsepower Rating (hp)

The horsepower rating will be indicated as a fractional horsepower (number less than 1.0) or a larger horsepower. Fractional horsepower numbers may be listed as fractions (1/2) or decimals (0.5).

10.11.6 Speed (rpm)

The motor speed will be indicated in rpm. This will be the rated speed for the motor, and it will not account for slip. The actual speed of the motor will be less because of slip. As you know, the speed of the motor is determined by the number of poles and the frequency of the AC voltage. Typical speed for a two-pole motor is 3600 rpm, for a four-pole motor is 1800 rpm, and for an eight-pole motor is 1200 rpm. The actual speed of the motor rated for 3600 rpm will be approximately 3450, for the 1800 rpm motor it will be approximately 1750, and for the 1200 rpm motor it will be approximately 1150. The actual amount of slip will be indicated by the *motor design letter*. The motor design letter will be explained in Section 10.11.14.

10.11.7 Volts

Voltage ratings for single-phase motors will be listed as 115 volts, 208 volts, or 230 volts. Three-phase motors have typical voltage ratings of 208, 240, 440, 460, 480, and 550. Other voltages may be specified for some special-type motors. You should always ensure that the power supply voltage rating matches the voltage rating of the motor. If the rating and the power supply voltage do not match, the motor will overheat and be damaged.

10.11.8 Amps

The amps rating is the amount of full-load current (FLA) the motor should draw when it is under load. This rating will help the designer calculate the proper wire size, fuse size, and heater size in motor starters. The supply wiring for the motor circuit should always be larger than the amps rating of the motor. The NEC (National Electrical Code) provides information to help you determine the exact fuse size and heat size for each motor application.

10.11.9 Frequency

The frequency of a motor will be listed in hertz (Hz). Typical frequency rating for motors in the United States is 60 Hz. Motors manufactured for use in some parts of Canada and all of Europe and Asia will be rated for 50 Hz. You must be sure that the frequency of the motor matches the frequency of the power supplied to the motor.

10.11.10   Service Factor (SF)

The *service factor* is a rating that indicates how much a motor can be safely overloaded. For example, a motor that has an SF of 1.15 can be safely overloaded by 15%. This means that if the motor is rated for 1 hp, it can actually carry 1.15 hp safely. To determine the overload capability of a motor, you multiply the rated horsepower by the service factor. The motor is capable of being overloaded because it is designed with ways of dissipating large amounts of heat.

10.11.11   Duty Cycle

The *duty cycle* of a motor is the amount of time the motor can be operated out of every hour. If the motor's duty cycle is listed as continuous, it means the motor can be run 24 hours a day and does not need to be turned off to cool down. If the duty cycle is rated for 20 minutes, it means the motor can be safely operated for 20 minutes before it must be shut down to be allowed to cool. The motor with this rating should be shut down for 40 minutes of every hour of operation to be allowed to cool.

Another way to specify the duty cycle of a motor is called the *motor rating*. The motor rating on the data plate refers to the type of duty for which the motor is rated. The types of duty include continuous duty, intermittent duty, and heavy duty, which includes jogging and plugging duty. Continuous duty includes applications where the motor is started and allowed to operate for hours at a time. The intermittent duty includes operations where the motor is started and stopped frequently. This type of application allows the motor to heat up because it will draw LRA more often than will a motor rated for continuous duty.

Motors that are rated for jogging and plugging are built to withstand very large amounts of heat that will build up when the motor draws large LRA during starting and stopping. Since the motor can be reversed when it is running in the forward direction for plugging applications, it will build up excessive amounts of heat. Motors with this rating must be able to get rid of heat as much as possible to withstand the heavy-duty applications.

10.11.12   Insulation Class

The *insulation class* of a motor is a letter rating that indicates the amount of temperature rise the insulation of the motor wire can withstand. The numbers in the insulation class are listed in degrees Celsius (°C). The table in Fig. 10–16 shows typical insulation classes for motors. The insulation class and other

Class	Temperature Rise °C
A	105
B	130
F	155
H	180

Figure 10–16 Insulation class for motors. This table indicates the amount of temperature the wire's insulation is rated.

temperature-related features of a motor will help to determine the temperature rise the motor can withstand.

10.11.13 Ambient Temperature Rise

The *ambient temperature* rise is also called the Celsius rise, which is the amount of temperature rise the motor can withstand during normal operation. This value is listed in degrees Celsius. A typical open motor can withstand a rise of 40°C (104°F) and an enclosed motor can withstand 50°C (122°F) rise. This means the motor should not be exposed to environments where the temperature is 104°F above the ambient. If the ambient is considered to be 72°F, it means the motor is limited to temperatures of 176°F. Another classification that will help to determine the amount of temperature a motor will be able to withstand is the type of insulation the motor winding has. The classes of insulation that are used with motors and the amount of temperature these classes can handle are listed in Section 10.11.14.

10.11.14 NEMA Design

The National Electrical Manufacturers Association (NEMA) provides design ratings that may also be listed as motor design. The motor design is listed on the data plate by a letter A, B, C, or D. This designation is determined by the type of wire, insulation, and rotor that are used in the motor and is not affected by the way the motor might be connected in the field.

Type A motors have low rotor circuit resistance and have approximate slip of 5–10% at full load. These motors have low starting torque with a very high locked-rotor amperage (LRA). This type of motor tends to reach full speed rather rapidly.

Type B motors have low to medium starting torque and usually have slip of less than 5% at full load. These motors are generally used in fans, blowers, and centrifugal pump applications.

Type C motors have a very high starting torque per ampere rating. This means that they are capable of starting when the full load is applied for applications such as conveyors, crushers, and reciprocating compressors like air-conditioning and refrigeration compressors. These motors are rated to have slip of less than 5%.

Type D motors have a high starting torque with a low LRA rating. This type of motor has a rotor made of brass rather than copper segments. It is rated for slip of 10% at full load. Normally, this type of motor will require a larger frame to produce the same amount of horsepower as a type A, B, or C motor. These motors are generally used for applications with a rapid decrease of shaft acceleration, such as a punch press that has a large flywheel.

These standards are set by NEMA, and a motor must meet all the requirements of the standard to be marked as a type A, B, C, or D. This allows motors made by several manufacturers to be compared on an equal basis according to application.

10.11.15   NEMA Code Letters

NEMA code letters use letters of the alphabet to represent the amount of locked-rotor amperage (LRA) in kVA per horsepower a motor will draw when it is started. These letters are listed in a table in Fig. 10–17. From this table you can see that letters in the front of the alphabet indicate low LRA ratings, and letters at the end of the alphabet indicate higher LRA ratings. It is important to remember that the number in the table is not the amount of LRA the motor will draw, but rather it is the number that must be multiplied by the horsepower rating of the motor.

10.12   Wiring Split-Phase Motors for a Change of Rotation and Change of Voltage

The wiring diagrams for the run and start windings of the split-phase motor were shown previously in Fig. 10–13. The terminals in the run and start windings can be reconnected in the field to change the rotation of the motor or to allow the motor to operate at a different voltage. Since the split-phase motor and the capacitor-start motor are very similar, these connections will be presented in Section 10.16, after you learn more about capacitor-start, induction-run motors.

10.13   Overview of Capacitor-Start, Induction-Run Open-Type Motors

The *capacitor-start, induction-run* (CSIR) motor is basically a split-phase induction motor that adds a capacitor with the start winding to create a larger phase shift between the start and run windings to start the motor. The capacitor-start, induction-run, open-type motor is generally used in applications where the load

NEMA Code Letter	Locked-Rotor kVA per hp
A	0–3.15
B	3.15–3.55
C	3.55–4.00
D	4.00–4.50
E	4.50–5.00
F	5.00–5.60
G	5.60–6.30
H	6.30–7.10
J	7.10–8.00
K	8.00–9.00
L	9.00–10.0
M	10.0–11.2
N	11.2–12.5
P	12.5–14.0
R	14.0–16.0
S	16.0–18.0
T	18.0–20.0
U	20.0–22.4
V	22.4 and up

Figure 10–17   Table with locked-rotor amperage (LRA) ratings. The ratings are listed as amount of kVa per horsepower. *(Courtesy of NEMA, National Electrical Manufacturers Association.)*

is connected directly to the shaft of the motor. For example, a capacitor-start, induction-run motor can be used to turn direct-drive fans and larger pump loads. Some capacitor-start, induction-run motors are used to drive open-type, belt-driven compressors. The torque for a capacitor-start, induction-run motor is larger than the split-phase motor. A typical capacitor-start, induction-run motor is shown in Fig. 10–18.

10.14   Theory of Operation for a Capacitor-Start, Induction-Run Motor

The capacitor-start, induction-run, open-type motor operates in a manner very similar to the AC split-phase motor explained in the previous chapter. It has a

Figure 10–18   A capacitor-start, induction-run open-type motor used for HVAC and refrigeration applications such as direct-drive fan motors and pump motors. *(Courtesy of GE Motors & Industrial Systems, Fort Wayne, Indiana)*

rotor, and a stator with a run winding and a start winding. A start capacitor is connected in series with the start winding to create a larger phase shift than the split-phase motor has. The start capacitor is generally rated from 50 to 90 μF (microfarads). The capacitance and voltage rating for the capacitor will be listed on its case. Fig. 10–19 shows a picture of two start capacitors. The start capacitor is so named because it is always connected in series with the start winding in the capacitor-start, induction-run motor to provide a larger phase shift, which will also create more starting torque. The start capacitor is also what gives the capacitor-start, induction-run motor its name.

The start capacitor has several physical features that make it easy to recognize when you are servicing equipment. The easiest way to recognize a start capaci-

Figure 10–19   The start capacitor is mounted in a black plastic case that has a round shape. *(Courtesy of North American Capacitor Company, Mallory.)*

tor is that it is generally mounted in a black plastic case. A second characteristic is that the black plastic case for the start capacitor is generally round in shape. From the picture of the start capacitors in Fig. 10–19, you can see that the start capacitor is mounted in a round black plastic case.

The start capacitor is made of two sheets of conducting foil called *plates* that are separated (sandwiched) with three sheets of insulating paper called the *dielectric*. Fig. 10–20a shows an example of the foil sheets and dielectric sheets before they are rolled up. Fig. 10–20b shows a diagram of the sheets rolled up so that they will fit into the round plastic case. A lead wire is soldered onto each piece of foil and brought out to the top of the capacitor to be used as the capacitor's terminals. Fig. 10–20c shows the electrical symbol for the capacitor. Notice that the symbol shows the capacitor conducting plates separated by the dielectric so current cannot pass directly from one plate to the other.

When voltage is applied to the capacitor in a circuit, current will flow as far as the plate on the left side of the capacitor. Since the capacitor plates are separated by dielectric paper, current cannot flow directly through the capacitor to the other plate, so the voltage must reverse and move backward through the circuit until it reaches the other capacitor plate. This process is referred to as *charging the capacitor and discharging the capacitor.* If the voltage that causes the current to flow is AC voltage, the AC sine wave will continually charge and discharge the capacitor, which keeps current continually flowing through the components in the circuit.

Figure 10–20a   Two sheets of conducting foil that are separated by sheets of dielectric paper. The dielectric paper acts as an insulator between the conducting foil. The two sheets of foil are called the plates. **b** The layers of conducting foil and dielectric paper are rolled up so that they will fit into the round black plastic case. **c** The electrical symbol for the capacitor shows the two conducting plates separated by the insulator.

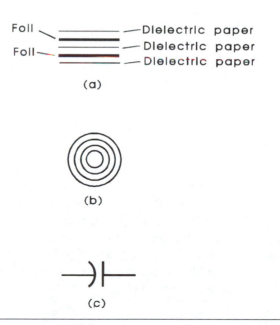

Since it takes time for the voltage to charge up the first plate, then reverse itself and discharge the plate, the capacitor will cause a phase shift of up to 90° between the voltage waveform and the current waveform in the circuit. It is important to remember that the capacitor will always cause the voltage waveform to lag the current waveform.

10.15   Applying Voltage to the Start Winding and Capacitor

Fig. 10–21 shows an electrical diagram of the capacitor-start, induction-run motor so that you can see that the start capacitor (SC) is connected in series with the centrifugal switch and the start winding. From this diagram you can see that when voltage is first applied to the leads of the motor, current will flow from L1 through the run windings and back to L2 or N. At the same time, voltage will also be applied to the first capacitor plate. Since current cannot flow through the capacitor to the other plate, the voltage must reverse itself and move back through the run winding until it finds its way back to the start winding at the other plate of the capacitor. Since the voltage must take the long way around the motor circuit to get to the other plate, it creates a phase shift between the current flowing in the run winding and the current flowing in the start winding. This process continues as long as the centrifugal switch remains closed and current flows through the start winding. Since the capacitor causes a larger phase shift than the split-phase motor can develop with just the start winding, the capacitor-start, induction-run motor will have more starting torque than the split-phase motor.

When the motor's rpm increases to 75–85% full rpm, the centrifugal switch will open and cause the current to stop flowing through the start winding. After the centrifugal switch opens the start winding circuit, current will continue to flow through the run winding so the motor will act as an induction motor when it is running. This is why the motor is called a *capacitor-start, induction-run motor.*

Figure 10–21 Electrical diagram of a capacitor-start, induction-run (CSIR) motor. Notice the start capacitor (SC) is connected in series with the centrifugal switch and start winding.

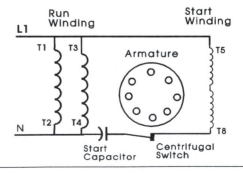

10.16 Wiring Split-Phase and Capacitor-Start, Induction-Run Motors for 115 Volts

Fig. 10–21 shows the electrical diagram for a capacitor-start, induction-run motor that is connected for 115 volts. This connection is also called the low-voltage connection. The diagram for the split-phase motor would be identical except it does not have a start capacitor connected to its start winding. When you must connect a split-phase motor or capacitor-start, induction-run motor for 115 volts, you should connect terminals T1, T3, and T5 together with the supply voltage wire L1, and you should connect terminals T2, T4, and T8 together with the neutral supply voltage wire. You should notice from the diagram that these connections place the two sections of the run winding in parallel with each other and with the start winding. You should also notice that the start capacitor is connected in series with the centrifugal switch. This means that each section of the run winding will have 115 volts applied to it and the start winding will have 115 volts.

10.17 Wiring a Capacitor-Start, Induction-Run Motor for 230 Volts

At times you may need to reconnect the capacitor-start, induction-run motor that you are working on for a higher voltage. Fig. 10–22 shows an electrical diagram of a capacitor-start, induction-run motor that is connected for 230 volts. This connection is also called the high-voltage connection. Again the diagram for the split-phase motor would be similar except it would not have the start capacitor. From

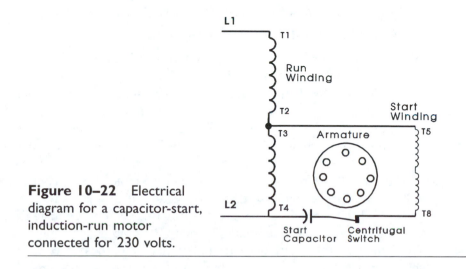

Figure 10–22 Electrical diagram for a capacitor-start, induction-run motor connected for 230 volts.

this diagram you can see that the two sections of the run winding are connected in series. This connection ensures that 115 volts is applied to each winding as a voltage drop. You should understand that the insulation on the wire that is used for the run and start windings is rated for 115 volts. Since the two run windings are connected in series, you are ensuring that each winding will only receive 115 volts, since the 230 volts will be dropped equally across each run winding. You should also notice that the capacitor is always connected to the centrifugal switch, so it will always be connected in series with the start winding.

It is important at this time to remember that the motor may be rated for 208 volts for its higher voltage rather than 230 volts. The diagram for the higher voltage is the same regardless if the motor data plate indicates the motor is rated for 230 volts or 208 volts. The important point to remember is that the motor must match the exact amount of incoming voltage that is used to supply the motor. This means that you should always measure the exact amount of voltage before you connect a motor.

You should also notice that the start winding is connected so that it only receives 115 volts. This means that the start winding is connected so that it is actually in parallel with the lower section of the run winding (T3-T4). One end of the centrifugal switch is connected permanently to one terminal of the start capacitor. The other terminal of the start capacitor is connected to terminal T4. This ensures that it always will be connected across the lower section of the start winding. You can see that if the motor is connected for 230 volts or 115 volts, one end of the centrifugal switch is always connected through the capacitor to terminal T4.

10.18   Wiring Split-Phase and Capacitor-Start, Induction-Run Motors for a Change of Rotation

When you are working in the field, you will find times when you must change the direction of rotation of the motor's shaft. Identify the rotation of the shaft by standing behind the motor (at the end opposite of where the shaft comes out of the end plate) and observe the direction of rotation. To change the direction of rotation of the motor shaft, you must change the direction current flows through the start winding. The easiest way to do this is to exchange T5 and T8 wires.

The capacitor-start, induction-run, open-type motor that is connected for high voltage can also be wired so that the direction of its shaft's rotation can be reversed. When the motor is wired for clockwise rotation, terminal T5 of the start winding is connected to terminals T2 and T3 of the run windings, and terminal T8 of the start winding is connected to terminal T4 of the run winding. Fig. 10–8b shows the electrical diagram of a high-voltage capacitor-start, induction-run motor connected for counterclockwise rotation. In this diagram you can see that terminal T8 of the start winding is connected to terminals T2 and T3 of the run winding, and terminal T5 of the start winding is connected to terminal T4 of the run winding.

10.19   Changing Speeds with a Capacitor-Start, Induction-Run Motor

The split-phase and capacitor-start, induction-run motors are generally not designed to have their speed changed. Some more expensive capacitor-start, induction-run motors have additional sets of poles provided so that their speed can be changed by changing the number of poles used in each run winding, but typically this type of motor cannot be reconnected for a change of speed. This means that if you need to change the speed of the motor, you would need to replace the motor with one of a different speed. It is also important to understand that if the motor is used to drive a belt-driven fan, the speed of a fan can be altered slightly by changing the size of the pulley on the motor.

10.20   Overview of Permanent Split-Capacitor, Open-Type Motors

The *permanent split-capacitor* (PSC) motor is similar to the split-phase induction motor and capacitor-start, induction-run motor in that it has a start winding and a run winding. However, it is different in that it does not use a centrifugal switch to remove power from the start winding when the motor reaches full speed. Instead of a centrifugal switch, the PSC motor uses a run capacitor that is connected in series with the start winding to limit the amount of current that is allowed to flow through the start winding when the motor reaches full speed. The run capacitor is mounted in a metal container so that it can dissipate the heat more easily, which allows it to remain in the circuit even after the motor is running at full speed. The permanent split-capacitor, open-type motor is generally used in applications where the load is connected directly to the shaft of the motor. For example, a permanent split-capacitor motor can be used to turn direct-drive fans such as squirrel-cage furnace fans or blade-type condenser fans, and combination evaporator/condenser fans in window air conditioners. The torque for a permanent split-capacitor motor is larger than that of a split-phase motor and about the same or slightly less than a capacitor-start, induction-run motor. A typical permanent split-capacitor motor is shown in Fig. 10–23. You should be able to see the run capacitor is mounted on the top of the motor. Fig. 10–24 shows a special PSC motor that has a shaft extending out of both ends of the end plates. This type of PSC motor is used in window air conditioners with a

Figure 10–23 Typical permanent split-capacitor, open-type motor used for HVAC and refrigeration applications such as direct-drive, squirrel-cage furnace fan motors and blade-type condenser fan motors. *(Courtesy of GE Motors & Industrial Systems, Fort Wayne, Indiana)*

blade-type fan mounted on one end to operate as the condenser fan and a squirrel-cage fan on the other end to operate as the evaporator fan.

10.21   Basic Parts and Theory of Operation for a PSC Motor

The basic parts of the PSC motor include the start winding, the run winding mounted in the stator, and the end plates that support the rotor with the motor

Figure 10–24 A special PSC motor that has a shaft extending out each end of the motor. This type of motor is used in window air conditioners to power the condenser fan and evaporator fan. *(Courtesy of GE Motors & Industrial Systems, Fort Wayne, Indiana)*

shaft on one or both ends. The permanent split-capacitor, open-type motor operates in a manner very similar to the AC split-phase motor. It has a rotor, and a stator with a run winding and a start winding. A run capacitor is connected in series with the start winding to create a larger phase shift than the split-phase motor has. The run capacitor is generally rated from 5 to 60 μF (microfarads). The capacitance and voltage rating for the capacitor will be listed on its case. Fig. 10–25 shows a picture of several examples of run capacitors. The run capacitor is so named because it remains in the motor circuit even after the motor is running at full speed.

The run capacitor has several physical features that make it easy to recognize when you are servicing equipment. The easiest way to recognize a run capacitor is that it is generally mounted in a metal container. A second characteristic is that the metal container for the run capacitor is generally oval or rectangular in shape. The run capacitor is made in a similar manner to the start capacitor where two sheets of conducting foil called *plates* are separated (sandwiched) by three sheets of insulating paper that are called the *dielectric*. The foil plates and dielectric are rolled up and then placed in the metal container and an electrical terminal is connected to each plate.

You should remember that when voltage is applied to the capacitor in a circuit, current will flow as far as the plate on the left side of the capacitor. Since the capacitor plates are separated by dielectric paper, current cannot flow directly through the capacitor to the other plate, so the voltage must reverse and move backward through the circuit until it reaches the other capacitor plate. This pro-

Figure 10–25 Examples of typical run capacitors. Notice that run capacitors are mounted in metal containers that have an oval or rectangular shape. *(Courtesy of North American Capacitor Company, Mallory.)*

cess is referred to as *charging the capacitor and discharging the capacitor*. If the voltage that causes the current to flow is AC voltage, the AC sine wave will continually charge and discharge the capacitor, which keeps current continually flowing through the components in the circuit.

Since it takes time for the voltage to charge up the first plate, then reverse itself and discharge the plate, the capacitor will cause a phase shift of up to 90° between the voltage waveform and the current waveform in the circuit. It is important to remember that the capacitor will always cause the voltage waveform to lag the current waveform.

The reason a centrifugal switch is not needed to remove the start winding from the circuit after the motor is running at full speed is that the CEMF that is generated by the rotor will be present across the start winding. When this counter EMF is present, it will oppose the applied voltage and the resulting current in the start winding will be determined by the difference of the applied voltage and CEMF. If the motor is running at near full rpm, the CEMF will be high and the amount of current in the start winding will be minimal. If the motor encounters a large load and its slip increases because the speed of its shaft slows down, the CEMF will decrease and the difference between the applied voltage and CEMF will become larger, allowing the start winding to draw extra current, which will help provide enough additional torque to get the motor shaft back to its original speed. In this manner, the run capacitor will allow the start winding to stay in the circuit and add current when additional torque is required to get the shaft to return to its normal speed. In this manner, the PSC motor is able to regulate its speed. This is an important feature for the PSC motor if it is used for moving air or pumping cooling water for a chiller or pumping refrigerant as a compressor.

10.22   Applying Voltage to the Start Winding and Capacitor of a PSC Motor

Fig. 10–26 shows an electrical diagram of a typical permanent split-capacitor motor. You can see that the start capacitor (RC) is connected in series with the start winding. When voltage is first applied to the leads of the motor, current will flow from L1 through the run windings and back to L2 or N. At the same time, voltage will also be applied to the first capacitor plate. Since current cannot flow through the capacitor to the other plate, the voltage must reverse itself and move back through the run winding until it finds its way back to the start winding where it reaches the other plate of the capacitor. Since the voltage must take the long way around the motor circuit to get to the other plate, it creates a phase shift between the current flowing in the run winding and the current flowing in the start winding. In the PSC motor this process continues as long as voltage is applied to the motor windings. Since the run capacitor remains in the start

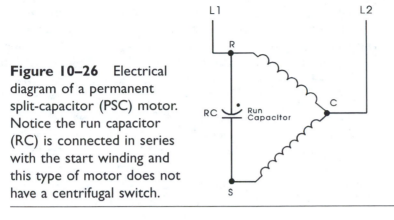

Figure 10–26 Electrical diagram of a permanent split-capacitor (PSC) motor. Notice the run capacitor (RC) is connected in series with the start winding and this type of motor does not have a centrifugal switch.

winding circuit at all times for this motor, it is called a *permanent split-capacitor motor.*

In the PSC motor the initial current flow into the motor windings will be rather large because the rotor is not turning at first. This current is called locked-rotor amperage (LRA). Since a run capacitor is mounted in series with the start winding, the current cannot flow directly through the capacitor. Instead the voltage will move as far as the first plate when L1 voltage is applied and then stop. At this point one plate of the capacitor will have a large positive charge, which will create a potential differential across the plates of the capacitor. This potential differential causes the voltage to move from the first plate, back through the run winding, and start winding until the voltage reaches the opposite plate of the capacitor. This causes a large phase shift between the current in the run winding and start winding, which is sufficient to cause the rotor to spin.

When the rotor begins to spin, it will begin to generate counter EMF, which will cause the current flow to diminish to a point of normal current flow called full-load amperage (FLA). The actual amount of current the motor draws will depend on the difference between the rated speed and the actual speed. This difference is called slip. The rotor will continue to rotate as long as voltage is applied to the windings.

10.23   Installation and Servicing Practices: Changing Rotation, Changing Voltages, and Changing Speeds for the Permanent Split-Capacitor Motor

When you are installing or servicing the permanent split-capacitor motor on HVAC or refrigeration equipment, you must be able to perform several important functions. These functions include reconnecting the permanent split-capacitor motor terminals to change its rotation, reconnecting its terminals so it will run on a different speed, and replacing the motor when it is rated for the wrong voltage. The reason you may need to change the rotation of the motor is if the fan blades the motor is driving are turning in the wrong direction or if the pump the motor is connected to is running in the wrong direction. You may need to reconnect the motor terminals to change the motor's speed if the motor is running a multispeed fan. Typically the motor uses its highest speed for air-conditioning applications and its lowest speed for heating applications. Sometimes you need to switch the low speed for an intermediate speed or a fast speed for the next slower speed if the fan is too noisy. If the PSC motor you are changing does not match the supply voltage, you would need to locate a motor that does because the PSC motor is not generally made with dual-voltage capability.

10.24   Wiring a Permanent Split-Capacitor Motor for a Change of Rotation

When you are changing the direction of rotation for the motor, you must change the direction current flows through the start winding. The easiest way to do this is to exchange the ends of the start winding wires (T5 and T8). Fig. 10–27a shows the terminal connection for wiring a permanent split-capacitor motor for clockwise rotation. You can see that terminal T5 (black wire) of the start winding is connected to the terminal of the run capacitor, and T8 (white wire) of the start winding is connected to the C terminal of the motor. Fig. 10–27b shows the diagram of the same motor connected for counterclockwise rotation. In this diagram you can see that terminal T8 of the start winding (white wire) is now connected to the terminal of the run capacitor, and terminal T5 of the start winding (black wire) is connected to the C terminal motor.

Some PSC motors provide a plug connection for the two leads of the start winding. When you need to reverse the start winding leads, you can simply unplug the start windings and reverse the plug and the motor will rotate in the opposite direction. The plug makes it fairly simple to reverse the direction of rotation of the PSC motor. If the PSC motor does not provide the two leads of the start winding exposed through the stator, you cannot change the rotation of the motor. Some original equipment motors are not designed to have their rotation changed, but most replacement motors have the ability to have their rotation changed. Fig. 10–28 shows an example of a diagram you would find on a PSC motor that indicates how to change the direction of rotation for the motor. In this diagram you can see that you need to remove the plug that has the start winding's black and white wire, attached to it and reverse it when it is plugged in again.

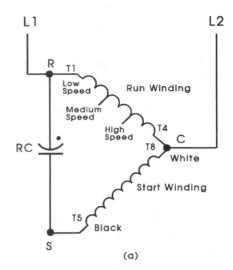

Figure 10–27a Electrical diagram for a 230 V permanent split-capacitor, open-type motor connected to run in the clockwise direction. **b** Electrical diagram of a 230 V permanent split-capacitor, open-type motor connected to run in the counterclockwise direction. Notice that terminals T5 and T8 have been reversed.

10.25   Changing Speeds with a Permanent Split-Capacitor Motor

At times you will be working with a PSC motor that needs to have its speed changed. For example, you may be working with a multispeed furnace fan motor and you will need to change the speed of the motor. The diagram on the motor's data plate will show which connections to change. Fig. 10–29 shows a

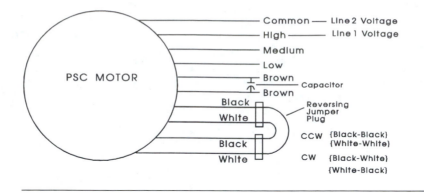

Figure 10–28 Electrical diagram that is shown on a PSC motor's data plate. This field diagram shows how the direction of rotation for the motor can be reversed by changing the start winding. This diagram shows the plug that must be reversed to reverse the start windings (black and white wires).

diagram of the multispeed motor connected for high speed. From this diagram you should notice that the run winding has multiple taps. When you use all of the segments of the run winding, the motor will run at its lowest speed, and when you use just one segment of the run winding, the motor will run at its highest speed.

The wire connected to the low-speed tap is generally colored red, the wire connected to the medium-speed tap is generally colored blue, and the wire connected to the high-speed tap is generally colored black. Fig. 10–30 shows a diagram of the PSC motor connected for medium speed and, if the motor is con-

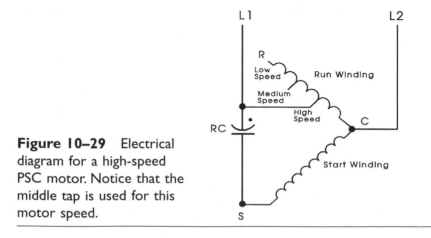

Figure 10–29 Electrical diagram for a high-speed PSC motor. Notice that the middle tap is used for this motor speed.

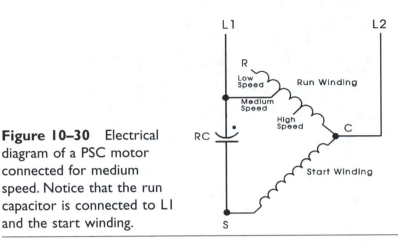

Figure 10–30 Electrical diagram of a PSC motor connected for medium speed. Notice that the run capacitor is connected to L1 and the start winding.

nected for high speed, the high-speed terminal would be connected to L1. Each of these diagrams shows the run capacitor connected between the point where L1 is connected to the speed tap that is used and the start winding.

10.26   Overview of Shaded Pole, Open-Type Motors

The physical appearance of the *shaded pole motor* is similar to the permanent split-capacitor (PSC) motor, but it is different in that it does not use a centrifugal switch to remove power from the start winding when the motor reaches full speed and it does not use any capacitors to create its phase shift. Instead of a capacitor to create its phase shift for starting, the shaded pole motor uses a *shading coil*. The shading coil is sometimes called the *shaded pole,* and this is how the motor gets its name. The shaded pole, open-type motor does not have as much starting torque as the other types of open motors, so it is generally used in applications where the smaller loads are connected directly to the shaft of the motor. For example, a shaded pole motor can be used to turn small direct-drive fans such as squirrel-cage furnace fans or blade-type condenser fans, and combination evaporator/condenser fans in window air conditioners. A typical shaded pole motor is shown in Fig. 10–31. You should notice that the shaded pole motor looks very similar to the PSC motor except it does not have a run capacitor. Fig. 10–32 shows a shaded pole motor that has a shaft extending out of both ends of the end plates. This type of PSC motor is used in window air conditioners, and a blade-type fan is mounted on one end to operate as the condenser fan, and a squirrel-cage fan is mounted on the other end to operate as the evaporator fan.

Figure 10–31 Typical shaded pole, open-type motor used for HVAC and refrigeration applications such as direct-drive, squirrel-cage furnace fan motors and blade-type condenser fan motors. *(Courtesy of GE Motors & Industrial Systems, Fort Wayne, Indiana.)*

Figure 10–32 A shaded pole motor that has a shaft extending out each end of the motor. The shaded pole motor, like the PSC motor with a double shaft, is used in window air conditioners to power the condenser fan and evaporator fan. *(Courtesy of GE Motors & Industrial Systems, Fort Wayne, Indiana.)*

10.27   Basic Parts and Electrical Diagram for the Shaded Pole Open Motor

The basic parts of the shaded pole motor are the rotor, the stator, and the end plates. The rotor is like the rotors in all of the other single-phase induction motors in that it is made of laminated pieces of steel pressed onto the squirrel-cage rotor. The shaded pole motor generally uses bushing-type bearings in its end plates. This means that it is very important to check the lubrication of the shaded pole motor several times a year. Fig. 10–33b shows an electrical diagram of the shaded pole motor. From this diagram you can see that the stator has only a run winding. Fig. 10–33a provides a diagram of the motor that shows the four poles of the run winding.

To really understand the shading coil, you need to see exactly what it looks like. Fig. 10–34a shows how the shading coil is physically mounted in the run winding, and Fig. 10–34b shows how the shading coil looks when it is mounted in each of the four poles of the motor. This diagram shows a large coil made of a single turn of a solid piece of copper that is inserted inside each pole of the run windings. This coil is called the shading coil and it will create an induced magnetic field when current is flowing through the run winding. The induced magnetic field creates a small phase shift that is large enough to give the motor enough starting torque to begin to run. In some cases the shading coil is called the start winding, since it provides a similar purpose to the start winding of other motors. You should notice from Fig. 10–34b that the shading coils are mounted only around the front part of each run winding pole.

The main parts of this type of motor that may have a problem are the run winding, which can develop an open circuit, and the bearings can become defective so that the rotor will not be able to rotate easily. Sometimes the shading coil will become loose and the motor's stator will become overheated. You will see how to test for each of these problems in the troubleshooting section of this chapter.

10.28   Theory of Operation for a Shaded Pole Motor

The shaded pole, open-type motor has a rotor and a stator. The stator has a run winding and a shading coil that is mounted around each run winding, which acts like the start winding. When voltage is applied to the run winding, current begins flowing through the winding, which creates a very strong magnetic field. The current flowing through the run winding also creates an induced current in the shading coil. Since the shading coil is made from a single-turn coil of copper, the magnetic field will be out of phase with the magnetic field in the run winding. This phase shift will be sufficient to cause the rotor to begin to turn.

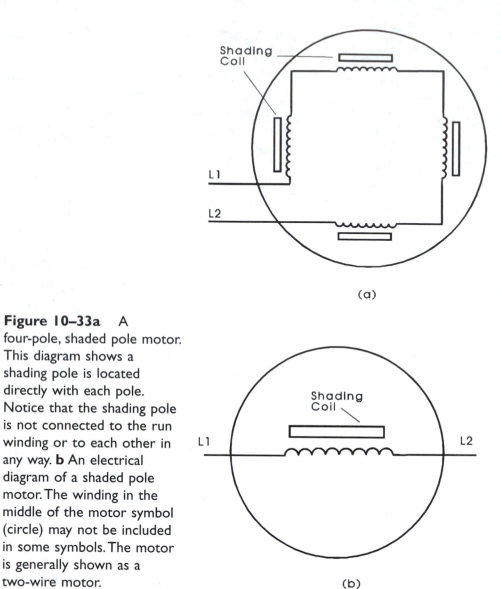

(a)

Figure 10–33a A four-pole, shaded pole motor. This diagram shows a shading pole is located directly with each pole. Notice that the shading pole is not connected to the run winding or to each other in any way. **b** An electrical diagram of a shaded pole motor. The winding in the middle of the motor symbol (circle) may not be included in some symbols. The motor is generally shown as a two-wire motor.

(b)

After the rotor comes up to near full rpm, the amount of phase shift between the run winding and shading coil becomes very small. At this point, the shading coil is no longer needed, and the motor is running on just the run winding, as do all of the other single-phase motors. The shading coil provides sufficient phase shift to allow the shaded pole motor to start.

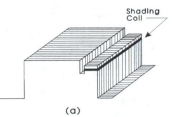

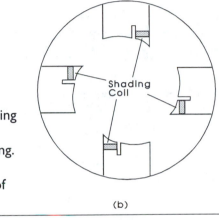

Figure 10–34a The physical location of a shading coil as you would find it mounted in the run winding. **b** The physical location of the shading coil on each of four run winding poles.

Fig. 10–34b shows a cut-away diagram of the stator of the shaded pole motor. From this diagram you can see that this motor has four poles that make up the run winding. A shading coil can be seen where it is pressed into the windings of each pole. It is important to understand that the shading coil is not electrically connected to the run winding or to other shading coils in any way. This means that when you test the run winding with an ohmmeter, you will only be able to determine the continuity of the run winding, and not the shading coil. This will not create any problems. Since the shading coil is made of a large piece of cop-

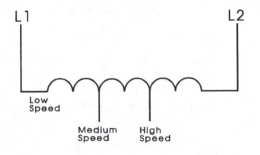

Figure 10–35 Electrical diagram for a 230 V shaded pole, open-type motor connected to run at low speed.

per, it will very seldom have any problems, since normal current flow cannot burn it out. It is also important to point out that the size of the wire in the run winding of the shaded pole motor is large enough to carry the locked-rotor amperage (LRA) for a sustained period of time if the motor's rotor stalls. This is important since the shaded pole motor is used for fans and it generally uses a bushing-type bearing in the end plate. This means that if the bushing ever runs dry and the rotor stalls, the motor's run winding will tend to survive without damage.

10.29   Installation and Servicing Practices: Changing Rotation, Changing Voltages, and Changing Speeds for the Shaded Pole Motor

When you are installing or servicing the shaded pole motor on HVAC or refrigeration equipment, you must be able to perform several important functions. These functions include reconnecting the shaded pole motor terminals so it will run on a different speed, and replacing the motor when it is rated for the wrong voltage. The shaded pole is generally not designed to have its direction of rotation changed. This means that if you need to change the direction of rotation of the shaft of a shaded pole motor, you will generally need to replace the motor with one that is turning in the correct direction.

You would need to reconnect the motor terminals to change the motor's speed if the motor is used in multispeed fan applications. Typically the motor uses its highest speed for air-conditioning applications and its lowest speed for heating applications. The shaded pole motor is generally not designed to have its windings changed if it does not match the supply voltage. Instead if you were replacing a motor that is connected to 230 volts, you must be sure to have a motor with the exact voltage rating. If the motor does not match the voltage rating of the supply voltage, you will need to order a new motor that matches the voltage rating.

10.30   Changing Speeds of a Shaded Pole Motor

At times you will be working with a shaded pole motor that needs to have its speed changed. For example, you may be working with a multispeed furnace fan motor and you will need to change the speed of the motor. The diagram on the motor's data plate will show which connections to change. Fig. 10–35 shows a diagram of the multispeed motor connected for low speed. From this diagram you should notice that the run winding has multiple taps. When you use all of the segments of the run winding, the motor will run at its lowest speed, and when you use just one segment of the run winding, the motor will run at its highest speed.

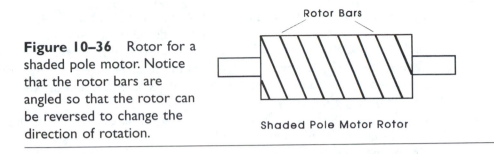

Rotor Bars

Shaded Pole Motor Rotor

Figure 10–36   Rotor for a shaded pole motor. Notice that the rotor bars are angled so that the rotor can be reversed to change the direction of rotation.

The wire connected to the low-speed tap is generally colored red, the wire connected to the medium speed tap is generally colored blue, and the wire connected to the high-speed tap is generally colored black. When you want to change the motor's speed, you would connect the L1 wire to the blue wire to make it run on medium speed, and you could connect L1 to the black wire if you wanted the motor to run on high speed.

When the L1 wire is connected to the red wire for slow speed, the motor is using all six poles in the run winding. When you use the medium-speed tap, you are using four poles, and when you are connected to the high-speed tap, you are using only two poles. You should remember that the higher the number of poles, the slower the motor will run.

10.31   Wiring a Shaded Pole Motor for a Change of Rotation

When you are working in the field, you will find times when you must change the direction of rotation of the motor's shaft. Since the shaded pole motor uses the shading coils as the start winding, it is not practical to open the motor and reverse the shading coils. This means that generally you must order the shaded pole motor as either a clockwise-rotation motor or a counterclockwise-rotation motor.

There is one other way to reverse the rotation of the shaded pole motor's shaft. This method is not used too frequently, but it will help you to get out of trouble if you absolutely must change the direction of rotation of the motor's rotor while you are servicing equipment in the field. This method involves removing the end plates of the motor and removing the rotor and replacing it in the stator housing to change to the opposite direction. This means that if the rotor shaft is coming out of the left side of the stator (when you are looking directly at the motor's data plate), you would remove the rotor and place it back into the stator so the shaft is sticking out of the right side of the stator.

The reason this method may work is that the rotor bars in the squirrel-cage rotor are slightly angled rather than being square with the outside of the rotor. This means that one end of each rotor bar will pass the field coil slightly before the other end. Fig. 10–36 shows the rotor bars angled for the shaded pole rotor. Since each rotor bar becomes magnetized when current is flowing in the run winding, one end of each rotor bar will be the north magnetic pole and the other end will be the south magnetic pole. When the rotor is switched end for end in the stator, this magnetic field relationship will be reversed, and the rotor will rotate in the opposite direction.

If you use this method to change the direction of rotation of the rotor, you should check the motor after it has run for several minutes to see if the stator is overheating. Some shaded pole motors were designed to have their rotors reversed, and if they are reversed they will cause the motor to overheat slightly.

Questions for This Chapter

1. Identify the main parts of an AC motor.
2. Explain how you can identify the run winding and start winding in the stator of an AC split-phase motor.
3. Explain how a magnetic field is developed in the rotor of an AC motor.
4. Explain how the squirrel-cage rotor acts like the bar magnets rotor in the AC motor.
5. Explain why the rotor rotates when a voltage is applied to the stator of an AC motor.

True or False

1. Torque is rotational force.
2. Slip is the difference between the actual speed of a motor and its rated speed.
3. The AC split-phase motor is called an induction motor because it uses a transformer with a capacitor to start.
4. The run winding is physically placed offset from the start winding in the stator of an AC motor to help it provide sufficient phase shift to start the rotor.
5. The start winding uses larger wire than the run winding.

Multiple Choice

1. The start capacitor is _____

    a. connected in series with the run winding of a PSC motor.
    b. connected in series with the start winding of a CSIR motor.
    c. connected in series with the start winding of the shaded pole motor.

2. The run capacitor is _____
    a. connected in series with the run winding of a PSC motor.
    b. connected in series with the start winding of a PSC motor.
    c. connected in series with the start winding of the CSIR motor.
3. The shaded pole motor _____
    a. uses a capacitor and a centrifugal switch.
    b. does not have any capacitors or centrifugal switch.
    c. uses a capacitor but does not need a centrifugal switch.
4. The permanent split-capacitor (PSC) motor _____
    a. uses a run capacitor and a centrifugal switch.
    b. uses a start capacitor and a centrifugal switch.
    c. uses a run capacitor but does not have a centrifugal switch.
5. The capacitor-start, induction-run (CSIR) motor _____
    a. uses a start capacitor that is connected in series with its start winding.
    b. uses a start capacitor that is connected in series with its run winding.
    c. uses a run capacitor that is connected in series with its start winding.

Problems

1. Calculate the speed of a two-pole, four-pole, six-pole, and eight-pole motor.
2. Draw the sketch of a typical data plate (name plate) for an AC motor and identify each of the items listed in Section 10.11 of this chapter.
3. Draw the electrical diagram of a split-phase motor wired for 115 VAC.
4. Draw the electrical diagram of a capacitor-start, induction-run motor that is wired for 230 VAC.
5. Draw the electrical diagram of a PSC motor that is wired for 230 VAC.

11 Single-Phase Hermetic Compressors

OBJECTIVES:

After reading this chapter, you will be able to:

1. Explain the theory of operation of a single-phase hermetic compressor.
2. Identify the main parts of the split-phase hermetic compressor.
3. Explain the operation of the coil and contacts of the current relay.
4. Identify the wiring diagrams of a split-phase; capacitor-start, induction-run (CSIR); permanent split-capacitor (PSC); and capacitor-start, capacitor-run (CSCR) compressors.
5. Explain the operation of the coil and contacts of a potential relay.

11.0 The Split-Phase Hermetic Compressor

Single-phase hermetic compressors are used in most residential air-conditioning and refrigeration units. The main difference between the open-type motors you studied in the previous chapter and a hermetic motor is that the hermetic motor is sealed and it must be cooled by the same refrigerant it is pumping. Since the hermetic motors are sealed and they have oil mixed with their refrigerant to lubricate their pistons and crankshafts, they cannot use a centrifugal switch to deenergize their start winding when the rotor reaches full speed. Instead the hermetic motor must have a relay or some other type of device to deenergize the start winding, and this device must be mounted on the outside of the hermetic compressor shell. The other major difference between the open motor and the hermetic motor is that the hermetic motor must have a terminal board that is specifically designed so that the terminals are mounted through the compressor shell. Inside the shell the terminals are connected to the run winding, start winding, and common point, and on the outside of the shell, the terminals are identified as R for run, S for start, and C for common. Fig. 11–1 shows a picture of several types of hermetic compressor motors you will find in residential air-conditioning and refrigeration equipment. You should remember that a motor is mounted inside each of the compressors, and the rotor of each motor is connected directly to the crankshaft of the compressor so that when the motor runs, the compressor piston pumps refrigerant. Some compressors use different methods of pumping the refrigerant such as a scroll-type compressor or centrifugal-type compressor. However, as far as the electric motor is concerned, they will all function as one of the four types of single-phase compressor motors: split-phase; capacitor-start, induction-run; capacitor-start, capacitor-run; or permanent split-capacitor.

The single-phase motor can be wired in any one of four ways to produce the required starting torque. The *split-phase* configuration is used for low-torque applications. The *permanent split-capacitor* (PSC) is used for medium-torque applications; the *capacitor-start, induction-run* (CSIR) and *capacitor-start, capacitor-run* (CSCR) configurations are used for high-torque applications.

These wiring configurations have become so common that they have become the names for the compressor when it is wired in that particular configuration. For example, when a capacitor is added to the single-phase compressor to give it more starting torque, it becomes a capacitor-start, induction-run motor and it will be called a capacitor-start, induction-run compressor. When a start capacitor and a run capacitor are added to the split-phase motor, it will become a capacitor-start, capacitor-run motor.

11.1   Basic Parts of a Split-Phase Hermetic Compressor

The basic parts of a split-phase hermetic compressor include the rotor, whose shaft is used to turn the compressor crankshaft, the stator that houses both the start winding and the run winding, and the starting switch that deenergizes the start winding after the motor is running at 75–85% full rpm. Fig. 11–2 shows a cut-away diagram of two split-phase hermetic compressors. The motors for each compressor are mounted in a metal container that is called the *shell.* The motor may be mounted so that the stator is in the top of the compressor shell and the rotor shaft is pointing downward so that the piston that pumps the refrigerant is in the bottom of the shell, or it may be mounted so the piston is in the top of the shell.

(a)

(b)

(c)

Figure 11–1   Examples of single-phase hermetic compressors. Each of these compressors has a single-phase motor mounted inside it to pump refrigerant. *(Courtesy of Copeland Corporation)*

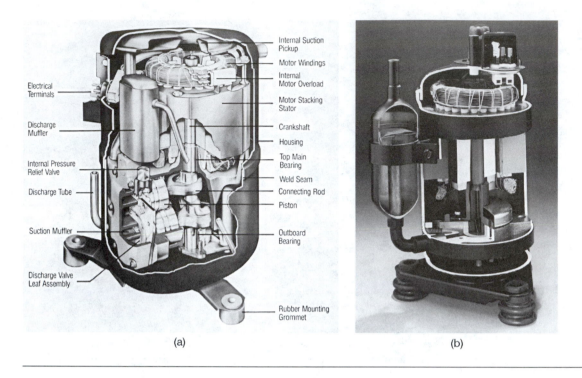

(a)   (b)

Figure 11–2 Cut-away pictures of two typical types of compressors. You should notice that the motor in both of these compressors is mounted in the top of the compressor, and the piston is located in the bottom of the compressor. In some compressors, the motor is mounted in the bottom of the compressor. *(Courtesy of Tecumseh Products Company; and courtesy of Carlyle Compressor Company, Division of Carrier Corp.)*

Another important part of the compressor is the way the wires are connected to the motor terminals. Since the motor is mounted inside the shell, a set of terminals is mounted through the shell in such a way as to create a seal that prevents refrigerant from leaking out of the compressor shell, yet provides a convenient electrical connection to the motor terminals inside the shell. Fig. 11–3 is a close-up cut-away picture of a compressor to show a typical terminal board on a compressor. From this diagram you should also notice that the motor in this compressor is upside down so that the piston is located in the top of the compressor and the motor is located in the bottom. The terminal board in this compressor is near the bottom, and you can see how the wires from the motor are connected to the terminal board inside the compressor, and three terminal pins are located on the outside of the compressor. The terminal pins make it easy to press on termi-

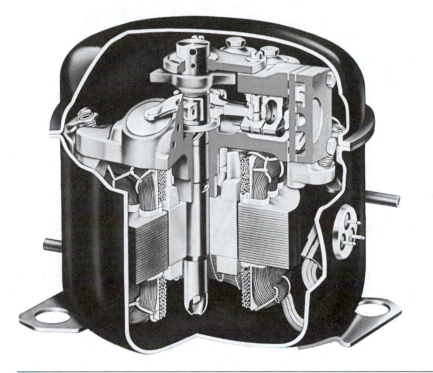

Figure 11–3   Cut-away view of a compressor that shows how the motor wires are connected to the compressor terminals that are mounted through the compressor shell. *(Courtesy of Tecumseh Products Company)*

nal connectors and the current relay directly. The *current relay* is used to take the place of the centrifugal switch and to disconnect the start winding when the motor reaches 75% full rpm.

The terminal connections for compressors have become somewhat standardized so that the current relay from any manufacturer will fit most any compressor. You will see in the next section that the current relay must be sized for the amperage draw for each compressor motor, but the terminal numbers are standardized. Fig. 11–4 shows an electrical diagram of the terminal board with the terminals identified. You should notice that the common terminal C is located at the top of the terminal board, and the start terminal S is located at the left side of the terminal board, while the main terminal M is located at the right side of the terminal board. It is very important to understand that the M terminal is called the run terminal R in the split-phase open motor. In some compressors the main terminal will also be identified as the run terminal.

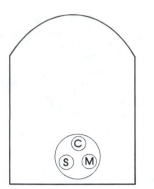

Figure 11–4 Terminal connections on a compressor. Notice that the C terminal is located to the top of the terminals, the S terminal is on the left, and the M terminal is on the right.

11.2   The Split-Phase Motor Inside the Compressor

The actual split-phase motor mounted inside the compressor shell is very similar to the open split-phase motor in that it has a run winding and a start winding. In some cases, the run winding is also called the main winding. Fig. 11–5 shows an electrical diagram of a split-phase compressor motor. From this diagram you can see that the run winding uses larger wire than the start winding, and the terminal end of the run winding is identified with the letter M because it is also called the main winding. The start winding is made of smaller-gauge wire than the run winding, and it is identified with the letter S. Notice that the point where the run and start windings come together is called the common terminal, and it is identified with the letter C. It is also important to remember that the universal symbol for a motor is a circle with the terminals identified.

Typically the run winding will have between 4 Ω to 20 Ω of resistance, and the start winding will have between 20 Ω and 200 Ω of resistance. You should remember that the start winding will have more resistance because it has more feet of smaller gauge wire than the run winding.

Figure 11–5 Electrical diagram of a split-phase hermetic motor. Notice that the start winding is found between the S and common terminal, and the run winding is found between the M and common terminal. You should remember that some motors have the run winding identified as the main winding.

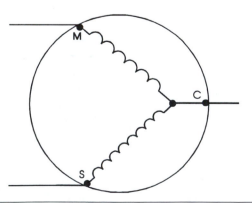

11.3   **The Current Relay**

The split-phase motor needs a means of deenergizing its start winding after the motor reaches 75–85% of its full rpm. In the open-type motor, the centrifugal switch is used, but it cannot be used inside the compressor shell because its open contacts may cause an explosion of the fumes from the refrigerant oil. It would also be difficult to check or replace the centrifugal switch if it is mounted inside the compressor shell, since the compressor shell is welded closed to create the hermetic seal when the compressor is manufactured.

Fig. 11–6 shows the current relay connected to the motor terminals. You should notice that the current relay coil is connected in series between L1 and the M terminal of the run winding, and the current relay contacts are connected in series between L1 and the S terminal of the start winding. The current relay contacts are normally open, until voltage is applied to the run winding of the motor. When voltage is applied to the run winding, the run winding will draw a large amount of current. Since the current relay coil is connected in series with the run winding, the large current flowing through the run winding must also flow through the current relay coil. The large current that flows through the current relay coil will cause it to pull its normally open contacts closed. When the contacts of the current relay are pulled closed, L1 voltage will be supplied to the start winding, which will begin to draw current. The current flowing through the start winding will cause sufficient phase shift to cause the rotor to begin to run. When the rotor is turning at nearly full rpm, the start winding and run winding

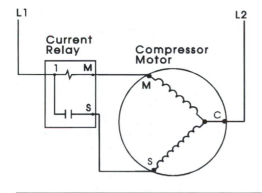

Figure 11–6   An electrical diagram that shows a current relay connected to a split-phase motor. The current relay has terminals identified as 1, M and S. The current relay coil is connected in series with the run winding (main winding) and the normally open current relay contacts are connected in series with the start winding.

will produce enough CEMF to allow the motor current to decrease to full-load amperage (FLA) levels. When the current returns to the FLA level, the current flowing through the current relay coil will not be strong enough to hold the current relay contacts in, and they will return to their normally open location. When the current relay contacts open, they will deenergize the start winding.

In the diagram the current relay looks as large as the compressor motor. In reality, the current relay is approximately 2 inches square, and the compressor may be as large as 18 inches tall. The important thing to remember about it is that it does the same job as the centrifugal switch in the split-phase motor. Fig. 11–7 shows a typical current relay mounted directly on the compressor motor M and S terminals. You can get an idea of the relative size of the current relay and compressor motor in this diagram.

11.4   Theory of Operation for the Split-Phase Compressor

You should remember from Chapter 10 that the split-phase motor gets its name because the motor uses the phase shift that occurs between the run and start winding currents to start the motor.

Air-conditioning and refrigeration manufacturers use the split-phase compressor in small residential air conditioners and refrigerators where a small amount of starting torque is required. Other equipment that uses the split-phase compressor includes dehumidifiers and window air conditioners.

11.5   Theory of Operation for the Current Relay

As you know, the start winding for a single-phase motor can only stay in the circuit for a few seconds during starting since it will draw a large current. A

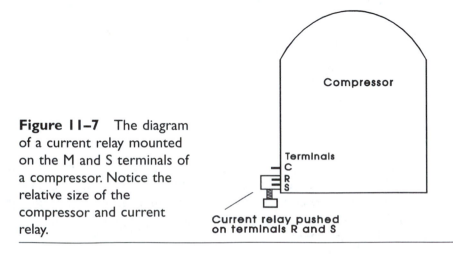

Figure 11–7 The diagram of a current relay mounted on the M and S terminals of a compressor. Notice the relative size of the compressor and current relay.

switch must be added to the start winding circuit to take the start winding out of the circuit when the motor reaches 75% of full rpm. This switch is called a current relay and it does the same job as the centrifugal switch in the open-type motor, where it deenergizes the start winding after the motor reaches 75% full rpm. Fig. 11–8a and Fig. 11–8b show an example of two types of current relays, and Fig. 11–8c shows the electrical diagram for the current relay.

You should notice the terminals are clearly marked on each of the current relays. The current relay in Fig. 11–8a has three terminals, and the current relay in

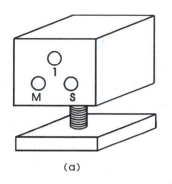

(a)

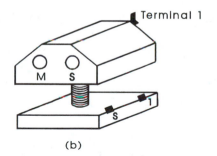

(b)

Figure 11–8a Example of a current relay used to disconnect the start winding in a split-phase hermetic compressor after the motor starts to run. **b** A second type of current relay. **c** A diagram of a current relay.

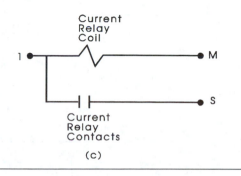

(c)

Fig. 11–8b has an M and an S terminal. It is important to understand that you must connect L1 power to the terminal identified as 1 and you must connect L2 for a 230 V single-phase compressor to the terminal identified as terminal C on the compressor motor. If the motor is a 115 V motor, the neutral power supply wire should be connected to the C terminal on the compressor.

Fig. 11–8c shows the electrical diagram of the current relay. Notice that this is a special relay that is used only for starting single-phase compressor motors. The wire in the coil of this relay is made of very large-gauge wire so that it can carry the same amount of current that the run winding will draw. You should also notice that the contacts of the current relay are normally open, and they will be pulled closed when the current flowing through the coil to the run winding becomes large enough.

11.6   Starting the Split-Phase Compressor with the Current Relay

The split-phase compressor used in window air conditioners or residential refrigerators is generally controlled by a line voltage thermostat. The line voltage thermostat is wired in series with the L1 supply voltage and terminal 1 on the current relay. Fig. 11–9 shows an electrical diagram of a line voltage thermostat connected in series with the current relay and compressor motor. Whenever the area being cooled becomes too warm, the thermostat contacts will close and pass voltage through the current relay to the motor windings.

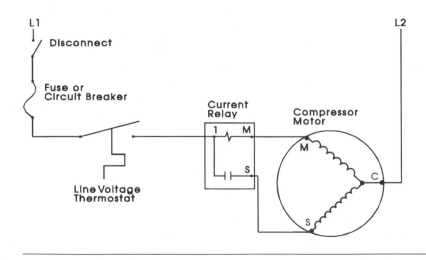

Figure 11–9   Electrical diagram of a line voltage thermostat that is used to control voltage to the compressor motor.

When voltage is first applied to the current relay and motor windings, the contacts of the current relay are open, so voltage can only reach the run winding, which will immediately begin to draw current. Since the rotor is not turning, the amount of current will quickly become very large, up to ten times the normal full-load amperage (FLA). This current is called the locked-rotor amperage (LRA) and it will remain high until the rotor begins to rotate. Fig. 11–10 shows the graph of the LRA and the FLA. From this graph you can see that the LRA gets very large as soon as voltage is applied to the motor windings. When the LRA is approximately 75% of its maximum, it will be large enough to energize the coil of the current relay and pull the current relay contacts closed.

When the normally open contacts of the current relay close, they will supply voltage to the start winding. When the voltage reaches the start winding, it will begin to draw current that will be out of phase with the current in the run winding. This phase shift will be sufficient to provide the starting torque required to get the compressor rotor to begin turning.

When the compressor rotor is turning at 75–85% of its full rpm, the start and run windings will produce enough CEMF to allow the current to drop to its FLA value. When the current flowing through the run winding drops to the FLA value, it will be too small to keep the current relay coil energized enough to keep the current relay contacts closed. At this point, the current relay contacts will return to their normally open condition and all voltage and current to the start winding will be deenergized. You should notice the point in the diagram in Fig. 11–10 where the contacts of the current relay are pulled in, and where they drop out.

11.7   Using the Wiring Diagram to Install the Permanent Split-Capacitor Compressor

The wiring diagram for the permanent split-capacitor compressor is shown in Fig. 11–11. From this diagram you can see all of the components of the system.

Figure 11–10 Graph that shows the locked-rotor amperage (LRA) and full-load amperage (FLA) of a split-phase compressor. Notice the point where the current relay contacts are pulled closed, and where they drop out.

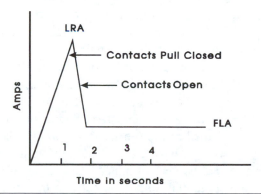

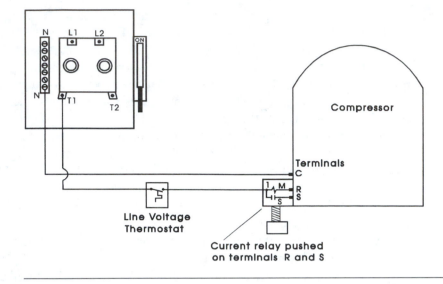

Figure 11–11   Electrical wiring diagram of a split-phase compressor. Notice the line voltage thermostat and current relay are also shown in this diagram.

The main difference between the wiring diagram and the ladder diagram is that the wiring diagram shows the location and outline of all of the components in the circuit. The wiring diagram will be used to install the system and ensure that it is connected correctly to the power from the disconnect box, and it will be used to locate components and their terminals.

For example, the current relay is shown in the wiring diagram as a square that is connected directly on the compressor. The wiring diagram also shows the locations of terminals on the current relay and the compressor as you would see them on the actual components. The wiring diagram also shows the location of wires that are used in the circuit between the disconnect switch, the line voltage thermostat, and the compressor.

When you are ready to install the split-phase compressor in an air-conditioning or refrigeration system, you will notice the diagram that is provided will look similar to the one shown in Fig. 11–11. Notice that the incoming power from the disconnect and line voltage thermostat to the compressor is connected to terminal 1 on the left side of the current relay. The L1 wire will provide power for both the run and start windings of the compressor from this point. Power is provided to the run winding through the coil of the current relay and power for the start winding is provided through the contacts of the current relay. The wire for L2 power is connected to terminal C on the compressor. This provides the L2 potential voltage directly to the compressor windings at terminal C. If the compressor is a 115 V compressor, the L2 wire would be replaced with a neutral wire.

You can see why it is necessary to use the wiring diagram for installation because you may become confused when you compare the ladder diagram to the actual wires. You should remember that the ladder diagram will provide detailed information that is useful during troubleshooting, and the wiring diagram will provide information for the installation of power wires and help you to locate individual components.

11.8   The Capacitor-Start, Induction-Run Hermetic Compressor

At times the starting torque from the split-phase compressor will not be sufficient to start the compressor. In these cases a start capacitor is added to give the motor more starting torque. The start capacitor is usually added at the time the compressor is manufactured, but sometimes it can be added to an existing split-phase compressor motor that is having a difficult time starting. When the start capacitor is added to the split-phase motor, it will be called a *capacitor-start, induction-run motor.*

Typical applications for the capacitor-start, induction-run motor include larger single-phase compressors for residential and commercial air-conditioning and refrigeration equipment. Fig. 11–12 shows a picture of a CSIR compressor. You should notice that the major difference between the CSIR compressor and the split-phase compressor is that the CSIR compressor has a start capacitor that is usually mounted on the compressor or very near it. You should remember that the start capacitor is assembled in a round, black, plastic case. Fig. 11–13 shows a diagram of the CSIR compressor. You should notice from this diagram that the CSIR compressor is basically a split-phase compressor with a start capacitor connected in series with the current relay contacts so that it is in series with the start winding.

11.9   Basic Parts of a Capacitor-Start, Induction-Run Hermetic Compressor

The basic parts of a capacitor-start, induction-run hermetic compressor are very similar to the split-phase hermetic compressor that was covered in the previous section, except a start capacitor and current relay are added. The start capacitor is identical to the start capacitors used in the CSIR open-type motors and the current relay for the CSIR compressor is similar to the one used on the split-phase compressor. Since the start capacitor needs to be connected in series with the current relay contacts and the motor start winding, an open needs to be cre-

Figure 11–12   A capacitor-start, induction-run (CSIR) compressor. Notice the start capacitor is mounted near the compressor. *(Courtesy of Copeland Corporation)*

ated in the circuit at the current relay. This can be accomplished by removing the jumper wire on the current relay. When the jumper is removed, the capacitor can be connected to the two terminals so it is in series with the contacts of the current relay. Fig. 11–14 shows the jumper on the current relay and the capacitor connected in place of the jumper. Fig. 11–15 shows the start capacitor connected to the current relay and the blue jumper removed.

11.10   The Operation of the Current Relay

The operation of the current relay in the CSIR compressor is similar to the operation of the current relay for the split-phase compressor, since it does the same job as the centrifugal switch as it deenergizes the start winding after the motor

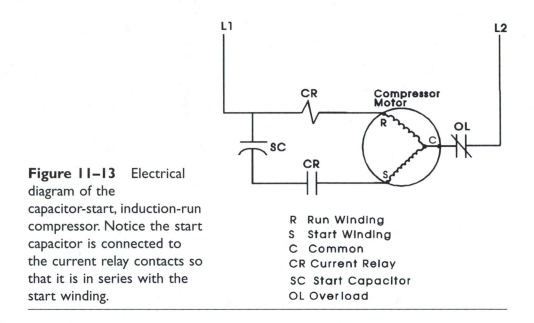

Figure 11–13 Electrical diagram of the capacitor-start, induction-run compressor. Notice the start capacitor is connected to the current relay contacts so that it is in series with the start winding.

R Run Winding
S Start Winding
C Common
CR Current Relay
SC Start Capacitor
OL Overload

reaches 75% full rpm. When voltage is first applied to the motor, current flows through the coil of the current relay into the run winding. Since the contacts of the current relay are normally open, current will only flow through the run winding at first. Since the current flowing through the run winding cannot get the rotor to turn, it will grow rather large to the LRA level. When the current reaches the LRA level, it will be sufficient to energize the current relay coil, which will pull its contacts closed. When the current relay contacts close, they will allow voltage to reach the start capacitor. After the first plate of the start capacitor charges, it will discharge and allow voltage to move to the opposite capacitor plate. This action will allow current to flow in the start winding, which will provide the phase shift between the run and the start windings. The phase shift is sufficient to allow the rotor to spin, which will generate CEMF. The increase in CEMF will cause the LRA to return to the FLA level, which will cause the cur-

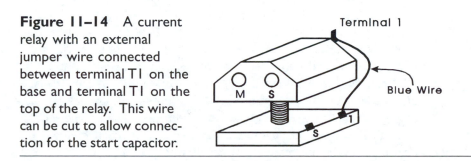

Figure 11–14 A current relay with an external jumper wire connected between terminal T1 on the base and terminal T1 on the top of the relay. This wire can be cut to allow connection for the start capacitor.

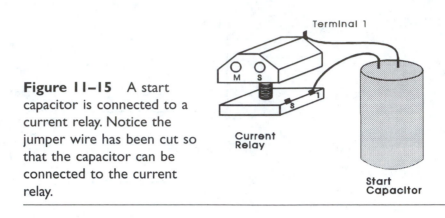

Figure 11–15 A start capacitor is connected to a current relay. Notice the jumper wire has been cut so that the capacitor can be connected to the current relay.

rent relay coil to reduce its magnetic field so that its contacts will return to their open position.

11.11 Thermal Overload Device

The thermal overload is used to sense the heat build-up by excessive current that the motor uses. It uses a set of normally closed contacts that is identified as OL. In most diagrams these contacts are located at the right of terminal C on the compressor. The overload is connected in series with the common terminal of the compressor so that all of the current in the compressor motor must go through it. Fig. 11–16a shows a picture of two types of thermal overload devices. Fig. 11–16b shows terminal connections for one type of overload and Fig. 11–16c shows the operation of the overload contacts in their open condition and closed condition.

Fig. 11–17a shows an electrical diagram of the thermal overload with its contacts in their normally closed position. Fig. 11–17b shows a diagram of the thermal overload after it has received excessive current and its contacts are open. From these diagrams you can see that the thermal overload consists of the thermal sensing element identified as point B and the overload contacts identified as point A. The contacts are made from copper and steel. These two metals are joined during a manufacturing process and they are called a bimetal element. The reason copper and steel are used is because they will expand at different speeds when they are heated. The heat causes the contacts to warp or change shape so that the contacts do not touch when the bimetal becomes hot.

When the current flows through the overload, the thermal sensing element begins to heat up. If the current is within limits, the amount of heat is not large enough to cause the normally closed contacts to open. When the current flowing through the overload becomes excessive, the thermal element produces excess heat that causes the bimetal overloads to trip (warp) to their open position.

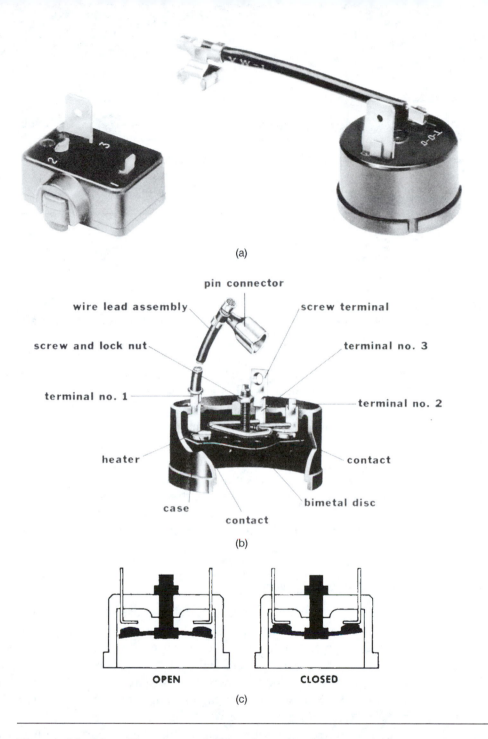

(a)

pin connector

wire lead assembly

screw terminal

screw and lock nut

terminal no. 3

terminal no. 1

terminal no. 2

heater

contact

case

bimetal disc

contact

(b)

OPEN

CLOSED

(c)

Figure 11–16a Two types of thermal overload devices. *(Courtesy of Therm-O-Disc Incorporated, Subsidiary of Emerson Electric)* **b** Terminal connections for an overload device. **c** The contacts of an overload in the open position and the closed position. *(Courtesy of Tecumseh Products Company)*

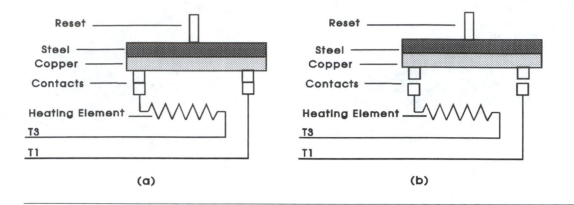

(a)   **(b)**

Figure 11–17a   Diagram of a thermal overload with its contacts closed. **b** Diagram of a thermal overload with its contacts open after the overload has been subjected to excessive current.

When the contacts trip to their open position, they will interrupt all current flowing through them, which in turn interrupts all of the current flowing through the motor. This protects the motor windings from becoming too hot due to excess current flow. When the motor has stopped running for a time and no current is flowing, the overload thermal sensing element will cool down and the bimetal will shrink, which will cause the contacts to go back to their normally closed position. When the contacts close, current will be allowed to flow through the overload so the compressor will start to run again. Some overload devices have a manual reset that forces a technician to physically go to the compressor and press the reset button. This makes the technician check out the cause of the overload condition.

11.12   The Hot Wire Relay

The hot wire relay is a starting switch that is used to deenergize the start winding of a hermetic compressor just like the current relay. The hot wire is not used in modern compressor applications because newer types of current relays and solid-state starting switches can do the job much better, but it is still important to understand how they work because you will see them on many older installations. Fig. 11–18 shows an electrical diagram of the hot wire relay, and Fig. 11–19 shows a sketch of a typical hot wire relay.

In the electrical diagram you can see that the hot wire relay has two terminals for incoming power that are identified as L1 and L2. The internal parts of the relay consist of a resistive wire that provides heat when current flows through it, and two sets of contacts that are made from bimetal. One set of contacts is iden-

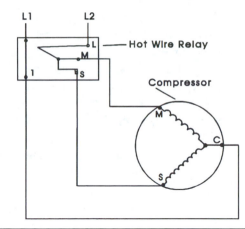

Figure 11–18 Electrical diagram of a typical hot wire relay. You should notice that its design is very similar to the current relay.

tified as the M contacts, and the other set is identified as the S contacts. The compressor's run winding is connected to the M contacts, and the compressor's start winding is connected to the S contacts. The common terminal C of the compressor can be connected to terminal 1, which is connected to L2. If a thermostat is used or other control devices, the C terminal can be connected to terminal 2 of the hot wire relay, and the thermostat can be connected between terminals 1 and 2 on the relay. This will ensure that the thermostat is connected in series with L2 and the common terminal of the compressor. If a starting capacitor is used, it would be connected between terminal S of the hot wire relay and terminal S of the compressor so that it is in series with the start winding just as when it is used with the current relay.

You can see that all of the current from L1 must pass through the resistive wire. After the current passes through the resistive wire, it splits and passes through either M contacts to go to the compressor's run winding, or through the

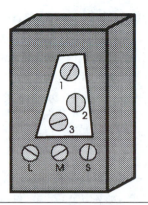

Figure 11–19 The hot wire relay is used on older compressors. You may see one on an older compressor application. The hot wire relay will be mounted on or near the compressor.

S contacts to the start winding. When voltage is first applied through the resistive wire it will produce a large amount of heat, which will be absorbed by the start bimetal switch and the run bimetal switch. After 2 to 3 seconds, the start bimetal switch will have absorbed enough heat to cause it to snap open. When this occurs, the start winding is deenergized from the circuit. You should remember this action is very similar to the action of the current relay that also deenergizes the start winding after 2 to 3 seconds.

After the start winding is deenergized, the total amount of current flowing through the resistive wire will be reduced, and the amount of heat it produces will not be large enough to cause the run bimetal switch to open. This allows the run winding to remain energized. If a problem occurs with the compressor that causes the run winding to draw excess current, the additional current will flow through the resistive wire, which will produce additional heat. The additional heat will cause the run bimetal switch to snap open and deenergize the run winding. In this way the hot wire relay is also used as an overload device. In its day, this was a great advantage to have the hot wire relay work as a starting switch and an overload device.

The only problem the hot wire relay causes is that it will not allow the compressor to be started quickly a number of times. Once the compressor has started, the hot wire has transferred a lot of heat to both the start and run bimetal switches. If the compressor is turned off and restarted in just a few minutes, the leftover heat may be sufficient to cause both the start bimetal switch and the run bimetal switch to open when the starting current is sensed again, which would not allow the compressor to run. This means that the hot wire relay must be allowed to cool down for several minutes anytime it has been running. It should be noted that you may consider replacing any hot wire relays that you find in the field with a current relay or solid-state starting device when you are troubleshooting.

11.13   The Solid-State Starting Relay

Since the advent of solid-state devices, newer starting switches have been designed to replace the current relay. The solid-state starting switch allows run and start current to flow through it when the motor is started, and then it deenergizes the start winding after the motor has started for 2 to 3 seconds. Fig. 11–21 shows an electrical diagram of the solid-state starting switch. In this diagram you can see that the power supply is connected to terminals x and X, the compressor S terminal is connected to the switch X terminal, and the compressor R terminal is connected to the switch X terminal. Fig. 11–20 shows a picture of the solid-state starting relay. If the solid-state starting relay is used with a CSIR compressor, the start capacitor will be connected in series with the start winding.

Figure 11–20 A solid-state starting relay. This starting relay is mounted near the compressor and has a starting capacitor with it. *(Courtesy of Watsco Components, Inc.)*

11.14   Overview of the Permanent Split-Capacitor Compressor

The permanent split-capacitor (PSC) compressor will operate like the open-type, permanent split-capacitor motor. The PSC compressor uses a run capacitor that is connected directly in series with the start winding without using a current relay or other starting relay. Fig. 11–22 shows a picture of a typical PSC compressor.

The PSC configuration becomes necessary for larger compressors. The split-phase and capacitor-start, induction-run motors can only be used on smaller compressors that are less than 1 horsepower. The reason for this is because compressors that are over 1 horsepower require larger currents that must travel through

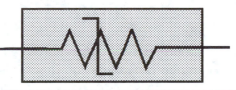

Figure 11–21   Electrical diagram of a solid-state starting relay.

(b)

(a)

Figure 11–22 A permanent split-capacitor (PSC) compressor is shown with its run capacitor. *(Courtesy of Copeland Corporation)*

the current relay coil and contacts to get to the motor windings. The size of wire in the current relay coil is not able to handle the amount of current that larger compressors draw in their run windings. If a current relay is used on these larger compressors, the large current would cause the current relay coil to burn open or become damaged.

For this reason, manufacturers have found that the PSC motor is a more reliable method of providing voltage to the start winding of the larger compressors and yet still get the effects of the phase shift caused by the capacitor and start winding. Fig. 11–23 shows a ladder diagram of the PSC compressor. From this diagram you can see that a capacitor is connected in series with the start winding of the compressor and no switches are used to disconnect the capacitor and voltage from the start winding. The PSC compressor, like the PSC open-type motor, does not provide a means to deenergize the start winding so the capacitor

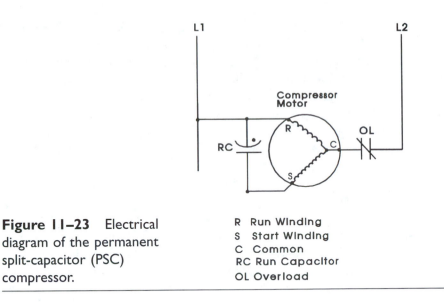

Figure 11–23 Electrical diagram of the permanent split-capacitor (PSC) compressor.

R Run Winding
S Start Winding
C Common
RC Run Capacitor
OL Overload

remains in the starting circuit permanently. This is the function that gives this wiring configuration for the compressor the name *permanent split-capacitor (PSC) compressor.* The word *permanent* indicates the capacitor always remains in the start winding circuit, and the word *split* indicates this is still a split-phase motor.

Since the capacitor remains in the start winding circuit at all times, the capacitor is placed in a steel case instead of plastic so it can dissipate more heat. The PSC compressor has become the most widely used compressor in air conditioners and refrigerators. Since there is no relay to disconnect the start winding after the compressor motor is started, there is one less component to fail. In the PSC compressor, the capacitor is the only component external to the motor used to start the compressor.

11.15 Basic Parts of the Permanent Split-Capacitor Compressor

The PSC compressor is basically a motor that has a run winding and a start winding just like the split-phase compressor and CSIR compressor. The major difference is that the PSC compressor motor is usually a larger horsepower than the split-phase compressor. The PSC compressor has the same terminal identification as the split-phase and CSIR compressors. The point where the run and start windings are connected in the compressor is called terminal C and it is the common point. The compressor will generally have an overload device connected to the common terminal and it will physically be mounted on the compressor where it can sense any excess heat from the compressor shell.

The other component used with the PSC compressor is the run capacitor. Fig. 11–22 showed a picture of the run capacitor with the compressor, and Fig. 10–26 in Chapter 10 shows pictures of several sizes of run capacitors. From these pictures you should remember that the run capacitor is housed in a metal container that is either oval shaped or rectangular shaped. Since the run capacitor remains in the circuit the entire time the motor is running, the metal container will help the capacitor to dissipate heat that builds up in it. You should notice that the run capacitor has a red dot or other method of indicating one of the terminals. It is important to be sure that the capacitor terminal that is identified is connected to the line side of the power supply and not to the compressor S terminal. The reason the identified terminal must be connected to the line-side voltage terminal is because this is the terminal connected to the capacitor plate that is nearest the metal container when the capacitor is inserted into the container. This means that if the capacitor plate would short out the metal container, it would cause the line-side fuse to blow. If this terminal is connected to the start winding of the compressor, the short circuit would cause the large current to flow through the compressor start winding before the fuse would detect the increase in current and open the circuit. This would cause damage to the compressor start winding.

11.16   Theory of Operation for the PSC Compressor

In the ladder diagram in Fig. 11–23 you can see that the run capacitor in the PSC motor is connected in series with the start winding. This means the phase shift that the run capacitor provides during starting will be similar to the phase shift in the capacitor-start, induction-run compressor. The run capacitor is usually rated between 5 μf and 50 μf. Since this rating is smaller than that of the start capacitor, the PSC motor will have a little less starting torque than the capacitor-start, induction-run compressor, but more than the split-phase compressor.

Once the PSC compressor starts and begins to run, the motor will again produce counter electromotive force (CEMF). The CEMF will build up to within a few volts of the applied voltage when the compressor is at full speed. The CEMF potential is present between the start and run windings. As long as the difference between the applied voltage and the CEMF is small, very little current will flow in the start winding. For example, if the applied voltage is 240 volts and the CEMF is 235 volts, the potential difference is 5 volts and the amount of current flowing in the start winding through the capacitor will be minimal. When the compressor loads up to pump a large quantity of refrigerant, the motor rotor will slow down slightly and the amount of CEMF will be reduced. When this occurs, the difference between the CEMF and the applied EMF will become larger and more current will flow through the capacitor. The phase shift will increase and cause more torque in the rotor shaft. You can see that this occurs at exactly the right moment to provide additional torque to move the larger load. The run capacitor is basically self-regulating since it will allow more current to pass as the difference in applied voltage and CEMF gets larger, and less current to flow when the voltage difference is small.

When the compressor is at full speed, the difference between the CEMF and applied EMF is small. The current in the start winding will be small, approximately 2 to 4 amps. This small current will not be enough to damage the compressor start winding and in effect it will be similar to having a current relay open its contacts and isolate the start winding.

Another effect of leaving the run capacitor in the start winding during the run cycle is that it helps to regulate the compressor speed. If the compressor starts to load up (that is, more refrigerant needs to be pumped), the compressor speed will tend to slow down slightly. When the rpm slows down, the CEMF becomes less, making a bigger voltage difference between the applied voltage and the CEMF. This causes the capacitor to allow more current to flow in the start winding, which gives the motor a little more torque, allowing the compressor to move the heavier load. As the load passes, the compressor again runs at full rpm and the current in the start winding returns to the small amount.

Some run capacitors actually have two separate capacitors mounted in the same container. One of the capacitors is smaller and it is used for the PSC condenser fan, and the second is a larger capacitor for the compressor. This type of run capacitor will have three terminals. One terminal is common for both capacitors, and a second terminal for the larger capacitor that is used for the compressor, and a third terminal for the smaller capacitor for the fan. The reason two capacitors are mounted in one container is that it saves space in the unit, and it saves time and wire when the system is manufactured. Fig. 11–24 shows a diagram of this type of capacitor used with the compresser and condenser fan.

11.17   Using the Wiring Diagram to Install a Permanent Split-Capacitor Compressor

The wiring diagram for the PSC compressor is shown in Fig. 11–25. From this diagram you can see all of the components of the system. The main difference between the wiring diagram and the ladder diagram is that the wiring diagram shows the location and outline of all of the components in the circuit. The wiring diagram will be used to install the system components and ensure that they are connected correctly to the power from the disconnect box. The wiring diagram will also be used to locate components and their terminals when you must take voltage measurements or other tests during troubleshooting. If a component

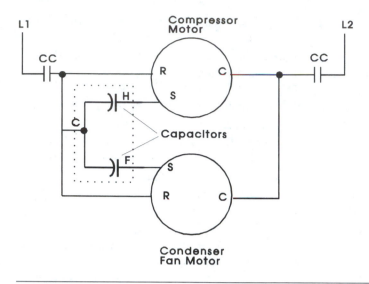

Figure 11–24 A special run capacitor that has two separate capacitors in one container. The terminals are identified for common, compressor, and fan. This capacitor is shown in a common, circuit with a compressor, and a fan.

on the wiring diagram is shown near the compressor in the wiring diagram, it means that the component is physically located near the compressor.

For example, the run capacitor is shown in the wiring diagram mounted near the compressor because that is where you would physically find it located. Its shape is shown as an oval because that is what the run capacitor physically looks like. The wiring diagram also shows the locations of terminals on the run capacitor and the compressor as you would see them on the actual components. The wiring diagram also shows the location of wires that are used in the circuit between the capacitor and the compressor. It is important to know the location of each individual wire in the circuit when you need to take voltage or current measurements.

When you are ready to install the PSC compressor in an air-conditioning or refrigeration system, you can use the wiring diagram that is provided with the system. In the diagram provided in Fig. 11–24 notice that the incoming power to the compressor is connected to the terminal on the left side of the run capacitor. This terminal on the capacitor is identified with a *red dot*. You should remember from Section 11.15 that it is important that the L1 voltage is only connected to the capacitor terminal with the red dot, and the compressor is connected to the other capacitor terminal. The L1 wire will provide power to both the run and start windings of the compressor from this point. The wire for L2 power is con-

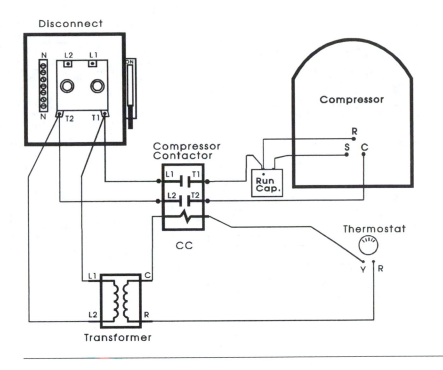

Figure 11–25   Wiring diagram for a PSC compressor.

nected to terminal C on the compressor. This provides the voltage directly to the compressor.

You can see why it is necessary to use the wiring diagram for installation because you may become confused when you compare the ladder diagram to the actual wires. The ladder diagram does not necessarily show where the actual wires are connected or located. You should remember that the ladder diagram will provide detailed information that is useful during troubleshooting about the sequence of operation, and the wiring diagram will provide information for the installation of power wires and help you to locate individual components.

11.18   Capacitor-Start, Capacitor-Run Compressor

Some air-conditioning and refrigeration compressors require very large amounts of starting torque. To give the single-phase compressor the most starting torque possible, both a start capacitor and a run capacitor are used. Since the compressor uses both a start capacitor and a run capacitor, it is called a capacitor-start, capacitor-run (CSCR) compressor. Fig. 11–26 shows a wiring diagram of a CSCR compressor.

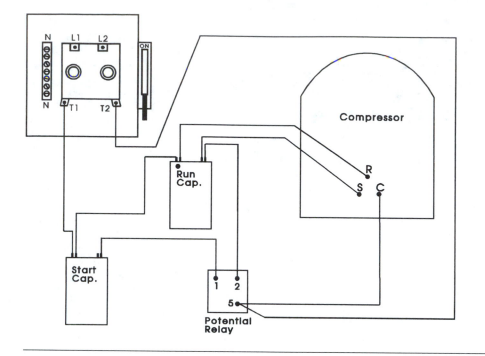

Figure 11–26   Wiring diagram for a capacitor-start, capacitor-run compressor. Notice the start capacitor, run capacitor, and potential relay with this compressor.

Equipment manufacturers will generally use the CSCR compressor on refrigeration equipment that uses an expansion valve and on air-conditioning systems when the permanent split-capacitor (PSC) compressor has trouble starting. In some cases the start capacitor is merely added to an existing PSC compressor to give it more starting torque. If the start capacitor is added to an existing PSC system, it is called a *hard start kit*. The start capacitor will supply the additional phase shift that provides sufficient starting torque to make the motor start more easily. The hard start kit will be explained later in this chapter. In either case, the CSCR compressor will have both a start capacitor and a run capacitor to provide the extra starting torque.

11.19   Components of the CSCR Compressor

Fig. 11–27 shows the ladder diagram of the capacitor-start, capacitor-run compressor. From this diagram you can see that the CSCR compressor has a start capacitor like the capacitor-start, induction-run (CSIR) compressor and it has a

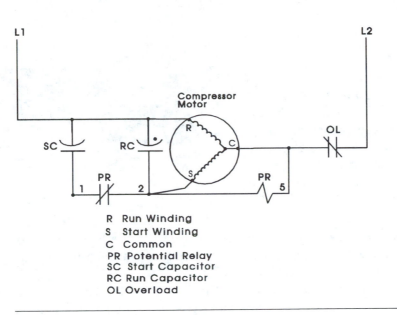

R   Run Winding
S   Start Winding
C   Common
PR Potential Relay
SC Start Capacitor
RC Run Capacitor
OL Overload

Figure 11–27   Ladder diagram of the capacitor-start, capacitor-run compressor. Notice that the wiring configuration for this compressor uses a start capacitor and a run capacitor.

run capacitor like the permanent split-capacitor (PSC) compressor. The CSCR compressor also has a potential relay that performs the function of disconnecting the start capacitor from the start winding when the motor is running at full speed, just like the current relay. The compressor for the CSCR configuration is usually 1 to 3 hp, which is larger than those used with the other wiring configurations, because it is the larger compressors that require the larger starting torque. If a compressor that is larger than 3 hp is required in the system, a three-phase compressor is generally used because it can provide even more starting torque than the CSCR compressor.

The motor used in the CSCR compressor is basically the same type of motor used in the other compressors in that it has a run winding and a start winding. This means the compressor will have terminals R, S, and C like the other compressors. The run winding is connected between terminals R and C and the start winding is connected between the S and C terminals.

You have studied the start capacitor in detail in the capacitor-start, induction-run compressor, and you have studied the run capacitor in the PSC compressor, so the only new component in the CSCR compressor that you need to learn about is the potential relay. Fig. 11–28 shows a picture of a typical potential relay. From the picture you can see that the potential relay is mounted inside a black rectan-

Figure 11–28 A typical potential relay with its cover removed. Notice that the relay looks like a black plastic cube with three screw terminals on it.

gular plastic case that is approximately 2.5 inches long. When you see a potential relay mounted near the CSCR compressor on an air-conditioning system, you will only see the black plastic case.

Fig. 11–29a shows a sketch of the plastic case with the screw terminals identified and Fig. 11–29b hows an electrical diagram of the potential relay's normally closed contacts and its coil. The three terminal screws on the potential relay are identified as terminals 1, 2, and 5. From the diagram you can see that the potential relay coil is connected to terminals 2 and 5, and the normally closed contacts of the potential relay are connected between terminals 1 and 2. Terminal 2 acts as a common point for both the contacts and the coil. The potential relay coil is made from a large number of turns of very fine wire, so its resistance will be approximately 10 kΩ to 20 kΩ. This is much higher resistance than resistance for the coil of a typical control relay. The potential relay is generally mounted close to the compressor or the start and run capacitors.

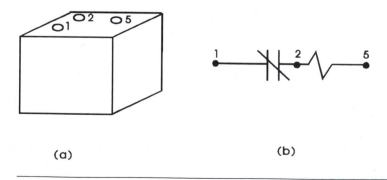

(a)   (b)

Figure 11–29a Potential relay with its terminals identified. **b** Electrical diagram of potential relay.

11.20   Theory of Operation for the CSCR Compressor

From the ladder diagram in Fig. 11–27, you can see that the CSCR compressor has the start capacitor in parallel with the run capacitor. Both capacitors are connected in series with the start winding. Connecting capacitors in parallel adds their capacitive values. The formula $C_T = C_1 + C_2$ is used for adding parallel capacitors, similar to adding resistors in series.

The effect of having a start capacitor with a large capacitance value (up to 600 µf) added to the value of the run capacitor (up to 75 µf) will cause a larger phase shift between the start and run winding voltages. You should remember that the larger the phase shift, the larger the starting torque and the easier it is to start the compressor against large loads. The CSCR compressor uses a potential relay to start the motor and deenergize the start winding. In effect, the potential relay replaces the current relay that is used in the split-phase compressor and the capacitor-start, induction-run (CSIR) compressor that was presented in earlier chapters.

In the diagram you can see the contacts of the potential relay are connected in series with the start capacitor just like the contacts of the current relay in the PSC compressor. From the diagram in Fig. 11–27, you will notice that when voltage is first applied, the potential relay contacts are normally closed, completing the circuit between the start capacitor and the start winding. As the compressor motor starts to turn, the CEMF begins to build. The CEMF will be present at terminals S and C. The coil of the potential relay (terminals 2 and 5) is connected to terminals S and C on the compressor. As the compressor reaches approximately 75% of full rpm, the CEMF will be strong enough to energize the potential relay coil and pull its contacts open. When the potential relay contacts open, the start capacitor is removed from the start winding.

The run capacitor is left in the start winding circuit, as in the PSC compressor. The combination of the start capacitor and run capacitor provides the phase shift during starting, which gives the CSCR compressor excellent starting torque, and the run capacitor remains in the circuit like the PSC compressor to give the CSCR compressor good running efficiency. The potential relay coil is connected in parallel with the S and C terminals on the compressor, which means it will last longer since it is not connected in series with the run winding like the current relay and it is not subject to large currents that tend to wear it out.

11.21   Hard Start Kits for Compressors

At times you will be working on a system that has a split-phase compressor or a PSC compressor that needs a little more torque to enable the compressor to start. When you encounter this problem in the field, you can add a starting capacitor to these compressors to provide them with more starting torque. If a start capacitor is added to the split-phase compressor, it technically becomes a capacitor-start, induction-run compressor. The start capacitor will provide additional phase shift, which develops the starting torque that allows the compressor to start and run. The start capacitor is connected in series with the normally open contacts of the current relay and the start winding. The parts that are used to accomplish this conversion are called a *hard start kit,* since the result is that the capacitor will help compressors that are having a hard time starting.

Recently a hard start kit has been designed that contains a capacitor and solid-state starting relay in one plastic case, which has two leads protruding out one end. The plastic case is basically a traditional start-capacitor case that has been extended so the solid-state relay can fit on top of the capacitor. Fig. 11–30 shows the picture and Fig. 11–31 shows the electrical diagram of this type of hard start kit. Since the solid-state starting relay is included in this kit, it can be connected

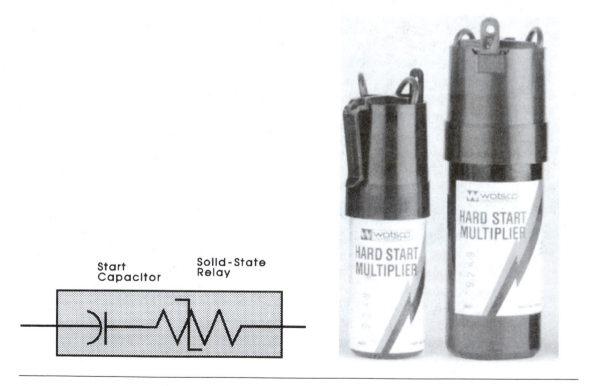

Figure 11–30 A solid-state hard start kit. Notice that this hard start kit has the capacitor and solid-state relay in a single plastic case. All you need to do is attach the two wires across the leads of the existing run capacitor in a PSC compressor. *(Courtesy of Watsco Components, Inc.)*

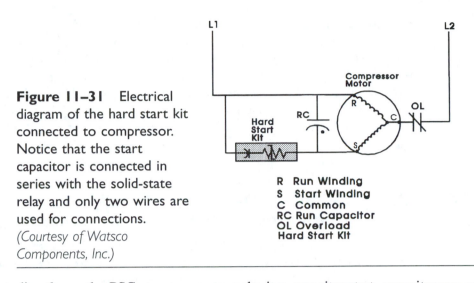

Figure 11–31 Electrical diagram of the hard start kit connected to compressor. Notice that the start capacitor is connected in series with the solid-state relay and only two wires are used for connections. *(Courtesy of Watsco Components, Inc.)*

R   Run Winding
S   Start Winding
C   Common
RC Run Capacitor
OL Overload
Hard Start Kit

directly on the PSC compressor to make it a capacitor-start, capacitor-run compressor. The solid-state starting relay takes the place of the potential relay. In Fig. 11–31 you can see the solid-state hard start kit connected to an existing PSC motor. After the hard start kit is connected, you can see the compressor now is technically a capacitor-start, capacitor-run compressor.

11.22 Using a Current Relay and Start Capacitor for a Hard Start Kit

The hard start kit is also used to add a start capacitor to a split-phase motor to provide additional torque. This is done in the field when the split-phase compressor needs additional starting torque. When the start capacitor is added to the current relay contacts and start winding, the compressor diagram will look similar to the CSIR one in Fig. 11–13. It is important to remember that the current relay can only be used on smaller single-phase compressors.

11.23 Adding a Potential Relay and Start Capacitor for a Hard Start Kit

If the PSC compressor is too large to use as a current relay, a potential relay can be used as part of the hard start kit. The potential relay must be selected so that it matches the CEMF of the compressor. You can usually use the model number of the compressor to help you select the potential relay and start capacitor to fit the compressor. When the potential relay and start capacitor are added to the PSC compressor, it technically becomes a capacitor-start, capacitor-run compressor.

It is also possible to use a hard start kit as a temporary repair to get a PSC or split-phase compressor to start that has not started for a number of months. Sometimes when a compressor has not started for a long period of time, the pistons become stuck in their cylinders. When this occurs, the compressor motor will hum when power is applied but the compressor will not start. When this occurs, you can place a start capacitor and potential relay on the PSC or split-phase compressor to get it started. After the compressor is started, you can remove the start capacitor and potential relay and allow the motor to run as it was originally designed.

11.24   Using the Wiring Diagram to Install the Capacitor-Start, Capacitor-Run Compressor

The wiring diagram for the CSCR compressor is previously shown in Fig. 11–26. From this diagram you can see all of the components of the system. The main difference between the wiring diagram and the ladder diagram is that the wiring diagram shows the location and outline of all of the components in the circuit. The wiring diagram will be used to install the system and ensure that it is connected correctly to the power from the disconnect box, and it will be used to locate components and their terminals.

For example, the potential relay is shown in the wiring diagram as a square because that is its physical shape. The wiring diagram shows the locations of terminals 1, 2, and 5 as you would see them on the actual potential relay. The wiring diagram also shows the location of wires and jumpers that are used in the circuit between the two capacitors, potential relay, and compressor. You can see that the start capacitor and run capacitors must have terminals that can allow two wires to be connected on each terminal.

When you are ready to install the CSCR compressor in an air-conditioning or refrigeration system, you can use the diagram provided. Notice that the incoming power to the compressor is connected to a terminal on the left side of the start capacitor. This wire will provide power to both the start and run windings from this point. The wire for L2 power is connected to terminal 2 on the potential relay, which also is connected to terminal C on the compressor. This provides the voltage directly to the compressor.

You can see why it is necessary to use the wiring diagram for installation because you may become confused when you compare the ladder diagram to the actual wires. You should remember that the ladder diagram will provide detailed information that is useful during troubleshooting, and the wiring diagram will provide information for the installation of power wires and help you to locate individual components.

11.25   **Controlling the CSCR Compressor**

Since the capacitor-start, capacitor-run compressor uses a large current to start and run, the CSCR compressor will normally be controlled by a relay or contactor, rather than being controlled by a thermostat (line voltage type) that is connected in series with the compressor (see Fig. 11–32). When the relay or contactor is used to control the compressor such as in a residential air-conditioning system, the thermostat will be a low-voltage type and it will be connected in series with the relay coil, which draws low current and uses low voltage (24 VAC) to become energized. The contacts of the relay or contactor are connected in series with the compressor motor windings and they will open or close to control the compressor.

In this type of circuit, when power is applied, it will move as far as the open contacts of the line voltage thermostat. When the room becomes too warm, the thermostat contacts will close and high voltage is applied to the run and start windings of the compressor, and a large amount of current will be pulled as the compressor starts. As you know, the large starting current is called locked-rotor current or locked-rotor amperage (LRA). After the compressor is running at full speed, you will notice that the current drops slightly to its full-load amperage (FLA) level. The amount of FLA current is listed on the name plate for the air-conditioning or refrigeration system or on the compressor so that you can compare it with any measurements that you take. If the measure of full-load current

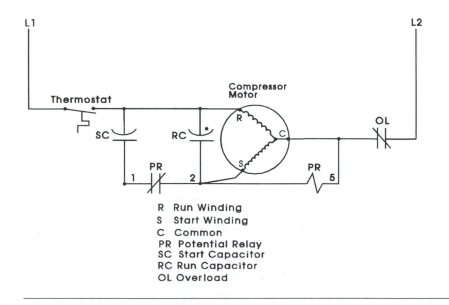

Figure 11–32   A CSCR compressor controlled by a line voltage thermostat.

is larger than the rating, the compressor will overheat and cause the overloads to open. The main reasons for high FLA are that the compressor is overloaded and it is pumping too hard, or the system could be low on refrigerant, and not enough refrigerant is coming back to the compressor to keep it cool.

11.26   Troubleshooting the Split-Phase Compressor

When a single-phase compressor fails to start in the air conditioner and refrigerator that you are working on, you must be able to troubleshoot the circuit and components quickly and accurately. The following procedures list the steps you should use to troubleshoot all of the four types of single-phase compressors. These steps will be broken into two parts. The first procedure should be used when power is applied to the compressor and it is trying to start and you can hear it humming. The second procedure should be used if the compressor does not try to start when power is applied. You should note that the same steps are used for the split-phase, CSIR, PSC, and CSCR compressors. The only difference is that you must take into account the extra capacitors and relays that some compressors use and be sure to test them also.

11.27   Troubleshooting Procedure 1

(The compressor tries to start and you can hear it humming.)

1. Put a clamp-on ammeter around the wire that is connected to L1. Apply power to the compressor to try to make it run by turning the disconnect to the closed position and switching the thermostat to the on position. Fig. 11–33 shows the correct location to place the clamp-on ammeter to check the current the motor is drawing as you try to start the system. If the current becomes excessive and the motor will not start, turn the power off immediately and go to part 2. It is important to always try to make the compressor run even though you know it has problems because sometimes the problem will be a loose connection. If the compressor runs at certain times and not at others, you should suspect loose connections.

2. If the compressor tries to run (makes a noise) but does not start, it is indicating that it is not drawing sufficient current in both the start and run windings, or the compressor piston is not able to turn due to mechanical problems. **(For this test, we will presume the problem is not mechanical.)**

    Since the motor is drawing some current but it will not start, you must test to see if the system has the proper amount of voltage available at the disconnect. If the proper amount of voltage is present, and the motor is drawing some amount of current but will not start, you

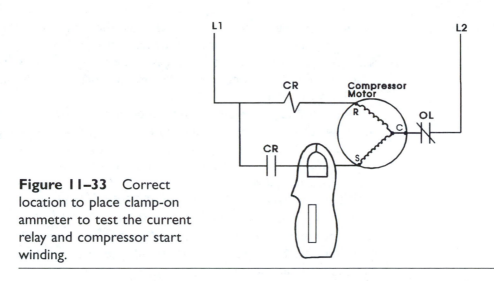

Figure 11–33 Correct location to place clamp-on ammeter to test the current relay and compressor start winding.

should suspect that sufficient current is not flowing in either the start winding or run winding. You can either test the start and run windings individually for current flow with an ammeter as shown in Fig. 11–33 or you can suspect one of the windings has an open and test each of them for continuity. Fig. 11–34 shows the correct location to place the

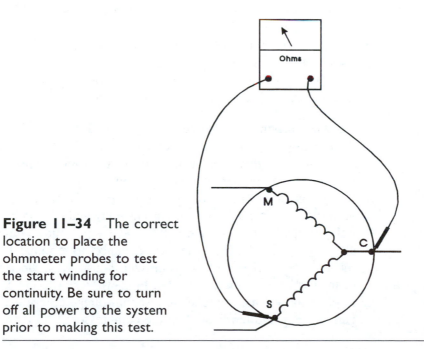

Figure 11–34 The correct location to place the ohmmeter probes to test the start winding for continuity. Be sure to turn off all power to the system prior to making this test.

ohmmeter probes to test the start winding after voltage has been turned off. If the start winding shows some amount of resistance (5–20 Ω), it should draw current. If the compressor is a CSIR or CSCR, you should also test any of the components that are in series with the run and start windings for continuity.

When you test the start winding to determine how much current it is drawing, you may need to remove the current relay from the compressor and put jumper wires between the current relay and compressor, so you have a place to clamp the ammeter around a wire. Be sure to place the current relay in an upright position so that it will operate correctly. If the current relay is operating correctly, the starting current should increase for 1 to 3 seconds and then return to zero after the current relay contacts open. If the current continues to be present in the start winding after 2 to 3 seconds, you should suspect the current relay contacts are welded closed.

11.28   Troubleshooting Procedure 2

(The compressor does not try to start when power is applied and it does not make a noise.)

Since the compressor does not try to start and no current is being drawn by the motor when power is applied, it suggests one of several possible problems is preventing any voltage from reaching the compressor, or the compressor may have an open in a wire that is common to both the run and start windings. These problems could include no power to the system, blown fuse, bad thermostat, bad wire between power source and compressor, open at terminal C in the motor, open in the thermal overload, or open in the neutral line that provides power to the compressor.

1.   Test the disconnect for proper voltage. Be sure to test for voltage at any point below the fuses at the terminals marked T1 and T2. If you do not have voltage at this point, you have lost the supply voltage.

2.   If you have the proper amount of supply voltage, you should next test for voltage at terminal 1 of the current relay. Place one probe of the voltmeter on terminal 1 of the current relay and one probe on the neutral terminal of the disconnect. If voltage is present, the thermostat and all the wires connecting from the disconnect to terminal 1 of the current relay are good. If no voltage is present, you have an open in this side of the circuit or in the thermostat. Fig. 11–35 shows the correct location to place the voltmeter probes to test for voltage at terminal 1 of the current relay and terminal C on the compressor.

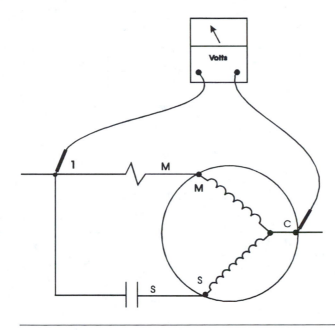

Figure 11–35 Electrical diagram that shows the correct location to place the voltmeter terminals to measure voltage at terminal 1 of the current relay and terminal C of the compressor.

3. If voltage is present at terminal 1 of the current relay and terminal C on the compressor, but the motor still does not hum or try to draw current, the problem must be in a part of the motor winding circuit that affects voltage to both windings of the motor. Turn off all power to the system and test the current relay and motor windings for continuity. You should also inspect the terminals of the compressor and current relay for dirty or loose connections. The problem you are looking for is an open in a part of the circuit that affects the run winding and the start winding. If either winding has a path for current, the compressor will try to start and you will hear it humming. Since the main symptom for this section is that the compressor does not make any noise when power is applied, the problem must be affecting both windings of the motor. This includes the current relay coil, the common terminal where the run and start windings of the of the motor come together, and the thermal overload. Fig. 11–36 shows the correct location to place the ohmmeter probes to test the run winding for continuity. Be sure to turn off all power to the motor circuit prior to making this test. Place the probes on

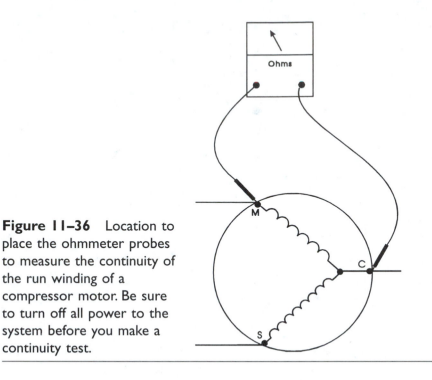

Figure 11–36 Location to place the ohmmeter probes to measure the continuity of the run winding of a compressor motor. Be sure to turn off all power to the system before you make a continuity test.

terminals R and C on the compressor terminals. You should also test the start winding in a similar fashion.

11.29   Testing the Current Relay

At times you will need to test a current relay that you suspect is malfunctioning. Use the following procedure to test your current relay. Fig. 11–37 shows the proper location to place the ohmmeter probes to test the current relay. (This procedure can also be used when you are in the field troubleshooting a system and you suspect the current relay is malfunctioning.)

1. Locate the current relay you are testing. Notice that the terminals are marked S, M, and 1.
2. The relay coil should be located between terminals 1 and M. (Refer to the diagram at the bottom of Fig. 11–37b.) Use an ohmmeter to test the relay coil. Set the meter R × 1. Put one probe on the M terminal and one probe on terminal 1.
   a. If the meter shows the coil has some resistance (2–100 Ω), the coil is good. Continue to step 3.

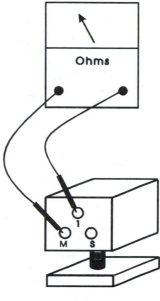

(a)

Figure 11–37a Location to place the probes of an ohmmeter when you are testing a current relay coil. Use the diagram at the bottom to locate the contacts at terminals 1 and M. **b** Diagram of current relay.

(b)

b. If the meter shows infinite (∞) resistance, the coil is open and the relay must be replaced.

3. The contacts are between terminals 1 and S. Use your ohmmeter to test the contacts. Place one meter probe on S and one probe on terminal 1. (*Note*: This relay may be position sensitive. The arrow on the side of the relay will show in which position the relay should be mounted. The arrow should point up when the relay is in the proper position.)

a. With the arrow pointing up, the contacts should be open. Set the meter to the highest resistance scale. Measure the resistance between terminals 1 and S on the relay. If the reading is infinity (∞), continue to step 3b. If the contacts show some small amount of resistance, the contacts may be welded or sticking, and the relay needs to be replaced.

b. You will need to also test the relay contacts to ensure that they have low resistance when they are closed. You can perform this test by turning the relay over so the arrow points down, which will cause the contacts to close by gravity. Repeat the resistance measurement with the contacts closed. This time the contacts should show less than 1 Ω of resistance. If the resistance in the contacts is more than 1 Ω, replace the current relay. Remember, a good current relay will show low resistance in its coil, zero resistance across terminals 1 to S with its contacts closed, and infinite resistance across terminals 1 to S when its contacts are open.

11.30   Troubleshooting Procedure 3

(The motor runs but draws excessive current.)

If the motor starts but it draws excessive current and becomes overheated and blows a fuse or causes the overload to open, it is a good possibility that the start winding is staying energized after the motor reaches full rpm. The problem occurs when the contacts of the current relay remain closed after the current in the run winding returns to FLA value. This can occur when the contacts become welded in the closed position, or if the wrong-size current relay is being used. If the current relay has been installed on the compressor for some time, and the compressor worked correctly at one time, you can rule out the possibility of the current relay being the wrong size and you should suspect the current relay is malfunctioning. If the current relay or compressor has been changed recently, you should suspect the wrong current relay is being used.

If the current relay has been working previously and the compressor is drawing excessive current after it starts, you should suspect the contacts are welded closed. You can test for this problem by removing the current relay and testing it for continuity, or you can use the clamp-on ammeter to test it. Remember during this test that you are trying to determine if the contacts of the current relay are open. If the continuity test shows the current relay contacts are closed when no current is flowing through the coil, the current relay is defective and should be replaced. If you use an ammeter to test the current relay, you should notice the current through the contacts to the start winding is rather large when the compressor is started, and returns to zero after 2 to 3 seconds when the compressor is started. If the current in the start winding does not return to zero after the compressor is running, the current relay is faulty and needs to be replaced.

11.31   Troubleshooting the CSCR Compressor

Troubleshooting the CSCR compressor is similar to troubleshooting the CSIR and split-phase compressors, since you can break the test into three parts: testing the circuit, testing the motor, and testing the capacitors and potential relay. The only major difference in testing the CSCR compressor is the potential relays and capacitor. If the problem is in the power supply voltage or the compressor windings, you can use the test provided in the preceding section.

11.32   Testing the CSCR Compressor Circuit for Current

The CSCR compressor circuit is much easier to troubleshoot than the CSIR or the split-phase compressor because the potential relay and capacitors are mounted remotely from the compressor, which means that you can easily isolate them and get the ammeter around the necessary terminals and wires to make tests. This section will show where to place the ammeter to measure current in the CSCR compressor. Fig. 11–38 shows the proper location to connect the clamp-on ammeter to test for FLA current. If the ammeter indicates the motor is drawing some

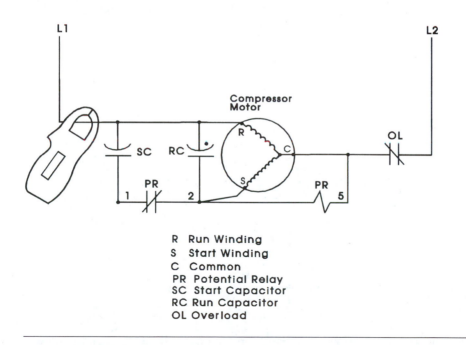

R Run Winding
S Start Winding
C Common
PR Potential Relay
SC Start Capacitor
RC Run Capacitor
OL Overload

Figure 11–38 Diagram that shows the clamp-on ammeter connected around the wire to measure the amount of full load current.

amount of current, it also means that the motor has voltage applied to its terminals and current is flowing through either the start winding or the run winding, or both. If the FLA test indicates no current flow, you should suspect the total loss of voltage somewhere between the power source at L1 or L2 or in the wire supplying this voltage to the compressor windings. Fig. 11–39 shows the proper location to place the clamp-on ammeter to test for current in the start windng and the run winding. It is important to determine if you do not have current flow through one of these windings so you will know in which part of the system to test for an open.

The previous test for current in the run winding and the start winding will indicate where the problem is located. If you have the proper amount of supply voltage and no current is flowing in the run winding, you can remove the wires from the compressor and test the run winding (R-C) for continuity, since this circuit must have an open. If the current test indicates that the start winding does not have any current flow, you can use the voltage test to test for the loss of voltage between the point where voltage is supplied to the compressor and the normally closed potential relay contacts and start winding. You can also turn off power and use a continuity test for this path. Remember you are looking for an open in this circuit. If you have no current flow in both the start and run winding

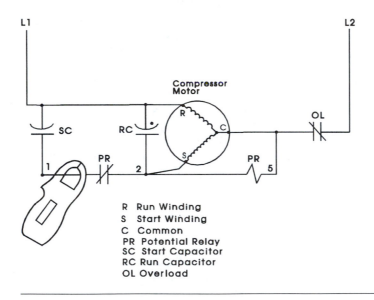

Figure 11–39 The clamp-on ammeter should be connected around the wire that is connected between the start capacitor and terminal 1 on the potential relay to measure the current in the start winding. The meter could also be connected around the wire between terminal 2 and terminal S on the compressor.

tests, you can suspect an open in the part of the circuit that is common to both windings, such as the overload, and you can use a continuity test to locate the problem. Remember if the overload is open, it is a symptom of overcurrent, and it is not causing the overcurrent.

11.33   Testing the Potential Relay Out of the Circuit

If you suspect the potential relay is malfunctioning, you can remove it from the circuit and test it for continuity. The potential relay has a high-resistance coil (approximately 10 kΩ to 20 kΩ) that is energized by the counterelectromotive force (CEMF) produced by the compressor at its C and S terminals. The coil is between terminals 2 and 5 of the relay and the normally closed contacts are between terminals 2 and 1. When you check the relay coil for continuity between terminals 2 and 5, be sure to set the meter on the R $\times$ 100 k scale. The relay should have very high resistance (approximately 20 kΩ). When you test the contacts, you should use the lowest resistance range (R $\times$ 1) and the meter should indicate that the relay has closed contacts between terminals 2 and 1. If you measure high resistance or infinite (∞) resistance, the contacts are dirty or burnt open. In either case, you should consider that the potential relay is not operating correctly and it should be replaced.

11.34   Testing the Potential Relay While It Is in the Circuit

At times you may need to test the potential relay while it is still connected to the compressor circuit. The major reason you would consider testing the potential relay is that the compressor is having trouble getting started. When this occurs, you can check the start capacitor, run capacitor, and potential relay with a clamp-on ammeter. For the first part of this test, clamp the ammeter around the wire that is between the S terminal on the compressor and terminal 2 of the potential relay to measure the amount of current flowing to the start winding of the compressor. You will need to refer to the wiring diagram in Fig. 11–38 or 11–39 to see the location of this wire. The reason you would use the wiring diagram instead of the ladder diagram is that when you look at the ladder diagram, it may not be apparent which wire to use because you will see that one end of the start capacitor and terminal 2 of the potential relay are both connected to the S terminal of the compressor through the wire that runs between the S terminal and terminal 2 of the potential relay.

If everything is operating correctly, the ammeter should indicate 10–20 amps is being used by the start winding for 1 to 5 seconds when it is trying to start and then the current falls off to zero. If the ammeter indicates 0 amps or very low amps (less than 2 amps) when power is first applied, the start capacitor, run capacitor, or potential relay is not operating correctly. If 0 amps is measured, this circuit must have an open in one of the wires, or one of the components has an open. If low current is measured, it means the wires are good, but one of the components is faulty.

Another problem occurring with the potential relay is that an open occurs in its coil, its contacts will remain closed, and the start winding will remain energized after the motor is started and running. The effects of the start winding remaining energized after the motor is running are that the start winding will continue to draw large amounts of current and the compressor motor will overheat to a point where the thermal overload will open and shut the motor off. When this occurs, you will be called because the compressor will only run for a short time and then turn off.

The best way to test for this type of problem is to place the clamp-on ammeter around the wire going to the S terminal of the compressor. When power is applied to the compressor, the current in this wire should show 10–20 amps for 1 to 5 seconds until the compressor is started and running at full speed. When the compressor is at full speed, the current in this wire should drop to 0 amps, since the potential relay coil should become energized by the CEMF and pull its normally closed contacts open. When the normally closed contacts of the potential relay open, all of the current in the starting circuit should go to zero. If the current does not go to zero, it indicates the contacts of the potential relay are still closed.

If the current test indicates the potential relay contacts will not open, you can turn all power to the system off, remove the relay, and replace it with a known good one and try the current test again. The reason you should replace the potential relay is because at this point all you can do is test its coil for continuity. If the coil has an open circuit, it is obviously faulty and you will need to change the relay. If the coil has the correct amount of resistance (10–20 kΩ), you can suspect the coil is good, but the contacts of the potential relay may be welded in the closed position so that they cannot open. In either case, you will need to replace the relay and try the current test again.

11.35   Troubleshooting the PSC Compressor

The PSC compressor can be tested during troubleshooting by using the CSCR diagrams and procedures and omitting the start capacitor and the potential relay, since this type of motor has only a run capacitor that remains in the ciruit at all times. The PSC compressor is perhaps the simplest type of compressor to troubleshoot, since it does not have a starting switch or relay. The start and run windings are identical to the windings in each of the other types of compressors, so you can use any of the preceding tests to test the windings for opens. The current and voltage tests for the start and run windings can also be used from the preceding tests.

Questions for This Chapter

1. Identify the windings and the terminal markings for any of the three types of single-phase hermetic compressor motors and explain why the windings will be similar in each of the four types of motors.
2. Explain the operation of the capacitor-start, capacitor-run hermetic compressor and its two capacitors and potential relay.
3. Use the graph of LRA and FLA in Fig. 11–11 to explain how the current relay operates.
4. Explain how you can test the run winding and start winding of a hermetic compressor.
5. Explain how the motor terminals are sealed in a hermetic compressor so refrigerant does not leak around them.

True or False

1. The run winding or the start winding of a single-phase compressor will draw current if it has an open.
2. The potential relay contacts are normally closed and are opened when the back EMF (CEMF) in the motor windings becomes large enough to energize the relay coil when the motor reaches 75% full speed.
3. The run capacitor in the PSC compressor remains in the circuit at all times and can help to regulate the speed of the compressor.
4. The contacts of the current relay are normally closed.
5. The thermal overload on the hermetic compressor can be opened by either excessive current draw or if the compressor becomes too hot.

Multiple Choice

1. The current relay has _____
   a. a coil with low resistance and normally closed contacts.
   b. a coil with high resistnce and normally closed contacts.
   c. a coil with low resistance and normally open contacts.
   d. a coil with high resistance and normally closed contacts.
2. The potential relay has _____
   a. a coil with low resistance and normally closed contacts.
   b. a coil with high resistance and normally closed contacts.
   c. a coil with low resistance and normally open contacts.
   d. a coil with high resistance and normally closed contacts.

3. The split-phase hermetic compressor motor _____
    a. uses a current relay that has its coil connected in series with the start winding and its contacts connected in series with the run winding.
    b. uses a current relay that has its coil connected in series with the run winding and its contacts connected in series with the start winding.
    c. uses a potential relay that has its contacts connected in series with the start windings.
4. The CSIR split-phase hermetic compressor motor _____
    a. uses a current relay that has its coil connected in series with the run winding and its contacts connected in series with the start winding.
    b. uses a current relay that has its coil connected in series with the start winding and its contacts connected in series with the run winding.
    c. uses a potential relay that has its contacts connected in series with the start windings.
5. The CSCR split-phase hermetic compressor motor _____
    a. uses a potential relay that has a run capacitor connected in series with the run winding and a start capacitor connected in series with the potential relay normally open contacts and with the start winding.
    b. uses a current relay that has its coil connected in series with the start winding and its contacts connected in series with the run winding.
    c. uses a potential relay that has its contacts connected in series with the start capacitor and start windings and a run capacitor connected between the run and start windings.

Problems

1. Draw a sketch of the split-phase hermetic compressor with a current relay connected to its windings.
2. Draw a sketch of a CSIR hermetic compressor with a current relay and start capacitor connected to its windings.
3. Draw a sketch of a CSCR hermetic compressor with a potential relay, start capacitor, and run capacitor to its windings.
4. Draw a sketch of a PSC hermetic compressor with a run capacitor connected to its windings.
5. Draw a sketch of any of the hermetic compressors and show where the ammeter would be placed to test current in the run winding and in the start winding.

12 Three-Phase Open Motors and Three-Phase Hermetic Compressors

OBJECTIVES:

After reading this chapter, you should be able to:
1. Explain the theory of operation of a three-phase motor.
2. Identify the main parts of the three-phase open motor.
3. Identify the main parts of the three-phase hermetic compressor.
4. Wire the three-phase open motor for high or low voltage.
5. Change the rotation of a three-phase open motor.
6. Wire the three-phase open motor for delta or wye configuration.

12.0   Three-Phase Motor Theory

When compressor or fan applications require larger motors than single-phase motors, three-phase motors are usually used. The three-phase motor can produce extremely large starting and running torque. In this section you will learn about how three-phase AC motors have some similarities to single-phase AC motors. Today all HVAC technicians must be able to work on commercial jobs that utilize three-phase motors and compressors. You must be able make small changes on three-phase motors in the field so they can operate at a different speed, different voltage, or at a different rotation. This chapter will make these field changes, installation, and troubleshooting of three-phase motors easy to perform.

The nature of three-phase AC voltage is that it has three independent sources of voltage that are 120° apart. Fig. 12–1 shows a diagram of three-phase voltage. Notice that this diagram shows three separate sine waves that are 120° apart. This natural phase shift in the voltage provides the necessary shift in the magnetic field when the voltage is applied to the stationary fields (stator) of the AC motor. This means that three-phase voltage can provide the phase shift required to start a motor naturally without any capacitors or start windings.

When the three-phase voltage is applied to the stator winding of a motor, the phase shift causes the magnetic field in the stator to actually rotate or move around the stator at the speed of the frequency of the AC voltage. The phase shift in the stator will also cause the squirrel-cage rotor in the three-phase motor to become magnetized. When the rotor is magnetized, its field will follow the rotating magnetic field in the rotor and cause the rotor to spin.

Since the three-phase voltage has a natural phase shift, it creates a very strong rotating magnetic field in the motor with the same natural phase shift. The strength of the magnetic field creates the strong torque at the shaft of the motor. This provides a means to start the motor under heavy loads such as the high pressure that a compressor must start against. The windings of a three-phase motor may be connected in a number of configurations to provide more or less starting torque and running torque, and the windings can be connected for high-voltage (480 VAC) applications and low-voltage (230 VAC) applications. The direction of rotation of the three-phase motor may be reversed by interchanging any two of the three supply voltage wires so that the phase relationship is reversed.

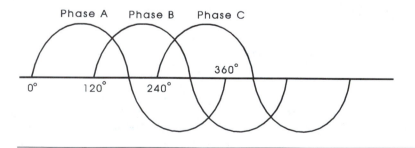

Figure 12–1   Example of three-phase voltage. Notice this voltage consists of three independent sine waves.

12.1   The Three-Phase Hermetic Compressor

In most air-conditioning and refrigeration systems, the compressor motor and pump are sealed in a container. When a compressor is sealed in this manner, it is called a hermetic compressor. The motor is sealed so that the system's refrigerant can be used to cool the motor windings without being lost to the atmosphere. Fig. 12–2 shows pictures of several types of three-phase hermetic compressors. The picture in Fig. 12–2a shows a picture of a full hermetic compressor. The three-phase motor for this compressor is completely sealed inside the hermetic shell. The compressor in Fig. 12–2b shows a semihermetic compressor. The three-phase motor in the semihermetic compressor is not sealed inside the shell; rather it is accessible much like an open motor. The compressor piston is the only part of the semihermetic compressor that is sealed inside the shell. In this section we will explain the wiring diagrams for only the full hermetic compressor. Since the semihermetic compressor uses an open-type motor, its diagram will be presented with the diagrams of the three-phase open motors.

As an air-conditioning and refrigeration technician, you will not open the hermetic compressor for maintenance. Instead the hermetic compressor is replaced as a unit if it fails. Since the hermetic compressor cannot be opened, you must rely on cut-away diagrams and pictures to learn what is inside the motor. You can also get some idea of the basic parts in a hermetic compressor by looking at the parts of a three-phase open motor. The basic parts of an open motor will be similar to those in the hermetic compressor.

12.2   Basic Parts of a Three-Phase Compressor Motor

The three-phase compressor motor consists of a stator with three separate windings that are mounted out of phase from each other. Fig. 12–3 shows a cut-away picture of the three-phase hermetic compressor. From this picture you can see that the windings are equally spaced around the stator and the wire in each winding is the same gauge size. This means that all of the three windings in the stator are equal. The rotor is also shown in the cut-away picture of the three-phase compressor. From this picture you can see that the rotor is a squirrel-cage rotor, which is similar to the rotors in other AC motors in that it is made of laminated steel plates that are pressed on the squirrel-cage frame. The steel plates allow the rotor to become magnetized easily and to give up its magnetic field easily. This allows the magnetic bars in the rotor to become magnetized positively and negatively as the rotor spins inside the stator. Since a three-phase motor has three separate voltages, the three-phase motor will have three separate and distinct magnetic fields in the rotor at the same time, which gives the rotor optimum torque.

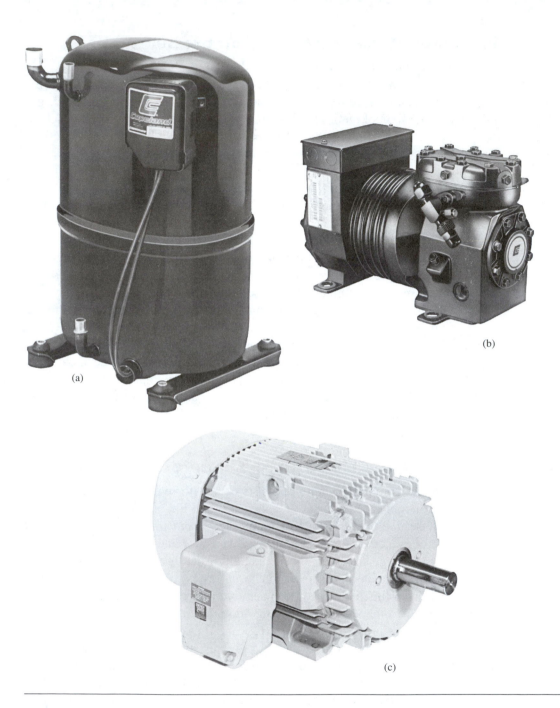

Figure 12–2a   Example of a three-phase hermetic compressor. **b** Example of a three-phase semihermetic compressor. **c** Example of a three-phase open motor. (**a** *Courtesy of Copeland Corporation,* **b** *GE Motors & Industrial Systems, Fort Wayne, Indiana*)

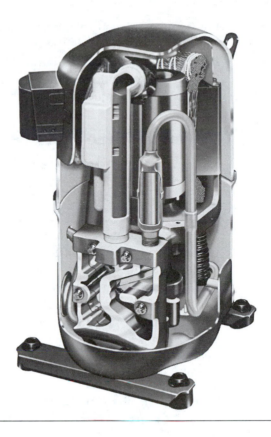

Figure 12–3   Cut-away picture of a three-phase compressor. *(Courtesy of Carlyle Compressor Company, Division of Carrier Corp.)*

12.3   Theory of Operation for the Three-Phase Compressor

The theory of operation for a three-phase compressor is easy to understand. When three-phase voltage is applied to the three windings of the motor, three separate magnetic fields are developed, which immediately begin to rotate through the stator windings. Since the magnetic field rotates around the stator, the three magnetic fields in the rotor of the compressor will "chase" the magnetic fields and cause the rotor to begin to spin. The speed of rotation of the magnet field is fixed and it is based on the number of windings and the frequency of the applied voltage. This means that the three-phase compressor motor will run at a constant speed unless the frequency of the applied voltage is varied. In some newer high-efficiency air-conditioning and refrigeration systems, a variable-frequency drive is used to control the speed of the compressor by providing less than 60 hertz at times when the amount of refrigerant that needs to be pumped is smaller.

You should remember that the nature of the phase shift in the three-phase voltage and the location of the three-phase windings in the stator of the compressor allow the compressor to develop the magnetic field in the rotor through induction so no extra components or circuits such as a current relay, potential relay, or capacitors are needed. As long as voltage is applied to the stator, a magnetic field will be provided in the rotor by induction and it will always try to cause the rotor to spin with the rotating magnetic field in the stator, which will allow the rotor to provide torque at its shaft that can be used to pump refrigerant.

12.4   Wiring a Three-Phase Hermetic Compressor

A diagram of a three-phase hermetic compressor is shown in Fig. 12–4 controlled by a motor starter and a thermostat. From this diagram you can see the motor terminals are identified as T1, T2, and T3. When you make the field wiring connections to these terminals, you can connect L1 to T1, L2 to T2, and L3 to T3. Since the motor windings are sealed inside the hermetic compressor, you cannot change the connections of the windings to change the speed of the motor or its torque. You can change its direction of rotation by switching L1 to T2 and L2 to T1. At times you may need to reverse the rotation of the three-phase hermetic motor to break it free if it has tied up after setting over the winter. Normally the compressor will pump refrigerant the same regardless of the direction the compressor motor is turning. Fig. 12–5 shows the location of terminals as you would find them on the side of a three-phase hermetic compressor.

12.5   Changing Connections in Open Three-Phase Motors to Change Torque, Speed, or Voltage Requirements

At times as a technician, you will need to make minor changes to open three-phase motors that are used to turn fans, water pumps, or open-type compressors

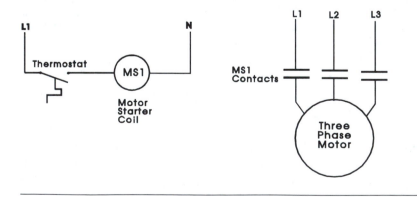

Figure 12–4   A three-phase motor controlled by a low-voltage thermostat and relay.

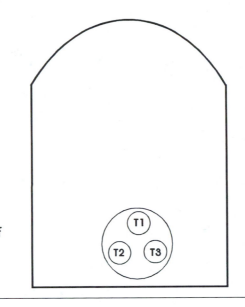

Figure 12–5   Location of the terminals on a three-phase hermetic compressor.

so that they will have more torque, run at different speeds, or be connected to a different voltage. Since these motors may cost several hundred dollars, it is important that you are able to make these changes in the field so a new motor does not have to be purchased and installed. After you learn about these field wiring changes, you will become confident in making these changes in the field.

12.6   Wiring a Three-Phase Motor in a Wye Configuration

As you know, the three-phase motor has three equal windings, and it does not need any starting switches or capacitors connected to any of its windings. Since the motor has three equal windings, they can be connected in a wye configuration or a delta configuration. Fig. 12–6 shows the three-phase motor windings for a six-lead motor connected in a typical wye configuration. The ends of each lead are numbered so that they can be changed or reconnected for different wiring configurations. The ends of the leads for the first winding are identified as T1 and T4, the second winding leads are T2 and T5, and the third winding leads are T3 and T6. You can see that terminals T4, T5, and T6 are all connected together at one point in the center of the motor called the *wye point*. The most important point to remember about connecting a three-phase motor in wye configuration is that it will have less starting torque than if the same motor leads were connected in a delta configuration. This means that the wye-connected motor will draw less current than the delta-connected motor.

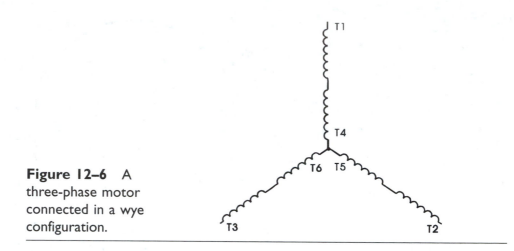

Figure 12–6 A three-phase motor connected in a wye configuration.

12.7 Wiring a Three-Phase Motor in a Delta Configuration

The six-lead motor shown in the Fig. 12–6 can also be connected in a delta configuration. Fig. 12–7 shows the motor connected in delta configuration. In this diagram you can see that the windings are connected in such a way that the three corners of the delta configuration are identified as T1, T2, and T3. In this configuration the incoming power supply is connected to T1, T2, and T3. You should remember that the delta-connected motor has more starting torque and draws more current than the wye-connected motor.

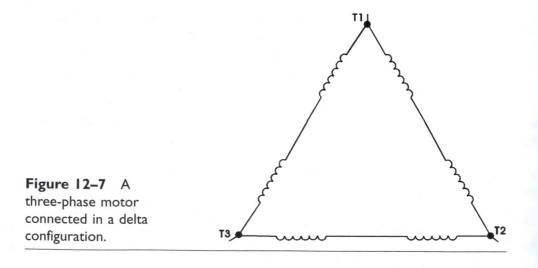

Figure 12–7 A three-phase motor connected in a delta configuration.

12.7.1 Wiring A Three-Phase, Six-Wire Open Motor for High Voltage or Low Voltage

A three-phase open motor is different than the three-phase hermetic motor in that the open-type motor has all of its terminal leads brought outside of the stator so the motor can be reconnected to operate with a high-voltage supply (480 VAC) or a low-voltage supply (240 VAC). The number of leads that are brought out of the stator for a three-phase motor may be six, nine, or twelve. If the motor has six leads brought out of the stator, you can use the diagram in Fig. 12–8a to connect the leads for low voltage (240 or 208 volts), and you can use the diagram in Fig. 12–8b to connect the leads for high voltage (480 volts). It is important to understand that if the motor is a six-lead, three-phase open motor, it is wired so it can be connected for high or low voltage.

12.8   Rewiring a Nine-Lead, Three-Phase, Wye-Connected Motor for a Change of Voltage

A number of three-phase open motors have nine leads brought out of the stator. The reason this motor has nine leads is that each of the three windings is broken into two pieces. The two sections that make up the three windings can be connected in a wye configuration or a delta configuration. Fig. 12–9a shows the nine leads connected in a wye configuration. From this diagram you can see that the two ends of the first half of the first winding are identified as T1 and T4, and the terminals for the second half of the first winding are identified as T7 and common. The common point is where the three windings are connected together to form the wye point.

The terminal ends for the first half of the second winding are identified as T2 and T5, and the terminals for the second half of the second winding are identi-

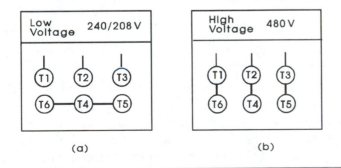

(a)   (b)

Figure 12–8a Connections for a low-voltage, wye-connected motor.
b Connections for a high-voltage, wye-connected motor.

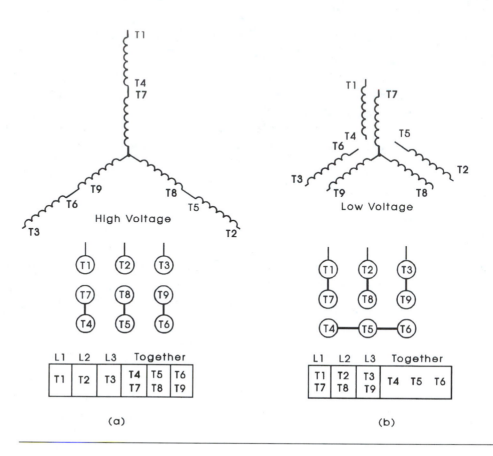

Figure 12–9a   Connections for a nine-lead, wye-configured, three-phase, open-type motor wired for high voltage (480 V). **b** Connections for a nine-lead, wye-configured, three-phase, open-type motor wired for low voltage (208 or 240 V).

fied as T8 and common. The terminal ends for the first half of the third winding are identified as T3 and T6, and the terminals for the second half of the second winding are identified as T9 and common. The diagram in Fig. 12–9a shows the correct way to connect the terminals of the wye-connected motor for high voltage (480 V) and Fig. 12–9b shows the correct way to connect the motor for low voltage (208 or 240 V). It is also important to note that if the three-phase open motor brings out nine leads, it is internally connected as a wye or a delta motor. If the motor is connected as a nine-lead, wye-connected motor and you need more starting torque and you want to change it to delta, you will not be able to make this conversion because the motor is internally connected so that it can only be wired as a wye-connected motor. This means that if you need a delta-connected motor, you will need to purchase a new nine-lead motor. This is a minor draw-back for the three-phase motor since you very seldom change the motor from delta to wye or wye to delta once the motor is installed.

12.9   Rewiring a Nine-Lead, Three-Phase, Delta-Connected Motor for a Change of Voltage

The nine-lead, three-phase, open-type motor can also be rewired for high and low voltage in the field. The nine-lead motor has each winding broken into two sections. Fig. 12–10 shows the nine leads connected in a delta configuration. From this diagram you can see that the two ends of the first half of the first winding are identified as T1 and T4, and the terminals for the second half of the first winding are identified as T1 and T9.

The terminal ends for the first half of the second winding are identified as T2 and T5, and the terminals for the second half of the second winding are identified as T2 and T7. The terminal ends for the first half of the third winding are identified as T3 and T6, and the terminals for the second half of the second wind-

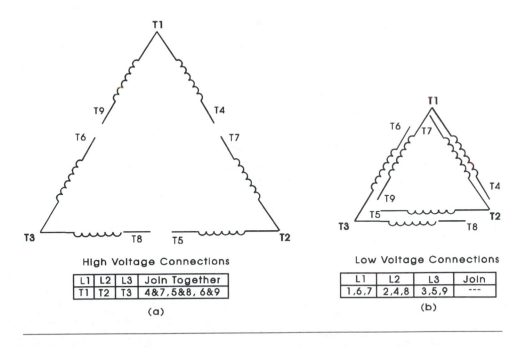

L1	L2	L3	Join Together
T1	T2	T3	4&7, 5&8, 6&9

(a)

L1	L2	L3	Join
1,6,7	2,4,8	3,5,9	---

(b)

Figure 12–10a Connections for a nine-lead, three-phase, delta-configured, open-type motor wired for high voltage (480 V). **b** Connections for a nine-lead, three-phase, delta-configured, open-type motor wired for low voltage (208 or 240 V).

ing are identified as T1 and T6. The diagram in Fig. 12–10a shows the correct way to connect the terminals of the delta-connected motor for high voltage (480 V) and Fig. 12–10b shows the correct connections for low voltage (208V or 240V). The important point to remember is that the windings are connected in series for high voltage, and they are connected in parallel for low voltage.

12.10   The Twelve-Lead, Three-Phase, Open-Type Motor

Some larger three-phase compressors can be connected as either a wye motor or a delta motor. In order for this change to occur, the motor must have either six or twelve leads brought out of the stator. If the motor has nine leads brought out, the motor cannot be changed from wye to delta in the field. For example, in some large refrigeration applications you must start the motor as a wye-connected motor so it does not draw too much LRA, and then reconnect it as a delta-connected motor after it is running so that it will provide sufficient running torque. This type of motor is called a wye-delta motor application, and the configuration is changed while the motor is running by a wye-delta motor starter. The wye-delta motor starter is a special motor starter that has two independent sets of three-phase contacts connected to it. The motor is connected in wye configuration on the first set of contacts, and it is connected as a delta-configured motor on the second set of contacts. When the motor is started, the coil for the first set of contacts is energized and the motor is connected as a wye-connected motor when voltage is first applied. When the motor reaches full speed, the relay of the second set of contacts is energized and the motor is rewired as a delta-configured motor. The motor for the wye-delta starting must be a six- or twelve-lead, three-phase, open-type motor. Fig. 12–11 shows a twelve-lead motor connected as a wye motor and Fig. 12–12 shows a twelve-lead motor connected as a delta-connected motor.

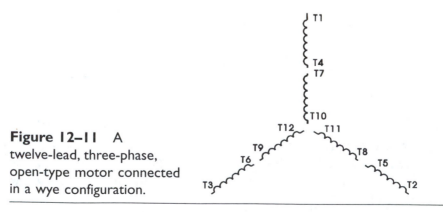

Figure 12–11 A twelve-lead, three-phase, open-type motor connected in a wye configuration.

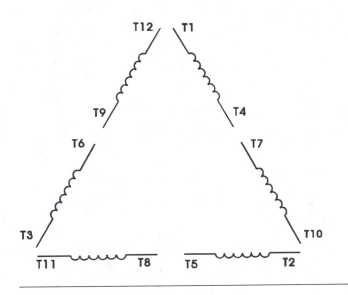

Figure 12–12   A twelve-lead, three-phase, open-type motor connected in delta configuration.

12.11   Wiring a Three-Phase, Open-Type Motor for a Change of Rotation

At times you will need to change the direction of rotation for the three-phase, open-type motor while it is connected to its application. This is perhaps the most common change you will make to the three-phase motor. As a technician, you will need to change the direction that a pump motor runs when it is turning the wrong way, or you may be requested to change the direction a fan motor is turning to ensure air is moving in the proper direction. In these cases all that you need to do is exchange any two of the supply leads. For example, you can change L1 and L2 so that L2 is connected to T1 and L1 is connected to T2. Fig. 12–13a shows the three-phase wye motor connected for clockwise rotation. You should notice that L1 is connected to T1 and L2 is connected to T2. In Fig. 12–13b you can see that the compressor is connected to operate in the counterclockwise direction. L1 and L2 are exchanged so that L1 is connected to T2 and L2 is connected to T1. Fig. 12–14a shows the delta-connected motor connected for clockwise rotation and Fig. 12–14b shows the motor connected for counterclockwise rotation.

12.12   Controlling a Three-Phase Compressor

The three-phase compressor is generally turned on or off with a relay or contactor. The voltage to the coil of the relay is controlled by a thermostat. When the

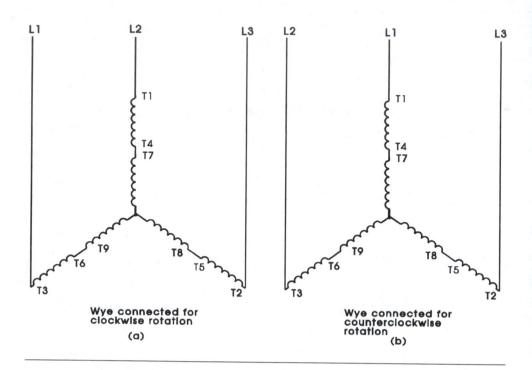

Wye connected for
clockwise rotation
(a)

Wye connected for
counterclockwise
rotation
(b)

Figure 12–13a A three-phase, wye-connected motor connected for clockwise rotation. **b** A three-phase, wye-connected motor connected for counterclockwise rotation. Notice T1 is now connected to L2 and T2 is now connected to L1.

contacts in the thermostat close, voltage is applied to the coil of the relay and it will draw current and become magnetized and pull the three sets of normally open contacts to their closed position. When the contacts close, three-phase voltage will be applied to the compressor and it will begin to run. When the temperature is satisfied at the thermostat, the thermostat contacts will open and deenergize the relay coil, and the three sets of contacts will open and turn the compressor motor off.

12.13 Controlling the Speed of a Three-Phase Motor

The speed of a three-phase motor can be changed by changing the number of poles in the motor windings, or by changing the frequency of the voltage supplied to the motor. Prior to the 1980s the only way that the frequency of the supply voltage could be easily changed was by changing the speed of the generator that produced the electricity. Since this is not practical, the preferred

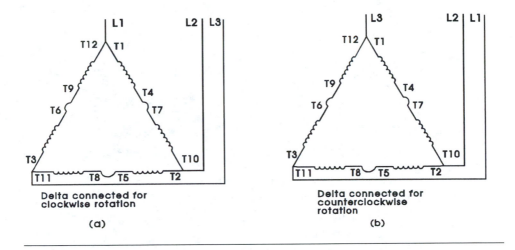

Delta connected for
clockwise rotation

(a)

Delta connected for
counterclockwise
rotation

(b)

Figure 12–14a   A three-phase, delta-connected motor connected for clockwise rotation. **b** A three-phase, delta-connected motor connected for counterclockwise rotation. Notice T1 is now connected to L2 and T2 is now connected to L1.

method of changing the speed of the three-phase motor was to reconnect the motor in the field so it would have a different number of poles. This basically allowed the fan motor or compressor to have two speeds.

In the 1980s more emphasis was placed on efficiency, so it became important to change the speed of three-phase fans and the speed of three-phase compressors so that they would only run at the speed necessary to provide the proper amount of cooling or heating. In the late 1970s and 1980s electronic technology began to produce products that could control larger amounts of voltage and current so that frequency of the voltage sent to a motor could be changed. Chapter 14 will go into detail about these devices. In this chapter all you need to know is that the variable-frequency drive is specifically designed to change the frequency of voltage that is used to power a three-phase or a single-phase motor. Since the frequency is changed, the speed of the motor can be adjusted. For example, if the motor needs to operate at a higher speed, the frequency would be increased above 60 Hz. If the speed needed to be decreased, the frequency would be lowered below 60 Hz.

Fig. 12–15 shows picture of a variable-frequency drive and Fig. 12–16 shows a block diagram of the variable-frequency drive. From the block diagram you can see that the drive is supplied with three-phase voltage. The first section of the drive is the rectifier section. In this section three sets of diodes are used to rectify the AC voltage to DC voltage. The second section of the drive is a filter

Figure 12–15 A three-phase, variable-frequency drive that is used to control the speed of three-phase fan motors and three-phase compressor motors. *(Courtesy of Rockwell Automation's Allen-Bradley Business)*

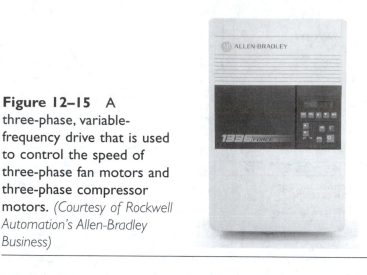

section. You can see that the voltage waveform is half of a sine wave as it enters the filter section and it changes to filtered pure DC voltage as it leaves the filter section. The final section of the drive is the transistor section. The transistors can be turned on or off by their triggering circuit so that the output waveform looks similar to the waveform of the original three-phase voltage that supplies this circuit. The main difference between the input voltage and the output voltage is that the frequency of the output voltage can be changed to any value between 1 Hz and 120 Hz. This means that the variable-frequency drive can create

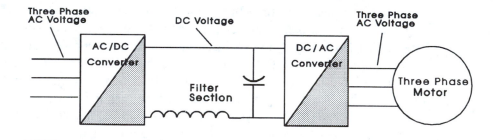

Figure 12–16 A block diagram of a variable-frequency drive. AC three-phase voltage enters the drive and is converted by the rectifiers to half-wave DC. The filter section converts the half-wave DC to pure DC. The transistors convert the pure DC back to variable-frequency AC. *(Courtesy of Rockwell Automation's Allen-Bradley Business)*

frequencies from 1% to 200% of 60 Hz, but the motor speed in practice is generally controlled between 75% and 125% of the rated motor rpm.

12.14   Troubleshooting the Three-Phase Compressor

At times in the field you will be expected to troubleshoot a three-phase motor that will not start, or a motor that is malfunctioning. When you are in the field, you can test the three-phase compressor and open-type motor for voltage, current, and to see if each of the windings has continuity. The following sections explain how to make each of these tests. The most frequent test for a three-phase motor is to test for supply voltage at the disconnect or fuse box and for voltage at each of the three motor terminals.

If the three-phase compressor fails to run when it is energized, you will need to make two basic tests. The first test involves testing each of the three lines that supply voltage to the compressor to ensure that three-phase voltage is flowing through the wires to the terminals of the motor. You can make this test by testing for voltage right at the terminals of the compressor motor. If the voltage supply for this compressor is 240 V three phase, you should measure 480 V between T1-T2, T2-T3, and T3-T1. If you have voltage at two of the three terminals, it indicates that you have lost one phase of the three-phase supply voltage and the most likely problem is a blown fuse. In some cases the compressor can continue to run when one of the phases is lost, but the compressor will not have sufficient torque to restart, and the internal overload will be tripped. When this occurs, you are likely to suspect the winding of the compressor is open and you may change a good compressor when the only problem is a blown fuse. Be sure that the voltage test indicates that you have the correct amount of voltage at each of the three tests you make at the terminals on the compressor. The voltage tests are similar to the test you learned about to determine if the voltage is present in a three-phase disconnect switch. Voltage tests for a three-phase, open-type motor should be similar to the tests for a three-phase compressor except some open-type motors will not have internal overloads.

If the correct amount of voltage is present at all three terminals and the motor will not run, you must suspect that one of the windings in the motor is open and it will have infinite (∞) resistance. (Remember that the compressor may also have an internal thermal overload that would give a similar symptom. Be sure to allow several hours for the overload to cool down before you make a decison about changing the compressor.) If you suspect the winding is open, you will need to turn the power off to the system and disconnect all of the wires from the compressor terminals and test each winding for resistance. If the three-phase motor is an open-type motor, you would use a similar test to see that the proper amount of voltage is present at its terminals. Fig. 12–17 shows the proper location to place a voltmeter to make these measurements.

12.15   Testing the Three-Phase Motor for Continuity

If the voltage test indicates the proper amount of voltage is present at the terminals but the motor will not start and only hums or does not make any noise, you should suspect one or more motor windings have an open and you must test the motor windings for resistance. Always remember that the voltage to the motor should be turned off for the continuity test, and the windings should be isolated if possible. The test for resistance should be between T1-T2, T2-T3, and between T3-T1. If any of these windings are open, the compressor must be removed and replaced, since it cannot be repaired while it is soldered into the refrigeration system. As stated before, some compressors have an internal thermal overload built into the windings and it may open if the internal temperature gets too hot. The motor windings may get too hot if the amount of refrigerant in the system is too low to provide cooling for the motor or if the motor is overloaded. Remember the hermetic motor is only cooled by the extra refrigerant brought back to the motor specifically to cool the windings. If the overload is open, it will not allow current to flow through the windings and hence the motor will not run. Since the overload may be the problem, it is always a

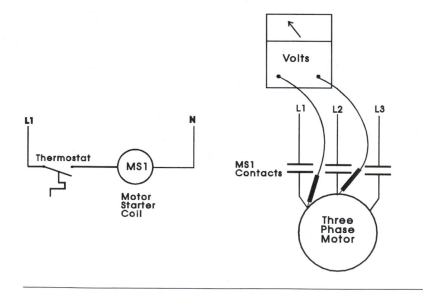

Figure 12–17   The proper location to place voltmeter terminals to test for three-phase voltage on a three-phase motor.

good practice to wait 1 to 2 hours for the compressor to cool down and retest the windings for continuity before you change it.

It is also important to understand that some compressors use the motor windings as a heater when the motor is not running. This is accomplished by connecting capacitors and resistors to allow a small amount of bleed current to flow through the windings to act as a small heater. This small amount of heat keeps the windings warm enough so that the refrigerant oil does not migrate away from the compressor to a part of the system that is warmer. A problem may occur when you test the compressor if you do not make sure that all of this circuit is isolated from the motor windings when you are testing for an open circuit. Fig. 12–18 shows the proper location to place the ohmmeter to make the continuity test.

12.16   Testing a Three-Phase Motor for Current

If the three-phase motor will start, you must check it for the proper amount of current to ensure that it is not overloaded. Fig. 12–19 shows the proper location to place the clamp-on ammeter around one of the three-phase wires that supplies voltage to the motor. Since each of the supply wires is connected in series with the motor windings, the current measurement will indicate how much current the motor is drawing through each winding. It is important to understand that the amount of current measured in each of the three wires should be nearly the same.

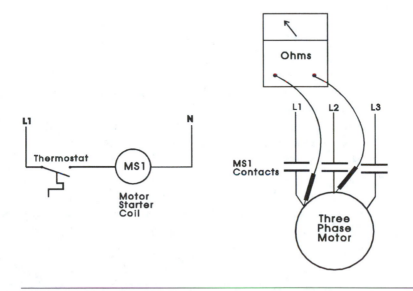

Figure 12–18   The proper locations to place ohmmeter terminals to test the three-phase motor windings for continuity to check for an open circuit.

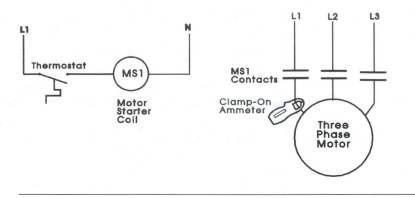

Figure 12–19   Proper location to place clamp-on ammeter to measure current in the motor winding.

If one of the leads is drawing excess current, it indicates the motor is beginning to fail. If all three of the measurements are larger than the data plate rating, the motor is being overloaded and you will need to determine the cause of the overload. Sometimes three-phase compressors will draw excess current when the refrigeration system is low on refrigerant and the motor is running exceedingly hot. The compressor could also draw excess current if it is pumping the refrigerant as a liquid instead of as a vapor.

Questions for This Chapter

1. Explain why the three-phase motor can start without using any start relays or capacitors.
2. Identify the main parts of a three-phase open motor and three-phase hermetic motor.
3. Explain why you would need to reconnect the terminals of a three-phase pump or fan motor to change its direction of rotation.
4. Explain why you would need to reconnect the terminals of a three-phase motor for a change of voltage.
5. The three-phase hermetic compressor you are testing for continuity indicates an open circuit. Explain why you should wait 2 hours or more before you make the decision to change the compressor.

True or False

1. The direction of rotation of a three-phase motor can be reversed by exchanging any two of its three terminals.

2. One of the windings in a three-phase motor is smaller than the other two so that a phase shift can be created to help the motor get started.
3. The three-phase motor has more starting torque than an equal size single-phase motor.
4. It is possible to change the internal winding connections on a nine-lead, three-phase, open-type motor so that it can operate on 480 VAC or 240 VAC.
5. It is possible to change the internal windings on a hermetic three-phase compressor so it will operate at a different voltage.

Multiple Choice

1. If you change motor terminal T1 to L2 of the supply voltage and terminal T2 to L1 of the supply voltage, the three-phase motor will _____
   a. not run correctly because it must be wired L1 to T1, L2 to T2, and L3 to T3.
   b. change the direction of its rotation.
   c. be able to run on 240 VAC instead of 480 VAC.
2. The three-phase motor does not require any current relays, potential relays, or a centrifugal switch because _____
   a. it has three equal windings instead of a start winding.
   b. it is a low-torque motor and these relays would cause its torque to become too large.
   c. these switches are mainly used to control for overcurrent conditions.
3. The phase shift that is required to create torque for a three-phase motor is _____
   a. created by adding capacitors to the motor windings.
   b. found naturally in the three-phase voltage that is used to provide power for the motor.
   c. created by making each of the windings in the three-phase motor slightly different.
4. The reason you need to know how to reconnect a three-phase motor for a change of voltage is _____
   a. because the voltage rating of the motor that you have in stock is rated 480 VAC and the motor that it is replacing is rated for 240 VAC.
   b. because the motor may need more speed.
   c. because some three-phase motors need DC voltage instead of AC voltage.

5. If the three-phase motor runs but draws excessive current, you should suspect _____

    a. a faulty centrifugal switch or current relay.
    b. the motor is running in the wrong direction and you should exchange any two leads.
    c. one of the motor windings is open, or one phase of the three-phase voltage is not supplying voltage.

Problems

1. Draw the sketch of a wye-connected, three-phase motor.
2. Draw the sketch of a delta-connected, three-phase motor.
3. Draw the connections that you would use to connect a wye-wired motor for low voltage and for high voltage.
4. Draw the connections that you would use to connect a delta-wired motor for low voltage and for high voltage.
5. Draw the electrical diagram of a nine-lead, three-phase motor connected for wye configuration and a delta configuration and explain why each is used.

13 Motor Starters and Overcurrent Controls

OBJECTIVES:

After reading this chapter, you will be able to:
1. Explain the operation of a motor starter.
2. Identify the main parts of a motor starter.
3. Identify the main parts of an overload and explain their operation.
4. Explain the operation of single-element and dual-element fuses.
5. Identify single-pole, two-pole, and three-pole circuit breakers and explain their operation.

13.0 Overview of Motor Starters

Motor starters are magnetically controlled devices that are usually used in larger commercial applications. The motor starter is a larger version of a relay or contactor and it is used to control larger motors. The motor starter also has *overcurrent protection* for motors built into it and relays and contactors do not. This overcurrent protection is called an *overload* and it is sized to trip if the amount of current drawn by the motor exceeds the designated limit.

13.1   Why Motor Starters Are Used in HVAC and Refrigeration Systems

In Chapter 9 you learned about relays and contactors that are used to start compressors and fan motors in residential air-conditioning and heating systems. These relays and contactors are designed to close their contacts and provide current to the motors in the system. Their main function is to turn on and off to provide current to these motors. The motors in these applications are protected against overcurrent by fuses or internal overloads that are built into the winding. Since the compressor is usually the largest motor in the system, it is also protected with an internal overload or an overload that is mounted externally directly on the shell of the compressor, so it does not draw too much current.

In larger commercial systems the compressor motors and fan motors are larger and are more expensive. These motors are protected by fuses or circuit breakers in the disconnect for short-circuit protection, and by the overloads in motor starters to protect against slow overcurrents. If an open motor has a problem such as the loss of lubrication or if the motor is overloaded, it will draw extra current, which would damage the motor if the overload is allowed to continue for any length of time. The overloads in the motor starter sense the excess current and trip the motor starter so that its contacts open and stop all current flow to the motor until a technician manually resets the overloads.

13.2   The Basic Parts of a Motor Starter

Fig. 13–1 shows a typical motor starter that has all of its parts identified. From this picture you can see that this is a three-pole starter that is used in three-phase circuits. The incoming voltage is connected at the top at the terminals identified as L1, L2, and L3, and the motor leads are connected to the bottom terminals that are identified as T1, T2, and T3. The three major sets of contacts are located in the top part of the motor starter and the overload assembly is mounted in the lower part of the motor starter. The coil is located in the middle of the motor starter and it has an indicator that shows the word *ON* when the coil is energized, and *OFF* when the coil is deenergized.

Fig. 13–2 shows a wiring diagram of a motor starter connected to a three-phase motor at the top of the diagram and a ladder diagram of just the control circuit that consists of the start push button, stop push button, the motor starter coil, and the motor starter overloads in the bottom diagram. It is easy to see the parts of the motor starter in the wiring diagram because they are all in the shaded area. You can see that the coil circuit uses smaller wire than the contact circuit. The motor starter has three sets of contacts that have a heater in series with it. This ensures that all of the current that flows to the motor must pass through a

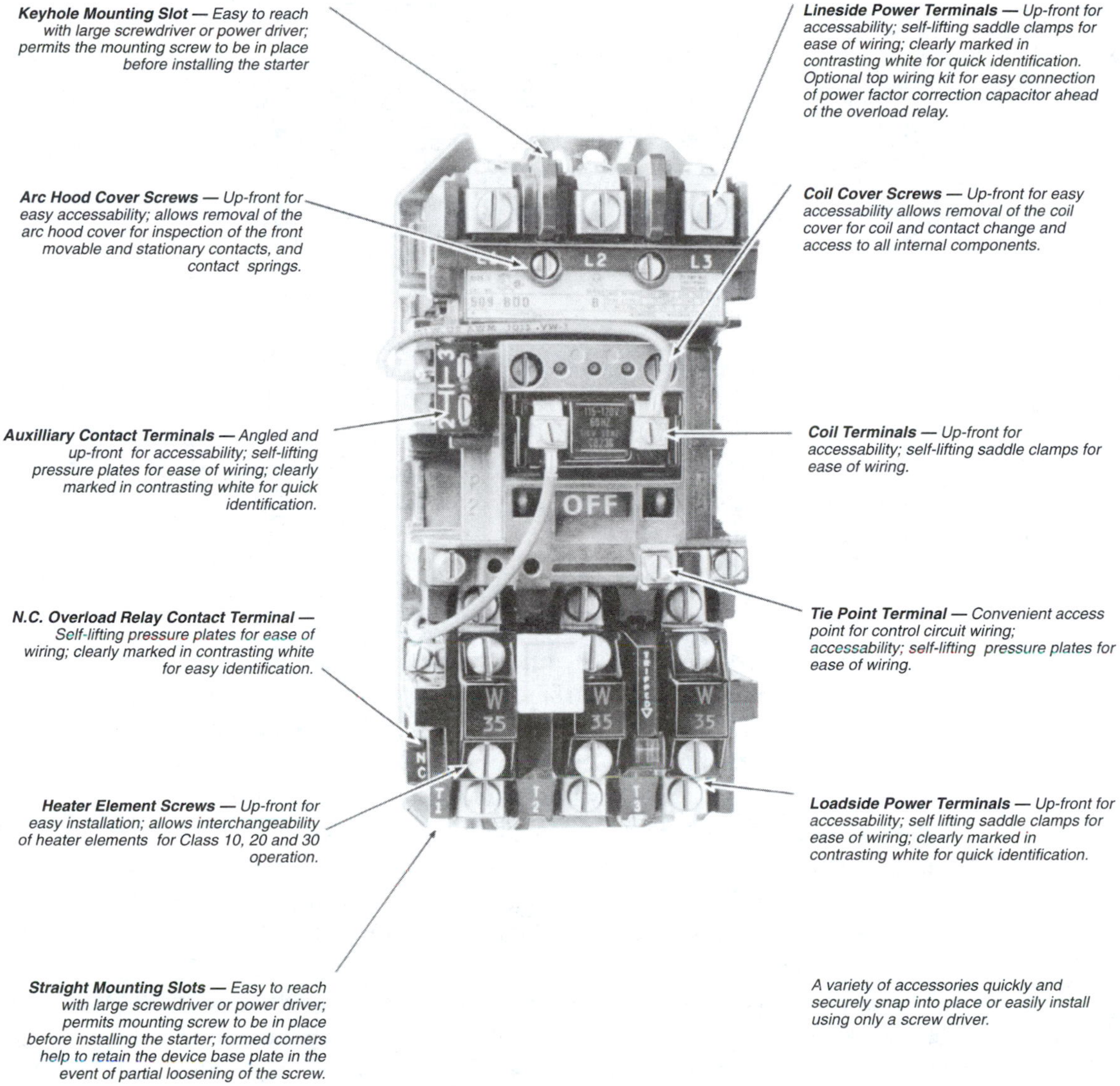

Keyhole Mounting Slot — *Easy to reach with large screwdriver or power driver; permits the mounting screw to be in place before installing the starter*

Lineside Power Terminals — *Up-front for accessability; self-lifting saddle clamps for ease of wiring; clearly marked in contrasting white for quick identification. Optional top wiring kit for easy connection of power factor correction capacitor ahead of the overload relay.*

Arc Hood Cover Screws — *Up-front for easy accessability; allows removal of the arc hood cover for inspection of the front movable and stationary contacts, and contact springs.*

Coil Cover Screws — *Up-front for easy accessability allows removal of the coil cover for coil and contact change and access to all internal components.*

Auxilliary Contact Terminals — *Angled and up-front for accessability; self-lifting pressure plates for ease of wiring; clearly marked in contrasting white for quick identification.*

Coil Terminals — *Up-front for accessability; self-lifting saddle clamps for ease of wiring.*

N.C. Overload Relay Contact Terminal — *Self-lifting pressure plates for ease of wiring; clearly marked in contrasting white for easy identification.*

Tie Point Terminal — *Convenient access point for control circuit wiring; accessability; self-lifting pressure plates for ease of wiring.*

Heater Element Screws — *Up-front for easy installation; allows interchangeability of heater elements for Class 10, 20 and 30 operation.*

Loadside Power Terminals — *Up-front for accessability; self lifting saddle clamps for ease of wiring; clearly marked in contrasting white for quick identification.*

Straight Mounting Slots — *Easy to reach with large screwdriver or power driver; permits mounting screw to be in place before installing the starter; formed corners help to retain the device base plate in the event of partial loosening of the screw.*

A variety of accessories quickly and securely snap into place or easily install using only a screw driver.

Figure 13–1   A typical motor starter with all of its parts identified. *(Courtesy of Rockwell Automation's Allen-Bradley Business)*

heater. If the current flowing to the motor is normal, the heater does not provide sufficient heat to cause the overload to trip. If the motor draws excess current, that current flowing through the heater will cause it to create excess heat that will trip the normally closed overload contacts. Since the normally closed over-

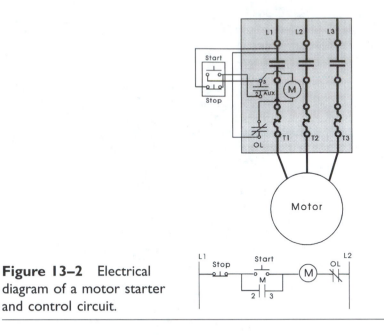

Figure 13–2 Electrical
diagram of a motor starter
and control circuit.

load contacts are connected in series with the motor starter coil, current to the
motor starter coil will be interrupted when the overload contacts open. A reset
button on the motor starter must be manually reset to set the overload contacts
back to their normally closed position.

The control circuit is shown as a ladder diagram at the bottom of this figure.
In this diagram you can see that the motor starter coil is energized when the start
push button is depressed. The manual push button is used for this example because it is easier to understand, but you should be aware that most motor starters
in HVAC and refrigeration applications are controlled by thermostats or other
types of controls.

The motor starter also has one or more additional sets of normally open contacts called auxiliary contacts. The auxiliary contacts are connected in parallel
with the start push button. These contacts serve as a seal-in circuit after the motor starter coil is energized. The start push button is a momentary-type switch,
which means that it is spring loaded in the normally open position. When the
start push button is depressed, current flows from L1 through the normally closed
stop push-button contacts and through the start push-button contacts to the motor starter coil. This current causes the coil to become magnetized so that it pulls
in the three major sets of load contacts and the auxiliary set of contacts. When
the auxiliary contacts close, they create a parallel path around the start push-
button contacts so that current still flows around the start push-button contacts to
the coil when the push button is released. Since the stop push button is con-

nected in series in this circuit, the current to the coil is deenergized and all of the contacts drop out when the stop push button is depressed.

13.3   The Operation of the Overload

The overload for a motor starter consists of two parts. The heater is the element that is connected in series with the motor and it has all of the motor current pass through it. The heater is actually a heating element that converts electrical current to heat. The second part of the overload device is the trip mechanism and overload contacts. The trip mechanism is sensitive to the heat, and if it detects excess heat from the heater it will trip and cause the normally closed overload contacts to open. Since the motor starter coil is connected in series with the normally closed overload contacts, all current to the coil will be interrupted and the coil will become deenergized when the overload contacts are tripped to their open position. When the coil becomes deenergized, the motor starter contacts will return to their open position and all current to the motor will be interrupted. When the overload contacts open, they remain open until the overload is reset manually. This ensures that the overloaded motor stops running and cool downs until someone investigates the problem and resets the overloads.

Fig. 13–3 shows a typical heating element for a motor starter overload device, which is called the heater. In the cut-away diagram for this figure you can see

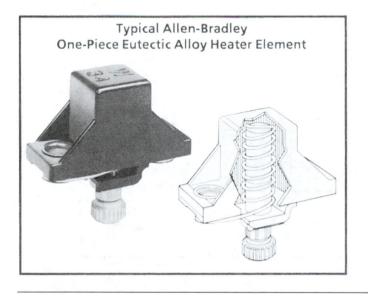

Figure 13–3   A typical heater assembly. Notice the ratchet that protrudes from the bottom of the heater. *(Courtesy of Rockwell Automation's Allen-Bradley Business)*

that the heater is actually a heating element that converts electrical current into heat as it passes through the heating element. You should also notice the knob protruding from the bottom of the heater. This knob has a shaft that is held in position inside the heater and teeth machined into the part that protrudes from the heater. The teeth are called the ratchet mechanism.

Fig. 13–4 shows the heating element mounted into the trip mechanism. The trip mechanism consists of the ratchet from the heater and a paw. The ratchet is actually the knob that protrudes from the bottom of the heater that has teeth in it. The paw has spring pressure that tries to rotate the ratchet. Since the heater holds the shaft of the ratchet tight with solder, the paw cannot move. When the heater becomes overheated, it melts the solder that holds the ratchet in place and allows it to spin freely. When the ratchet spins, it allows the paw to move past it, which in turn allows the normally closed contacts to move to their open position.

After the overload condition has occurred and the overload contacts have opened, the motor starter will deenergize and the motor will stop running. When the motor stops running, the heating element is allowed to cool down. After the heating element cools down for several seconds, the reset button can be depressed, which moves the paw back to its original position, and the normally closed overload contacts move back to their closed position. If the motor continues to draw excess current when it is restarted, the excess current will cause the heater to trip the overload mechanism again. If the motor current is within specification, the heaters will not produce enough heat to cause the overload mechanism to trip.

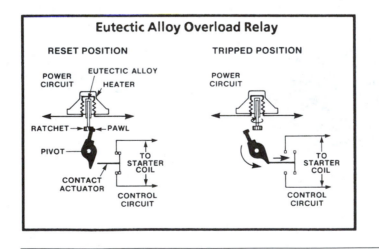

Figure 13–4   Diagram of the ratchet and paw mechanism that trips and allows the normally closed overload contacts to open. *(Courtesy of Rockwell Automation's Allen-Bradley Business)*

Since the overcurrent condition must last for several minutes to cause the overload mechanism to trip, the overloads allow the motor to draw high locked-rotor amperage (LRA) during the few seconds the motor is trying to start without tripping. If the motor has an overcurrent condition while it is running, the overloads allow the condition to last several minutes to see if it clears up on its own before the motor is deenergized. If the problem continues, the overloads will sense the overcurrent and trip to protect the motor. Fig. 13–4a shows the overload contacts in their normally closed position, and Fig. 13–4b shows the overload contacts in their tripped or open condition.

13.4   Exploded View of a Motor Starter

It may be easier to see all of the parts of the motor starter in an exploded view picture. Fig. 13–5 shows an exploded view picture of the motor starter. At the far left in this picture you can see that the main contacts of the motor starter are much larger than those in a traditional relay. The coil is shown in the middle of the picture. The coil has two square holes in it that allow the magnetic yoke to be mounted through it. The magnetic yoke and coil are mounted in the movable contact carrier. When the coil is energized, it pulls the magnetic yoke upward, which causes the movable contacts to move upward until they make contact with the stationary contacts that are shown at the far left. The overload mechanism is shown at the far right in this figure. It is mounted at the lower part of the motor starter, and all current that flows through the contacts must also flow through the overload mechanism.

13.5   Sizing Motor Starters

At times you will need to select the proper size motor starter for an application. The size of motor starters is determined by the National Electrical Manufacturers Association (NEMA). The ratings refer to the amount of current the motor starter contact can safely handle. The sizes are shown in the table provided in Fig. 13–6, and you can see the smallest size starter is a size 00, which is rated for 9 amps and is sufficient for a 2 hp motor connected to 480 volts three-phase, or for a 1 hp single-phase motor connected to 240 volts. You can see that the size 1 motor starter is rated for 27 amps, which is sufficient for a 10 hp three-phase motor connected to 480 volts or a 3 hp single-phase motor connected to 240 volts. A size 00 motor starter is about 4 inches high, and a size 2 motor starter is approximately 8 inches high. You will typically use up to a size 3 or 4 motor starter to protect compressor motors for commercial air-conditioning and refrigeration systems. The size 4 starter would protect motors up to 100 hp. It is important to understand that the overload heaters for the motor starter can also

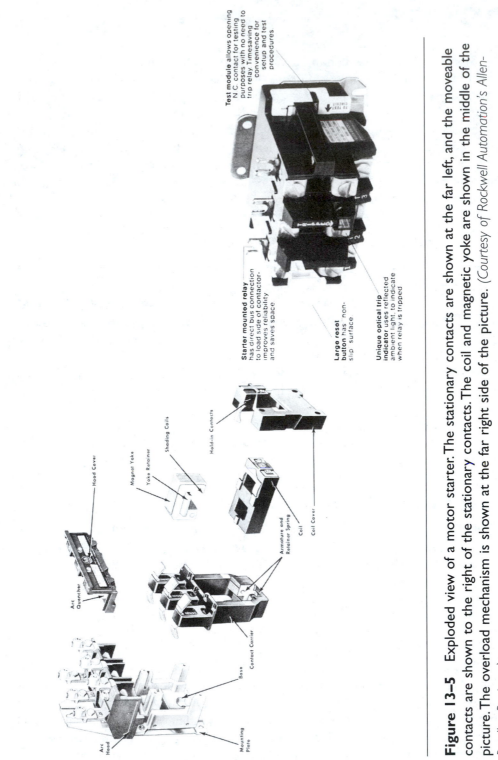

Test module allows opening N.C. contact for testing purposes with no need to trip relay. Timesaving convenience for setup and test procedures.

Starter mounted relay has direct bus connection to load side of contactor-improves reliability and saves space.

Large reset button has non-slip surface.

Unique optical trip indicator uses reflected ambient light to indicate when relay is tripped.

Hood Cover

Magnet Yoke

Yoke Retainer

Shading Coils

Hold-in Contacts

Armature and Retainer Spring

Coil

Coil Cover

Arc Quencher

Contact Carrier

Base

Arc Hood

Mounting Plate

Figure 13–5 Exploded view of a motor starter. The stationary contacts are shown at the far left, and the moveable contacts are shown to the right of the stationary contacts. The coil and magnetic yoke are shown in the middle of the picture. The overload mechanism is shown at the far right side of the picture. (*Courtesy of Rockwell Automation's Allen-Bradley Business*)

RATINGS FOR NEMA FULL VOLTAGE STARTERS

NEMA Size	Continuous Ampere Rating	Maximum Horsepower Rating Full Load Current Must Not Exceed "Continuous Ampere Rating"			
		Motor Voltage			
				50 Hz	
		200V	230V	380V 415V	460V 575V
00	9	1-1/2	1-1/2	2	2
0	18	3	3	5	5
1	27	7-1/2	7-1/2	10	10
2	45	10	15	25	25
3	90	25	30	50	50
4	135	40	50	75	100
5	270	75	100	150	200
6 ❷	540	150	200	300	400
7 ❷	810	–	300	600	600
8 ❷	1215	–	450	900	900
9	2250	–	800	1600	1600

Figure 13–6 The NEMA sizes for motor starter. Notice the smallest motor starter is a size 00, and the largest is a size 9. *(Courtesy of Rockwell Automation's Allen-Bradley Business)*

be purchased for a specific current rating. This means that each motor starter can have a heater that is rated specifically to the amount of current the motor that is connected to it draws.

13.6   Solid-State Motor Protectors

Solid-state motor protectors provide a feature similar to motor starters in that they protect compressor motors against problems. For example, when the compressor is supplied with three-phase voltage, the solid-state motor protector provides protection against the loss of any phase of voltage, low voltage on any phase, high voltage on any phase, reversal of any phase, unbalance voltage between any phase, and short cycling. Short cycling is a common problem that occurs when a compressor is turned off and turned on again after a few seconds before the pressure in the system has a chance to equalize. This can occur if someone sets the thermostat up and then back down too quickly, or if power is interrupted due to a power outage such as occurs during a lightning storm, and then immediately comes back on. If the compressor is allowed to try to start before its pressure is equalized, it will draw large locked-rotor current and possibly not start.

Figure 13–7 A solid-state motor protector. This device protects the motor against low voltage, loss of phase, and short cycling. *(Courtesy of SymCom, Inc.)*

Fig. 13–7 shows a picture of a solid-state motor protection device, and Fig. 13–8 shows a diagram of the location of the motor protection device in a compressor starting circuit. In the diagram you can see that the motor saver has a set of normally open contacts that are connected in series with the motor starter coil.

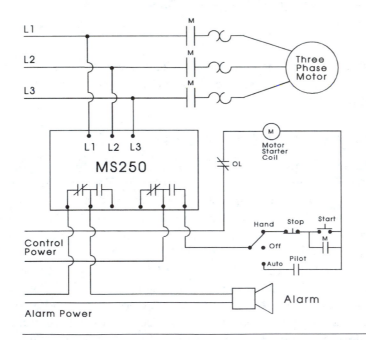

Figure 13–8 Electrical diagram of a motor saver that is connected to the coil of the motor starter or contactor. *(Courtesy of SymCom, Inc.)*

The motor starter samples the voltage at L1, L2, and L3 from the three-phase voltage supplied to the compressor. If the supply voltage has a problem (low voltage, phase loss, and so on), the motor saver circuit will sense the problem, will open its contacts, and cause current to the motor starter to be interrupted. A second circuit provides contacts for an audible alarm. When the motor saver trips, the alarm will indicate the problem exists.

13.7   Overcurrent Controls

Overcurrent controls include thermal overloads used in motor starters, thermal overloads used in compressors, circuit breakers, and fuses. In this section you will learn about thermal overloads for compressors, solid-state overloads for compressors, circuit breakers, and fuses. These devices are used to protect the compressor motor and other motors in the system from drawing too much current and overheating. You will find that the heat that damages compressors can come from overcurrents or from other conditions that cause the compressor to overheat when the current is within normal limits. These conditions occur when the amount of refrigerant in a system is low and insufficient refrigerant is pumped back to the compressor to help keep it cool. Since these overheating problems can occur when the motor current is normal, additional means of detecting overheating must be used.

13.8   Thermal Overloads

Thermal overload is specifically designed to detect excessive heat that comes from overcurrent or from other physical conditions. It is mounted directly on the dome of a compressor to sense the temperature in the compressor, and it is connected in series with the compressor run and start windings to detect excessive current. Fig. 13–9 shows a series of diagrams that shows how the thermal overload operates. Fig. 13–9a shows the contacts when the overload senses too much current, and Fig. 13–9b shows the overload when its contacts are closed.

All of the current flowing through the compressor motor windings will flow in terminal C1 at the bottom of the diagram, through the heating element, through the contacts, and out terminal C2. Since all of the current for the motor must flow through the heating element, the overload must be sized for the exact amount of current the motor will draw safely. When the motor draws current in excess of this limit, the heater will produce extra heat that will cause the bimetal contacts to warp to the open position. When the overload is allowed to cool, the bimetal will snap the contacts to the closed position.

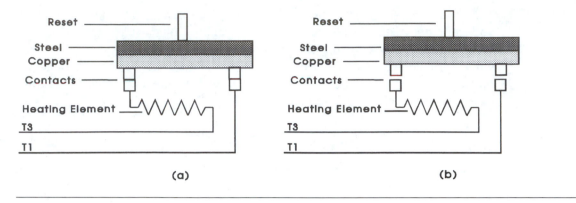

Figure 13–9a Electrical diagram of thermal overload with contacts open. **b** Electrical diagram of thermal overload with contacts closed.

13.9   Fuses

Fuses perform a function similar to an overload, except the fuse uses an element that is destroyed when the overcurrent occurs. The fuse provides a thermal sensing element that is capable of carrying current. When the amount of current becomes excessive, the heat that is generated is sensed by the fuse element, which will melt when the temperature is high enough.

Fuses are available in a variety of sizes and shapes for different applications. Fig. 13–10 shows a plug-type fuse and a cartridge-type fuse. Each fuse is sized for the amount of current it will limit. When the amount of current is exceeded, the fuse link melts and opens the fuse. Fig. 13–11a shows examples of a single-element fuse in the process of blowing and Fig. 13–11b shows dual-element fuses blowing. The single-element fuse provides protection at one level. This type of fuse is generally used for noninductive loads such as heating elements or lighting applications. The dual-element fuse provides two levels of protection. The first level is called slow overcurrent protection and it consists of a fusible link that is soldered to a contact point and attached to a spring. When a motor is started, it will draw locked-rotor amperage (LRA) for several seconds. This excess current will cause heat to build in the fuse and this heat is absorbed in the slow overcurrent link. If the motor starts and the current drops to the FLA level, the link will cool off and the fuse will not open. If the LRA current continues for 30 seconds, the amount of heat generated will cause the solder that holds the slow overcurrent link to melt. When the solder melts, the spring will pull the link open and interrupt all current flowing through the fuse. The slow overcurrent link will allow the fuse to sustain overcurrent conditions for short periods of

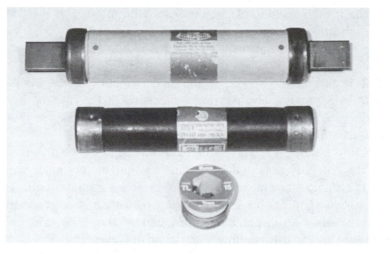

Figure 13–10   Plug-type and cartridge-type fuses. *(Courtesy of Copper Industries, Bussmann Division)*

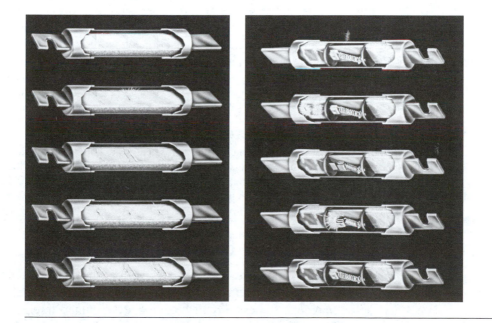

Figure 13–11a   Single-element fuses in the process of blowing.
b Dual-element fuses in the process of blowing. *(Courtesy of Bussmann)*

time, and if the condition clears, the fuse will not open. If the condition continues to exist, the fuse will open.

The second type of element is called the short-circuit element. This element will open immediately when the amount of current exceeds the level of current the link is designed to handle. In the dual-element fuse, the short-circuit link is sized to be approximately five times the rating of the fuse. A short circuit is by definition any current that exceeds the full-load current rating by five times. (Some manufacturers use the rating of ten times.)

The single-element fuse has only a short-circuit element in it. These types of fuses are generally not used in circuits to start motors since the motor draws LRA. If a single-element fuse is used to protect a motor circuit, it must be sized large enough to allow the motor to start, and then it is generally too large to protect the motor against an overcurrent condition of 20%, which will eventually damage the motor if it is allowed to occur for several hours.

13.10   Fused Disconnect Panels

The cartridge-type and screw-base fuses are generally mounted in a panel called a *fused disconnect*. The fused disconnect is normally mounted near the equipment it is protecting and it serves two purposes. First, it provides a location to mount the fuses and, second, it provides a means to disconnect the electrical supply voltage to a circuit. A three-phase disconnect is shown in Fig. 13–12. The fuse disconnect has a switch handle that is used to disconnect main power from the circuit and the fuses so that you can safely remove and replace or test the fuses.

SAFETY NOTICE!!

It is recommended that you always you use plastic fuse pullers to remove and replace fuses from a disconnect to protect you from electrcial shock even when the fuse disconnect switch is in the off position. It is important to remember that even though the switch handle is open, line voltage is still present at the top terminal lugs in the disconnect. You should never use metal pliers or screwdrivers to remove fuses.

13.11   Circuit Breakers and Load Centers

The *circuit breaker* is similar to the thermal overload in that it is an electromechanical device that senses both overcurrent and excess heat. Some circuit breakers also sense magnetic forces. Circuit breakers are mounted in electrical panels called load centers. The load center can be designed for three-phase circuits or for single-phase circuits. The single-phase panel provides 240 VAC and 120 VAC circuits. Fig. 13–13 shows a typical load center without any circuit breakers mounted in it.

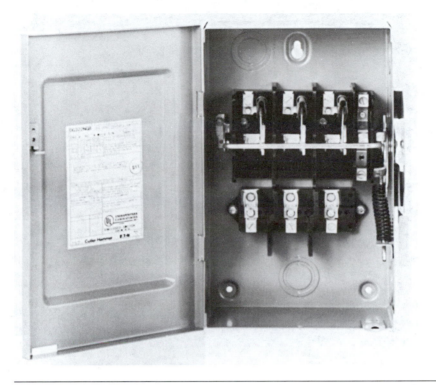

Figure 13–12 A three-phase fused disconnect. *(Courtesy of Challenger Electrical Equipment Corp.)*

The circuit breakers are manufactured in three basic configurations for single-phase 120 VAC applications that require one supply wire, 240 VAC single-phase applications that require two supply wires, and three-phase applications that require three wires. The three-phase circuit breakers can only be mounted in a load center that is specifically manufactured for three-phase circuits. The two-pole and single-pole breakers can be mounted in a single-phase or three-phase load center.

The operation of the circuit breaker is similar to the thermal overload in that it senses excessive current that will trip its circuit after a specific amount of time. The main difference between the circuit breaker and the thermal overload is that the circuit breaker is mounted in the load center to protect both the circuit wires as well as the load. The circuit breakers are sized to total current rating for the wire and all of the loads that are connected to the wire. In some cases this means that the circuit breaker is sized for the current flowing to the compressor, condenser fan, and evaporator fan. For instance, if the compressor draws 20 amps and the condenser fan and evaporator fan both draw 5 amps each, the circuit

Figure 13–13   A load center that is used to mount circuit breakers. The load center is sometimes called a circuit breaker panel. *(Courtesy of Challenger Electrical Equipment Corp.)*

breaker would be rated for 30 amps. A problem can occur if the compressor and condenser fan are turned off and the evaporator fan is running to circulate air and has an overload that doubles its current to 10 amps. In this case the circuit breaker will not be able to detect the additional 10 amps in the evaporator fan as an overload because the total circuit current is below the 30 A total. In these circuits it may be necessary to use a circuit breaker to protect the entire circuit, with overloads at each motor to protect them individually from overheating. This means that the main job of the circuit breaker is to protect the circuit against short-circuit conditions and protect the entire circuit against overcurrent conditions rather than protecting individual motors against overcurrent. This is why many circuits have a combination of circuit breakers and overload devices. Fig. 13-14 shows examples of single phase and 3 phase circuit breakers.

Figure 13–14   Single-pole, two-pole, and three-pole circuit breakers for load centers. *(Courtesy of Challenger Electrical Equipment Corp.)*

Questions for This Chapter

1. Explain the differences and similarities between a motor starter and a contactor.
2. An overload on a motor starter has contacts and a heater. Describe the function of each.
3. Identify the parts of a thermal overload that are used to protect a compressor and explain each part's operation.
4. Discuss the differences between a single-element fuse and a dual-element fuse.
5. Explain the difference between slow overcurrent and short-circuit current.

True or False

1. The heater is the part of the motor starter overload that opens and interrupts current flow.
2. The overload contacts on a motor starter are normally open and interrupt current when an overload condition is detected.
3. The main job of the circuit breaker is to protect the circuit against short circuits, and the entire circuit against overcurrent rather than protect individual motors against overcurrent.

4. A fused disconnect provides a location to mount fuses for the system and also provides a means to disconnect main power from the circuit.

5. An overload device may be mounted internally in a compressor or externally to detect heat on the compressor shell.

Multiple Choice

1. A motor starter ＿＿＿＿＿＿

   a. is just a larger version of a relay since it has only a coil and contacts.
   b. has a coil and contacts like a relay or contactor and it also has overloads to provide overcurrent protection.
   c. is more like a circuit breaker than a relay in that it can detect excessive heat build-up on the outside of the compressor motor though its heaters.

2. A dual-element fuse ＿＿＿＿＿＿

   a. can detect slow overcurrent and large short-circuit currents.
   b. can be used two times before it must be replaced.
   c. has two elements so it can be used in single-phase circuits to protect both L1 and L2 supply voltage lines.

3. The overload device on a motor starter ＿＿＿＿＿＿

   a. has a heater to detect excess current and overload contacts to interrupt the control circuit current.
   b. has a heater to detect excessive current and overload contacts that directly interrupt the large load current to the motor.
   c. has a heater that interrupts control circuit current and overload contacts that directly interrupt the large load current to the motor.

4. The motor starter has auxiliary contacts that ＿＿＿＿＿＿

   a. are connected in parallel with a start switch to seal in the circuit when it is operating correctly and ensure the coil remains deenergized if the overload contacts trip.
   b. are connected in series with a start switch to seal in the circuit when it is operating correctly and ensure the coil remains deenergized if the overload contacts trip.
   c. can switch current to an auxiliary motor such as a condenser fan at the same time the main contacts supply current to the compressor motor.

5. Heaters for motor starters are available in different sizes so that
   _____

   a. they can provide the correct amount of current protection to match the motor that is connected to the motor starter.
   b. they can match the voltage rating of the fuses used in the circuit.
   c. they can match the temperature rating of the ambient air the motor will operate in.

Problems

1. Draw a sketch of a motor starter that has its contacts connected to a three-phase motor and its coil controlled by start and stop push buttons. Be sure to show the heaters and the overload contacts in the proper circuits.
2. Draw a sketch of a thermal-type overload and explain the function of each of the parts.
3. Draw a sketch of a single-element fuse and a dual-element fuse and explain how they are different.
4. Draw a sketch of a thermal overload device connected to the outside of a compressor and explain how it protects the compressor against overheating.
5. Use the table shown in Fig. 13–6 to select a motor starter for a single-phase 7½ hp compressor that draws 27 amps at 230 volts.

14 Thermostats and Heating Controls

OBJECTIVES:

After reading this chapter, you will be able to:

1. Identify the major parts and circuits of a heating thermostat.
2. Identify the major parts and circuits of a cooling thermostat.
3. Explain the basic operation of a gas valve.
4. Explain the operation of an oil burner control.
5. Explain the operation of a fan and limit switch.

14.0 Overview of Thermostats and Heating Controls

The heating thermostat and the cooling thermostat are used to control the temperature in a conditioned space by energizing and deenergizing the furnace system and the air-conditioning system. These controls use bimetal elements and other methods to sense the temperature and to activate a set of contacts for control. Some thermostats have a fan switch to control the operation of the furnace fan in manual operation.

The furnace can use electricity, gas, or oil as its main fuel. Each of these furnaces must have special controls to operate safely. For example, a gas furnace must have a gas valve with a pilot light to ignite the fuel safely. A fuel oil furnace must have a pump, nozzle, and main burner control to atomize the fuel oil so that it will ignite and burn. An electric furnace must have stage controls to energize each bank of electric heat in stages. All of these types of furnaces need a control called a *high limit* to limit the highest temperature the system can reach if a problem arises such as a furnace fan failure. Other controls such as a *fan control* are also needed to energize the fan when the furnace is producing heat. If air conditioning is added to the furnace, additional cooling controls may be added to ensure the fan runs when the system is in the cooling mode and when it is in the heating mode.

14.1   Sensing Temperature

The oldest types of sensing elements used in modern times are temperature sensors. Early scientists found that different types of metals expanded at different rates when they are heated. When two dissimilar metals are bonded together and heated, one of the metals will expand faster than the other and cause the two strips to curl at one end if the other end is mounted securely. Fig. 14–1a shows an example of two strips of metal bonded together, and Fig. 14–1b shows the bonded strips curled on one end when they are bonded. Since two strips are used, this element is called a *bimetal strip.*

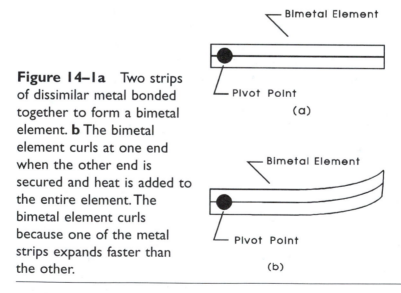

Figure 14–1a   Two strips of dissimilar metal bonded together to form a bimetal element. **b** The bimetal element curls at one end when the other end is secured and heat is added to the entire element. The bimetal element curls because one of the metal strips expands faster than the other.

14.2   Using the Bimetal Element to Sense Temperature

Fig. 14–2a shows a set of electrical contacts mounted to one end of the bimetal element. One half of the contact is mounted stationary, and the second half of the contact is attached directly to the end of the bimetal element. When the element is cold, the contacts are in the normally closed position, and when the element is heated, the end of the bimetal will curl and cause the contacts to open. Fig. 14–2b shows the contacts when they are open. Contacts are also designed so that they are in the open position when the bimetal strips are cool, and in the closed position when the strip is heated. Fig. 14–2c shows two sets of contacts mounted on the bimetal element. One of the sets is in the closed position when the element is cool. This set of contacts would control the heating system. The second set of contacts will become closed

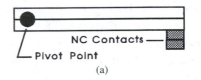

(a)

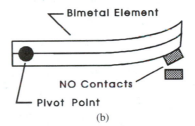

(b)

Figure 14–2a A set of normally closed electrical contacts is mounted on a bimetal element. **b** A set of normally open electrical contacts is mounted on a bimetal strip. **c** A set of normally open and a set of normally closed contacts are mounted on a bimetal element.

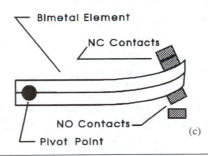

(c)

when the bimetal strip is heated, and it will control the air-conditioning system.

14.3   Amplifying the Movement of the Bimetal Element

The movement of the bimetal element is limited, so additional methods may be required to help the element become an effective temperature control switch. Fig. 14–3a shows a small permanent magnet attached to the bimetal element so that when it moves close to the point where the contacts will be closed when the element is heated, the magnets will make the contacts snap to the closed position. Fig. 14–3b shows the bimetal element in the shape of a spiral. When the element is lengthened and twisted into a spiral, it provides more area to sense the change in temperature and amplifies the movement caused by one metal expanding faster than the other. The spiral also provides a means of moving larger sets of contacts. Fig. 14–4a shows a set of mercury contacts sealed in a glass tube and mounted to the end of the bimetal spiral. Several sets of terminals are mounted in each end of the glass tube. The mercury is in a liquid form at room temperature, so gravity will move it like water to the end of the tube that is down. Since mercury is a conductor, the two sets of terminals that are covered by the mercury will act like a set of closed contacts. Fig. 14–4b shows the spiral expanded when heat is applied. Notice that the expansion of the bimetal element causes the mercury to move to one end. The contacts on this end of the tube are

Figure 14–3a A set of permanent magnets is mounted to the bimetal element to make the contacts snap into the closed position when the element moves the magnets near each other. **b** The bimetal element is shaped in a spiral to amplify the movement when it is heated or cooled.

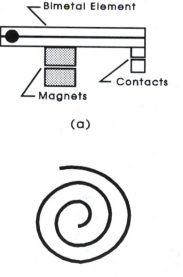

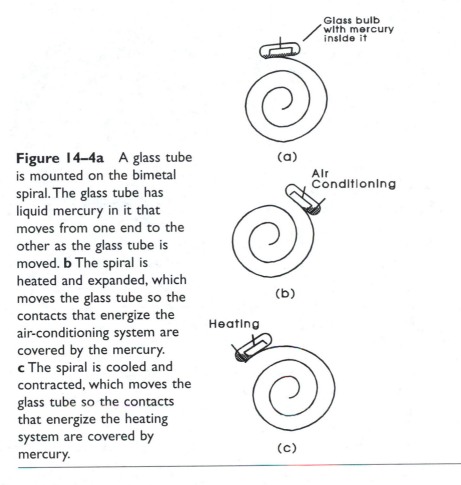

Figure 14–4a   A glass tube is mounted on the bimetal spiral. The glass tube has liquid mercury in it that moves from one end to the other as the glass tube is moved. **b** The spiral is heated and expanded, which moves the glass tube so the contacts that energize the air-conditioning system are covered by the mercury. **c** The spiral is cooled and contracted, which moves the glass tube so the contacts that energize the heating system are covered by mercury.

used to energize the air-conditioning system. Fig. 14–4c shows the spiral when it is cooled down, and you can see the glass tube is positioned so that the mercury moves to the opposite end of the tube and covers the contacts used to energize the heating system.

14.4   The Heating Thermostat

The bimetal element is the part of the thermostat that senses a change in temperature, but the thermostat has several other parts that make it operational. It is important to understand that all thermostats operate in a similar manner and most have adopted a standard terminal identification so that you can install and troubleshoot nearly any brand of thermostat. Fig. 14–5 shows a picture and Fig. 14–6 shows a diagram of a heating thermostat. The majority of thermostats are rated

Figure 14–5   A typical
heating thermostat. *(Courtesy
of Honeywell)*

to switch 24 VAC as control power. You will find some thermostats designed specifically for electric heating systems that are rated to switch 230 VAC or 110 VAC. The thermostats in the following examples are rated for 24 VAC.

Power is brought into the thermostat through terminal R. Terminal W is connected to the contacts that control heating and terminal Y is connected to the contacts that control cooling. The letters R, W, and Y represent the color of wires (red, white, and yellow) that are used to connect the thermostat to the heating and cooling system. A selector switch allows the thermostat to be set to control the heating cycle, cooling cycle, or to be turned off so that neither heating nor cooling is available. Some thermostats have a position for the selector switch that is called *auto,* which allows both the heating and cooling to be controlled simultaneously. This function is useful where it is cool at night and a home or building requires heating, and it may be warm enough during the day to require the air-conditioning system to run.

14.5   The Heat Anticipator

When a heating thermostat is used to control a gas or oil furnace, the heat exchanger inside them stores heat energy when the fuel is being burned. When the fuel source is turned off, the heat exchanger can continue to release the stored heat for several minutes as the furnace fan blows air over it. This excess heat may cause the room to become warmer than the thermostat setpoint. One way to keep this excess heat from causing the room to overheat is to provide an extra source of heat to the bimetal element in the thermostat to cause it to turn off 1 to 2 minutes prematurely.

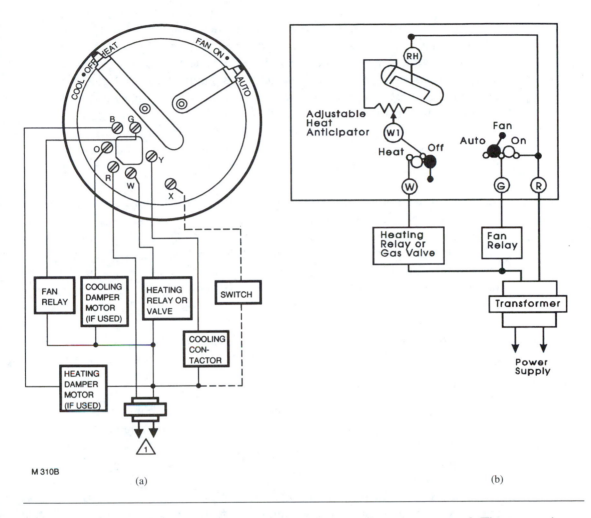

M 310B

(a)   (b)

Figure 14–6a   Electrical diagram of a subbase for a heating thermostat. **b** The internal diagram of a heating thermostat. *(Courtesy of Honeywell)*

The *heat anticipator* is a wire-wound variable resistor that is placed near the bimetal to provide the extra source of heat to help prevent the thermostat from allowing the system to overheat the room. The electrical resistor that is used as the heat anticipator is connected in series with the R-W circuit, so that anytime the heating circuit is closed to provide voltage to the W terminal, the current flowing through this circuit to the coil of the gas valve will also flow through the resistor. The current flowing through the resistor will cause it to heat up slightly. This heat is transmitted directly to the spiral bimetal element, causing it to open the heating circuit before the room temperature reaches the setpoint. When the thermostat heating circuit opens before the room temperature reaches setpoint, the gas valve will be deenergized and the fuel will no longer be provided to be burned. The residual heat that is built up in the furnace heat exchanger will then be released and moved into the room by the furnace fan.

For example, if the thermostat is set for 72°F, and the room temperature drops to 70°, the thermostat's heating contacts will close and energize the gas valve. When the gas is ignited, it will begin to heat up the furnace heat exchanger, and the fan will move this heat into the conditioned space. The bimetal element in the thermostat will sense the heat being added to the space as the room warms up to 71°. During this time the current flowing to energize the gas valve is also flowing through the heat anticipator, which causes it to add approximately 0.5°. When the room reaches 71.5°, the thermostat bimetal element will actually sense 72° because the heat anticipator is adding 0.5°. This will cause the heating contacts to open and deenergize the gas valve when the room is actually 71.5°. The fan will continue to run for several minutes and continue to remove the residual heat that has built up in the heat exchanger. This heat will add the last 0.5° and bring the room to exactly 72°F.

Since the heat anticipator is a variable resistor, it can be adjusted to add more or less heat to the bimetal element, which will lengthen or shorten the time the gas valve is energized. This allows the resistor to anticipate how much heat is left in the heat exchanger, so that the room temperature will be controlled to the exact temperature for which the thermostat is set. You should be aware that the variable resistor (heat anticipator) can be adjusted to any amount of resistance, but if the amount of resistance is set too low, the current flowing through the resistor to the gas valve may be too large for the wattage rating of the resistor, causing a burn spot to develop on the wire.

Numbers are provided on the heat anticipator resistor adjustment to keep the resistor from burning. These numbers refer to the amount of current that should flow through the resistor when it is moved to each setting. Use an ammeter to measure the amount of current flowing through the circuit to the gas valve, and then set the heat anticipator at the corresponding setting. The words *longer* and *shorter* are also marked on the heat anticipator adjustment so that you can extend or shorten the amount of time the gas valve is energized. You will need to use the *trial-and-error* method to find the precise setting to match the furnace to the room that is being heated. Fig. 14–6b shows the location of the heat anticipator in the heating thermostat. The heat anticipator only has current flowing through it when the circuit between R-W is made, and the gas valve is drawing current. The adjustment on the heat anticipator can also be used to adjust the thermostat for fuel oil and electric heating systems.

14.6   The Thermostat Subbase

The thermostat mounts to a subbase that is screwed to the wall. The subbase has the terminals for all of the wires and it also supports the fan switch and heat-off switch. The subbase is mounted to the wall and leveled so that when the thermostat is mounted to it, the thermostat will also be level. Fig. 14–7 shows a typical subbase.

14.7   Cooling Thermostat

Fig. 14–8 shows the diagram of a typical cooling thermostat. The cooling thermostat can be used exclusively for cooling or it can be added to the heating thermostat. The cooling thermostat contacts are connected between terminals R and Y. The cooling thermostat also needs to anticipate the changes in temperature in the conditioned space. The cooling anticipator is a fixed resistor and it is wired so that it will draw current and add heat to the bimetal element when the air-conditioning contacts in the thermostat are closed. The exact function of the cooling anticipator is to provide extra heat to the bimetal element when the air-conditioning system is not running to cause the thermostat to close the cooling contacts a few minutes prior to the room temperature warming to the thermostat setpoint. This extra time allows the air conditioning to come on and begin cooling before the room gets too uncomfortable.

14.8   Heating and Cooling Thermostats

Fig. 14–9 shows a diagram of a combination heating and cooling thermostat. This type of thermostat has a Heat-Off-Cool switch that allows you to set the thermo-

Figure 14–7   A subbase for a thermostat. *(Courtesy of Honeywell)*

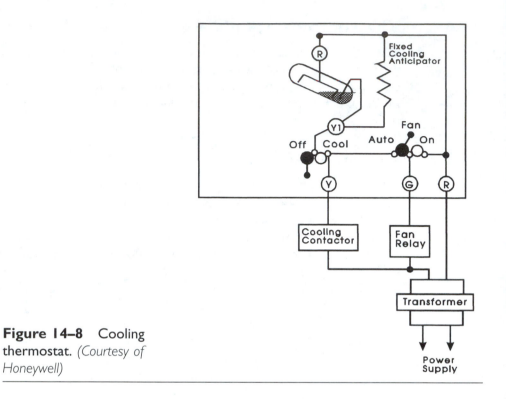

Figure 14–8 Cooling thermostat. *(Courtesy of Honeywell)*

stat for heating, cooling, or off. A fan switch is provided to allow settings for auto so that the fan will run when the heating or cooling system is called, and an on position that causes the fan to run continually. The electrical diagram shows that this thermostat provides two stages of heating and a single stage of cooling. In some heating and cooling thermostats each stage of heating has its own independent contacts mounted in a glass bulb with mercury, and the cooling part of the thermostat has its own mercury bulb. These bulbs are interconnected with a mechanical link so that when the heating setpoint is adjusted to 70°, the cooling is mechanically set to 75° so that the thermostat does not try to energize the heating and cooling equipment at the same time.

In other types of heat-cool thermostats, a single glass bulb has a center contact that makes continuity through the mercury to the heating contact when the bulb is tipped one way, and to the cooling contact when the bulb is tipped the other way.

14.9   The Fan Switch

The fan switch is used to provide a means of energizing the furnace fan on demand. This switch provides voltage from terminal R to terminal G when it is in

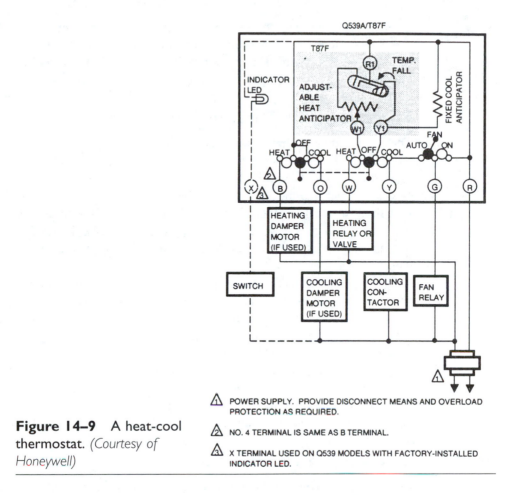

Figure 14–9 A heat-cool thermostat. *(Courtesy of Honeywell)*

⚠ POWER SUPPLY. PROVIDE DISCONNECT MEANS AND OVERLOAD PROTECTION AS REQUIRED.

⚠ NO. 4 TERMINAL IS SAME AS B TERMINAL.

⚠ X TERMINAL USED ON Q539 MODELS WITH FACTORY-INSTALLED INDICATOR LED.

the on position. The voltage at terminal G is used to energize the coil of a fan relay. Fig. 14–10 shows the electrical diagram of the fan switch in the thermostat. You can see when the switch is set to the auto position, voltage will be supplied to terminal G anytime the heating thermostat contacts close. When the switch is in the on position, voltage is supplied to terminal G at all times.

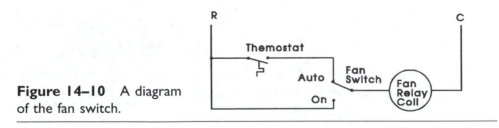

Figure 14–10 A diagram of the fan switch.

The fan switch can be used to cause the fan to run continuously for systems that have an electronic air cleaner. The fan switch can also be used to energize the fan on days where it is too warm for comfort, but not warm enough to use air conditioning.

14.10   Programmable Thermostats

The programmable thermostat includes a time clock with a standard-type thermostat. The timer allows the thermostat to have more than one temperature setting. For example, a heating thermostat can be set for 65°F at night, and the time clock can energize the 72°F setting that will bring the temperature back up the first thing in the morning. If no one is in an office building over the weekend, the lower temperature can be selected by the time clock during the weekend, and it can bring the temperature back to the higher temperature the first thing Monday morning.

Newer programmable thermostats have a built-in microprocessor (computer chip) to drive a digital display and allow input from a digital keypad. This type of thermostat allows multiple settings for different times of each day and for different days of the week. You should remember that these types of thermostats require both the R and C terminals to bring the "hot" and "common" of the 24 V power supply to the thermostat to run the clock and other parts of the thermostat such as LED lamps that require power. The remainder of the thermostat will include circuits for heating (terminal W), for cooling (terminal Y), and for the fan (terminal G) just like every other thermostat. Fig. 14–11 shows a programmable thermostat.

14.11   Line Voltage Thermostats

Line voltage thermostats are heavy-duty thermostats that are constructed to switch 230 VAC or 120 VAC on or off. The current rating of these thermostats is large

Figure 14–11
Programmable thermostat.
(Courtesy of Honeywell)

enough to operate electric heating elements that may draw 10–20 amps. They may also be used to control some refrigeration systems or commercial air-conditioning systems. Since the line voltage thermostat can control these large currents, a relay is not needed. Fig. 14–12 shows one type of line voltage thermostat. You should notice that this thermostat has a capillary bulb that is connected by a 4- to 6-foot capillary tube. The bulb is partially filled with refrigerant and it is placed where the temperature is to be measured. The heat in the space where the bulb is located will cause the refrigerant to expand and activate the switch contacts. It is important to understand that since the thermostat has 230 VAC or 120 VAC, it should be mounted in an electrical box.

14.12   Thermocouples

When you service heating and air-conditioning equipment, you will find a wide variety of sensors that measure temperature. The most common sensors include the thermocouple, RTD, and thermistors. The thermocouple is manufactured by placing two dissimilar metals such as iron and constantan together and welding them at the tip. When the tip is heated, the two different metals produce a potential voltage that is approximately 10 millivolts (0.010 volt). The actual amount of voltage a thermocouple will produce depends on the type of metal in the thermocouple and the amount of heat that is applied.

Thermocouples are classified by the type of metal in them. Fig. 14–13 shows a table of common thermocouples and the temperature and voltage range for each.

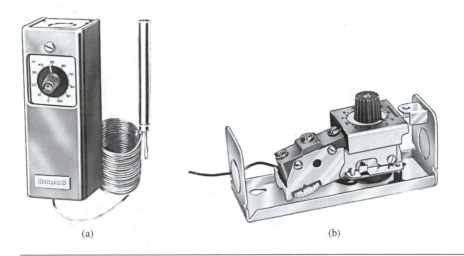

(a)   (b)

Figure 14–12a   A line voltage thermostat. **b** A line voltage thermostat with its cover removed. *(Courtesy of Honeywell)*

Types of Thermocouples

Type	Material	Range
J	Iron-constantan	−328°F to +1432°F
K	Chromel-alumel	−328°F to +2472°F
T	Copper-constantan	−328°F to +752°F
E	Chromel-constantan	−328°F to +1832°F
S	Platinum-rhodium-platinum	+32°F to +3232°F

Figure 14–13   Thermocouple table that shows types of metals and temperature ranges for each type of thermocouple.

You should notice that each type of thermocouple is identified by a letter such as J, K, T, E, or S. The most common thermostat that you find in furnaces is the J type. You can see that the temperature range for the J-type thermocouple is −328°F to +1432°F. This thermocouple will produce 0 millivolts at 32°F and approximately 44 millivolts at 1432°F.

14.13   Resistive Temperature Detectors

Some newer systems use resistive temperature detectors (RTDs) as sensors to determine temperature. The RTD increases its resistance as temperature increases. Several types of RTDs are commonly used including platinum and nickel. The platinum RTD is the most common type used and it is made by placing coils of platinum wire on a ceramic core. Fig. 14–14 shows a table of typical RTDs and their temperature ranges. You can test an RTD by placing an ohmmeter on the two leads of the RTD and using the temperature chart for RTDs. For example, you should measure 138 Ω when the RTD is at 110°F, and you should measure 230 Ω when the RTD is at 350°.

Some motor-starting circuits use an RTD in the start winding circuit like a current relay. At normal temperatures, the RTD has low resistance, which allows

Figure 14–14   Types of RTDs and temperature ranges

Type	Range
Platinum	−364°F to +1382°F
Nickel	−76°F to +356°F
Copper	−100°F to +300°F
Tungsten	−100°F to +5000°F
Balco	+32°F to +400°F

current flow to the start winding. When the motor is started, the current flow through the RTD to the start winding causes the RTD to warm up and increase its temperature. The increase in temperature reduces the current flow in the start winding just like the contacts of a current relay. The main problem with this type of motor-starting device is that it must cool down completely before the resistance is low enough for the motor to start again.

14.14   Thermistors

A *thermistor* is a thermal resistor made from semi-conductor P and N type material. Several types are available, and one has a positive temperature coefficient, which means its resistance increases as the temperature increases, and the other type has a negative temperature coefficient, which means its resistance decreases as the temperature increases. You will find thermistors used in compressors' motor windings as a temperature sensor and as temperature sensors for other devices like heat pump ambient temperature sensors. The main problem with thermistors it that their resistance is not linear over their entire temperature range, so they are usually designed into a circuit that senses temperature over a very narrow range, such as sensing the over temperature of a compressor winding. You can test a thermistor by changing its temperature and measuring the amount of change in its resistance. If the thermistor does not show a change of resistance when its temperature is changed, it is open and must be replaced.

14.15   Gas Furnace Controls

Gas furnaces have changed radically in the past 20 years to meet changing efficiency standards. Early gas furnaces used a standing pilot light that burned a small flame continually. The pilot light was available to ignite gas for the main burner anytime the thermostat energized the main gas valve. Modern furnaces do not use standing pilot lights; rather, they ignite the pilot light just prior to igniting the main burner. In some furnaces an electric spark is used to ignite the pilot light, and in other furnaces, a glow coil is used in place of the pilot light to ignite the furnace.

Older furnaces used natural draft vents, which included a long vent pipe that usually went from the furnace and extended through the roof several feet. These furnaces were designed to use excess heat from the combustion process to keep the vent pipe warm enough to make the exhaust gases rise through them quickly. This meant that a lot of heat was wasted to make the exhaust gas move. Some newer furnaces incorporate a combustion blower to assist in removing exhaust gases. This allows the furnace to remove more heat from the exhaust gas and use it.

14.16   **The Pilot Assembly and Thermocouple**

The function of the pilot assembly and thermocouple is to establish a pilot flame that is used to ignite the main burner. The thermocouple is used as a safety device to indicate the pilot light is burning correctly. Fig. 14–15 shows a pilot and thermocouple. The pilot light is a small orifice that gas flows through. This gas must be ignited with a match or other source of fire when the flow of gas is turned on to the pilot. The gas for the pilot comes through the main gas valve. The pilot circuit in the gas valve is controlled by a solenoid that is energized by the small voltage produced by the thermocouple. Since the thermocouple produces a small amount of voltage that is not sufficient to pull the armature of the solenoid open initially, the solenoid must be opened manually by depressing the valve by hand. After the solenoid is opened manually, the small voltage is able to keep the coil energized so that it keeps the pilot valve open.

14.17   **The Gas Valve**

The gas valve that uses a standing pilot is shown in Fig. 14–16 and a cut-away diagram of the gas valve is shown in Fig. 14–17. In the cut-away diagram you can see that it has a main valve that is operated by a solenoid coil, and a pilot valve that is operated by a separate solenoid. The main valve is spring loaded in the closed position, and when its solenoid coil receives voltage from the thermostat, it becomes energized and pulls the valve open. The pilot valve is controlled by voltage from the thermocouple.

The body of the gas valve has one inlet on the left side and two outlets on the right side. The small outlet supplies gas to the pilot assembly, and the large outlet supplies gas to the main burners. You should notice that the flow of gas to the main valve must pass through the pilot valve. This means that if the pilot valve is not open, gas cannot flow to the main valve, so the pilot valve acts as a safety device for the main valve. The electrical voltage that energizes the pilot solenoid comes from the thermocouple and it is very small. To activate the pilot valve, you must depress the pilot valve manually and hold it down, which pro-

Figure 14–15   A typical pilot assembly for a gas-fired furnace. *(Courtesy of Honeywell)*

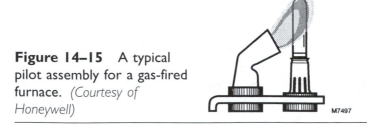

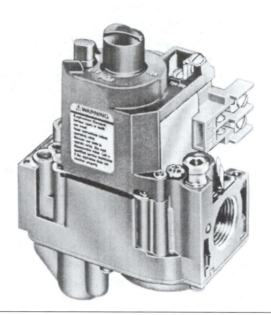

Figure 14–16   A typical
gas valve. *(Courtesy of
Honeywell)*

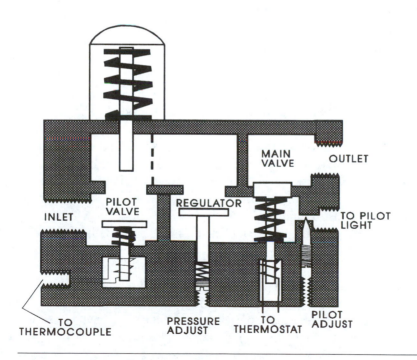

Figure 14–17   Cut-away view of a typical gas valve. *(Courtesy of Honeywell)*

vides the *pull-in* function for the solenoid valve. After the valve is pulled in, the small voltage from the thermocouple is sufficient to "hold in" the solenoid. If the pilot light fails, the temperature at the thermocouple will drop to a point where the voltage cannot hold the pilot valve open, and a spring forces the valve closed. This action creates a fail-safe control for the gas valve and pilot assembly so that the gas valve does not open and allow gas to flow into the combustion chamber of the furnace if the pilot light goes out.

14.18   Direct-Spark, Ignition-Type Gas Valves

Newer gas valves use a pilot light that does not stay on continuously. Instead the pilot light is ignited by a series of sparks generated by a direct-spark ignition module. After the pilot flame is established, it will heat a flame sensor that tells the main valve it is safe to open. In this type of gas valve assembly, the 24 VAC signal from thermostat terminal W starts the pilot ignition cycle, and heat from the pilot flame heats a sensor that allows the main valve to open and provide fuel to the main burner. Fig. 14–18 shows the electrical diagram of the direct-spark gas valve, and Fig. 14–19 shows a picture of a direct-spark ignition module, a spark igniter, and a flame rod for electronic ignition.

14.19   Igniter and Flame Rod

The main parts of the direct-spark ignition gas valve are the spark igniter and flame rod. In Fig. 14–19c you can see the igniter looks like a large spark plug with a gap between its main electrode and ground. The shell of the igniter is made of ceramic material so that it can withstand the 30,000–40,000 V electrical voltage that is supplied to it to produce the spark at the gap. The control module uses an oscillator signal to provide three to four sparks per second. The control module also provides a voltage signal to the coil of the pilot solenoid so that pilot gas can flow to the pilot assembly where the spark from the igniter can cause it to start burning.

After the pilot flame is established, it will provide heat to the flame rod temperature sensor. When the flame rod senses sufficient temperature, it sends a signal to the control module that the pilot flame is established and it is safe to allow the main valve to open so that gas can flow to the main burner. When the main gas comes into contact with the pilot flame, it ignites and establishes the main fire. When the main burner has operated for several seconds, it produces enough heat to cause the fan switch to close and energize the fan motor. The furnace fan moves air across the heat exchanger and into the conditioned space. When the desired temperature is reached, the thermostat is satisfied, and it opens its contacts between R-W, which deenergizes the direct-spark ignition control

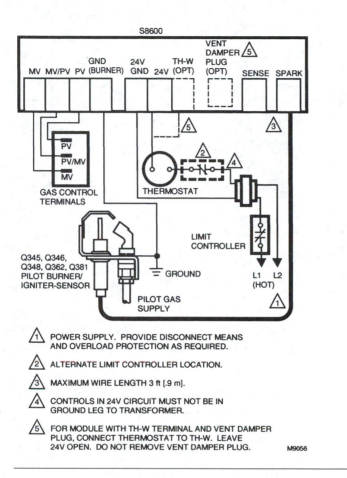

POWER SUPPLY. PROVIDE DISCONNECT MEANS
AND OVERLOAD PROTECTION AS REQUIRED.

ALTERNATE LIMIT CONTROLLER LOCATION.

MAXIMUM WIRE LENGTH 3 ft [.9 m].

CONTROLS IN 24V CIRCUIT MUST NOT BE IN
GROUND LEG TO TRANSFORMER.

FOR MODULE WITH TH-W TERMINAL AND VENT DAMPER
PLUG, CONNECT THERMOSTAT TO TH-W. LEAVE
24V OPEN. DO NOT REMOVE VENT DAMPER PLUG.

M9056

Figure 14–18   Electrical diagram of a direct-spark ignition system. *(Courtesy of Honeywell)*

module. This deenergizes the main gas valve and pilot valve so that they close and shut off the flow of gas. When the thermostat calls for heating again, the pilot ignition process starts again.

14.20   Oil Burner Controls

Fuel oil furnaces have been used for many years, and the oil burner controls have not changed significantly during this time. This means that if you understand the main parts of this system, you will be able to troubleshoot newer and older systems. The oil burner is shown in Fig 14–20b, and the oil burner control is shown in a picture in Fig. 14–20a.

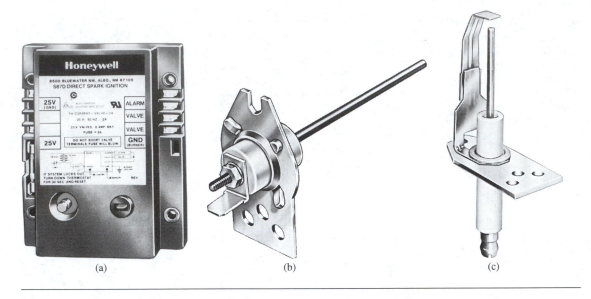

(a)   (b)   (c)

Figure 14–19a   A direct-spark ignition module for electronic ignition. **b** The flame rod for electronic ignition. **c** The spark igniter for electronic ignition. *(Courtesy of Honeywell)*

The main parts of the system are the primary transformer, the igniter electrodes, the cadmium sulfide cell (CAD cell), the primary controls, the oil pump motor, and the limit switches that shut down the system in the event of overtemperature. The basic function of these parts is to compress the fuel oil to a high pressure with a pump so that when the oil passes through a small nozzle, it will atomize completely so that it can burn easily. When fuel oil is in a liquid form, it does not burn well. When it is atomized into a vapor, it ignites easily and burns completely. The ignition system provides a continuous spark across a gap that is created by placing two electrodes close to each other at the tip of the pump nozzle so that the atomized fuel comes into contact with the spark that jumps across the electrodes.

14.21   The Ignition Components

The ignition system uses a step-up transformer. The primary side of the transformer is connected to a 120 VAC source, and the secondary side of the transformer produces a voltage in the range of 10,000 volts. The voltage is applied to the electrodes where it will jump the gap between the electrodes to create a high-voltage spark. The gap in the electrodes is placed directly in the spray pattern of the nozzle so that the atomized fuel can be ignited by the spark. The control module provides supply voltage anytime the pump is running. You should also

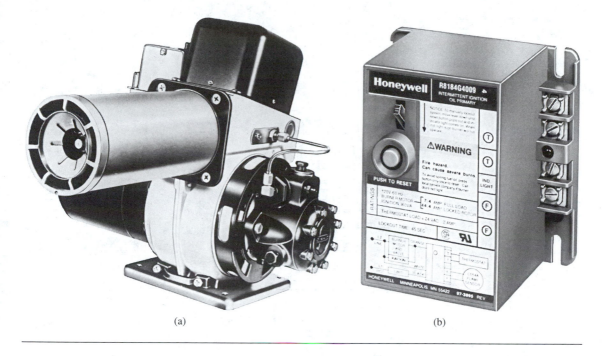

(a)   (b)

Figure 14–20a   An oil burner. **b** An oil burner control. *(Courtesy of R.W. Beckett Corporation, Honeywell)*

notice that the spark from the electrodes will occur at the frequency of the 60 Hz supply voltage.

14.22   The Primary Control and CAD Cell

The primary control contains the safety circuit for the oil burner and relays to energize the various parts of the system. See Fig. 14–21. The main problem that occurs with the fuel oil furnace is that the burner flame is interrupted, and fuel continues to be pumped into the burner chamber. If flame is established later, the excess fuel that is in the chamber will ignite and burn uncontrolled. This excess fuel may also vaporize, which can cause a dangerous explosion. The cadmium sulfide cell (CAD cell) is a sensor that changes its resistance as light strikes its surface. Since the flame of burning fuel oil produces light, the CAD cell can determine if the flame is occurring safely. If the flame diminishes or goes out completely, the CAD cell can detect this condition and change its resistance. This change in resistance causes the main burner control to deenergize relays that control voltage to the pump motor. You can see in the electrical diagram that the burner control has a bilateral switch to provide a signal to the base of the triac. You should remember from the chapter on electronic devices that the triac is a

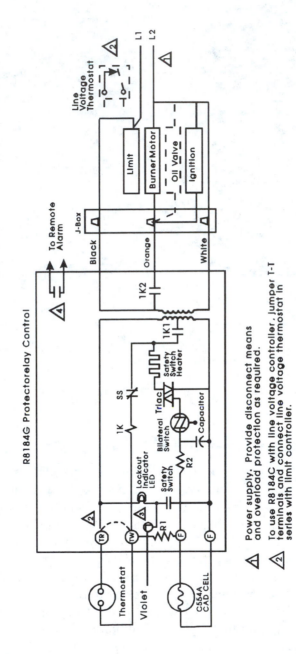

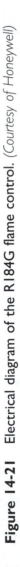

Figure 14-21 Electrical diagram of the R184G flame control. (*Courtesy of Honeywell*)

device that switches AC voltage. The triac provides a current path through its main terminals when a voltage signal is applied to its base.

14.23   The Operational and Safety Circuit

When the thermostat calls for heating, a 24 VAC signal is sent to terminal T on the burner primary control. This 24 V signal originates from the secondary of the control transformer that is inside the primary control. When the thermostat calls for heating, it closes its contacts, which completes the circuit to the coil of the 1K relay. The 1K relay closes its contacts to supply 110 VAC to the pump and to the ignition system. When the pump runs, fuel is pressurized so that it comes through the nozzle as a fine mist. When the oil mist comes into contact with the sparks at the electrodes, it will become ignited and a flame will develop. The flame is directed into a ceramic fire box and its heat will begin to warm the heat exchanger. The furnace fan blows air across the outside of the heat exchanger into a warm air duct and through registers into the room that is being heated.

The burner control requires a safety circuit to ensure that the flame is established when the pump begins to run, and remains burning as long as the pump runs. Fig. 14–22 shows the safety circuit highlighted on the diagram. From this diagram you can see that the circuit uses a transformer that has a center-tapped secondary winding. The primary side of the transformer is energized by 120 VAC that also powers the burner motor and furnace fan. The secondary voltage of the transformer is 24 VAC. This voltage is switched on and off through the R-W terminals of the thermostat. When the thermostat calls for heating, the voltage is supplied to the coil of the 1K relay coil. The contacts of the 1K relay control voltage to the burner motor in the high-voltage side of the circuit, and also interlock its own relay coil on the secondary side of the circuit. Voltage is also provided to the heating element of the safety switch SS1 through terminals MT1 and MT2 of the triac. When a flame is established, the light from the flame causes the resistance in the CAD cell to lower, which changes the voltage that is used to provide a signal to the bilateral switch. Since the bilateral switch provides the pulse for the triac, it will turn the triac off when a flame is detected.

When the triac is off, it interrupts the current flow to the heater of the safety switch SS1. Fig. 14–23 shows the current path when a flame is detected and everything is normal.

If a flame is not detected, the resistance in the CAD cell remains high and the bilateral switch will receive sufficient voltage through R1 and R2 to provide a pulse to the gate of the triac. When the gate of the triac is pulsed, it completes the circuit through the main terminals of the triac, which allows current to flow to the heating element of the SS1. The heat from the element of SS1 will cause

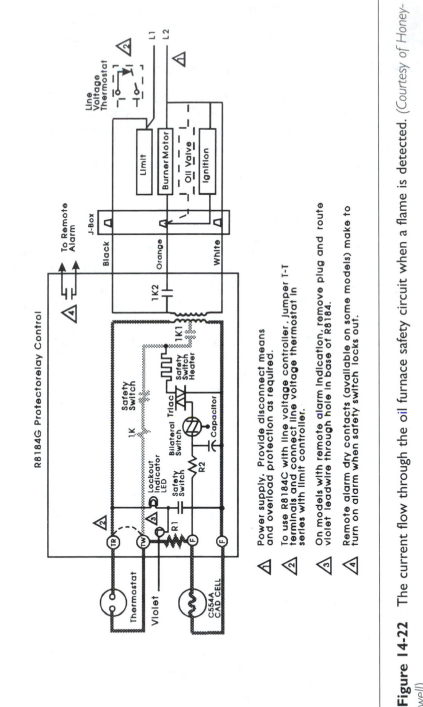

Figure 14-22 The current flow through the oil furnace safety circuit when a flame is detected. *(Courtesy of Honeywell)*

340

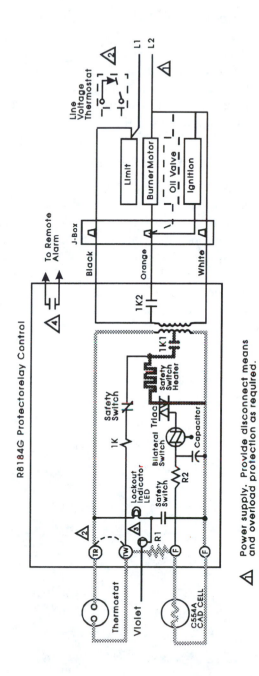

Figure 14–23 Current path through the safety control when a flame is not detected and the heating element of the SS1 switch is energized. *(Courtesy of Honeywell)*

the normally closed SS1 contacts to open after 15 to 20 seconds. When SS1 "trips" to the open position, it activates an external button to indicate the system has tripped. When contacts of SS1 open, the secondary voltage to the thermostat and the coil of 1K is interrupted. The coil of 1K is deenergized, which opens the 110 VAC circuit to the pump motor. The SS1 safety switch must be manually reset to get the pump to operate again. It is important to check the furnace to determine the reason the loss of flame occurred.

14.24   Fan and Limit Controls for Furnaces

Each type of furnace needs a control to protect it against overheating. If the furnace fan fails to operate correctly, the fire in the heat exchanger will make the furnace overheat and could possibly cause a fire. The high-limit switch is used to sense the overtemperature and open its contacts. The contacts for the switch are usually wired in series with the voltage that is supplied to the gas valve or oil burner pump or other source of heat. When the high-limit contacts open, they deenergize the source of heat and cause the furnace to cool down safely. The limit is usually a snap-action switch that must be manually reset. The manual reset function causes someone to come to the furnace to depress the reset button. This person must also check the furnace to learn the cause of the overtemperature. Fig. 14–24 shows an example of a high-limit switch with a reset button. Fig. 14–25a shows a high-limit switch that automatically resets when the overtemperature condition returns to normal. Fig. 14–25b shows a high-limit switch

Figure 14–24 A typical high-limit switch with reset. *(Courtesy of Honeywell)*

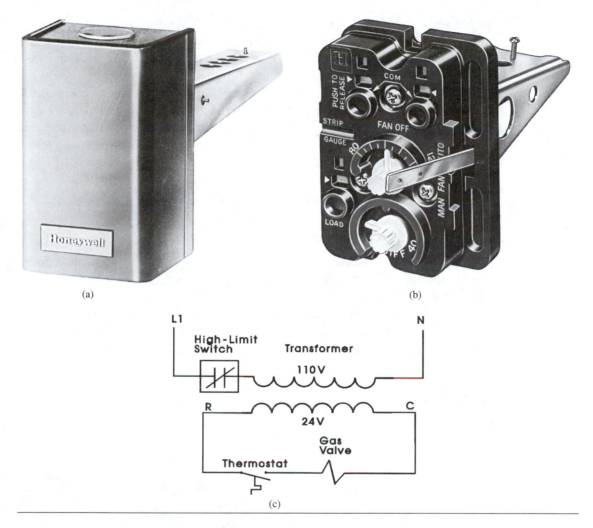

(a)   (b)

(c)

Figure 14–25a   A high-limit switch that automatically resets itself when the overtemperature condition returns to normal. **b** A high-limit switch with its cover removed. **c** Diagram of high-limit switch connected in a circuit. *(Courtesy of Honeywell)*

with its cover removed. Fig. 14–25c shows the high-limit switch connected in a circuit.

14.25   The Fan Switch

The furnace also needs a temperature control that controls the fan. When the gas valve or oil burner is first energized, the heat exchanger is cool. If the fan is energized at this time, it would blow cool air into the conditioned space. The fan control senses the temperature in the heat exchanger and energizes the fan when the temperature reaches approximately 110°F. When the source of heat is deenergized, the fan continues to run to allow the residual heat that is built up in the furnace to be moved into the room. The fan switch allows the heat exchanger to cool down to approximately 100°F before its contacts open and deenergize the fan. Fig. 14–26 shows a picture of a fan switch. The turn-on and turn-off temperatures for the fan switch are adjustable on some fan switches, and they are fixed for others. Some fan switches use a time-delay element in addition to the temperature-sensing element to provide sufficient time delay for the fan to cool the heat exchanger at the end of a cycle. The time delay can also be used to provide time delay at the beginning of the cycle as well.

14.26   Fan and Limit Control

The fan and limit switch can be operated off of the same temperature-sensing element. Fig. 14–27 shows an electrical diagram of a fan and limit switch. This control has two separate switches that are activated by two separate cams that are mounted to the spiral bimetal element. The spiral bimetal element is placed into the top of the heat exchanger where it can sense the warmest temperature. When the heat exchanger begins to warm up, the spiral bimetal element begins to turn and the cam for the fan switch moves across the fan switch and activates it. This causes the normally open fan switch contacts to close and energize the fan. The fan moves cool air from the conditioned space over the heat exchanger

Figure 14–26 A typical fan switch. *(Courtesy of Honeywell)*

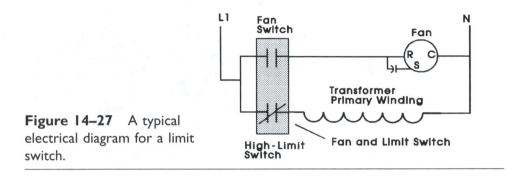

Figure 14–27 A typical electrical diagram for a limit switch.

and keeps the heat exchanger at a constant temperature. When the source of heat is deenergized, the heat exchanger cools down and the spiral bimetal element returns to its original position, allowing the fan switch to return to its open position.

If the fan failed to operate correctly, the temperature in the heat exchanger would continue to increase, and the spiral bimetal element would continue to turn. When the temperature of the heat exchanger reaches approximately 130°F, the cam for the limit switch on the bimetal element would cause the limit switch part of the fan and limit to activate. When this occurs, the normally closed contacts of the limit switch open and deenergize the source of heat for the furnace. When the furnace cools down, the bimetal element returns to its normal position, the limit contacts return to their closed position, and the heat source is energized again.

The furnace would continue to cycle on the high temperature until the temperature exceeded the fixed high-temperature limit or until someone noticed that the furnace is cycling on the fan and limit switch.

14.27   **Air Flow Switches**

Air flow switches detect the flow of air in heating and air-conditioning ducts or in locations near the condenser fan to indicate the condenser fan is operating correctly. Fig. 14–28 shows an example of this type of switch and it shows the electrical symbol for the switch. The air flow switch is sometimes called a sail switch because the element that senses air flow is in the shape of a sail. The sensing element is also called a paddle. The switch is usually used to detect critical air flow in larger systems that might be damaged by an evaporator fan not running or a condensor fan not running. If an evaporator fan does not move sufficient air across the evaporator in a commercial air-conditioning system, the evaporator fins will develop frost and freeze up. This condition will cause the refrigerant in the low side of the system to partially evaporate, and liquid refrig-

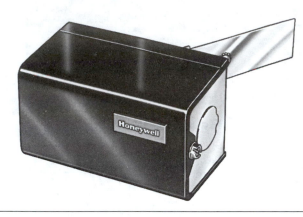

Figure 14–28   Air flow
switch. *(Courtesy of
Honeywell)*

erant may reach the compressor where it will damage the pistons and valves. The lack of air could be caused by slipping belts or a faulty fan motor. The flow switch can detect the low air flow and shut the compressor off.

If the condenser fan is not moving sufficient air across the condenser coil, the system will develop high pressure, which can also damage the compressor. The flow switch can detect the low air flow and shut the compressor off before it is damaged. You may also find a liquid-type flow switch in some water lines for boilers or water-cooled evaporators. The flow switch determines if the proper amount of liquid flow is available.

The electrical part of the switch is a single-pole, single-throw switch, or a double-pole, single-throw switch. You can test the switch contacts with a continuity test. Be sure the flow element is free to move far enough to activate the switch contacts. It is also important to ensure the switch is mounted in the air flow so that the flow is directed on the paddle. The switch can be designed to open the electrical contacts when flow is detected or it can be designed to close the electrical contacts when flow is detected. You should notice the abbreviation for the flow switch is FS.

14.28   Humidistat

Another type of control that is used in heating systems is called a humidistat. The humidistat is used to detect the amount of humidity that is in the air in the conditioned space. It is important to have a specific amount of humidity in the air during heating season. The outdoor conditions during the winter usually have lower humidity, and the process of warming the air indoors usually removes humidity. The humidistat is mounted on the wall or in the air duct like a thermostat so that it can sense the amount of humidity in the air. If the humidistat senses that the humidity is low, it closes its electrical contacts and energizes a

Figure 14–29   Residential
humidistat. *(Courtesy of
Honeywell)*

humidifier that puts moisture back into the air. Fig. 14–29 shows a typical humidistat.

Some commercial applications such as press rooms where your local newspaper is printed must control humidity very closely so that electrostatic sparks do not occur. When the humidity gets low, the air is more conducive to allowing sparks to jump from the paper to the other parts of the system. These sparks contain large amounts of static voltage that can injure personnel or damage electronic equipment in the press room. Other applications that require humidity control include textile mills where the amount of moisture in the product must be controlled, and the amount of moisture in the air must be controlled to prevent explosions. Another application where the amount of humidity is controlled is in manufacturing plastic pellets that are the raw material used in plastic injection molding machines. The moisture in the raw plastic pellets must be controlled very closely so that the moisture is not released into the finished product where it will create surface defects. As a technician who understands these controls, you will be requested to troubleshoot humidity control systems.

Questions for This Chapter

1. Explain how two strips of metal that form a bimetal element sense a change in temperature.
2. Explain how a spiral bimetal element reacts to changes in temperature.
3. Explain the function and operation of the heat anticipator.
4. Explain the operation of a thermocouple and where you would find a thermocouple.
5. Identify the components of a direct-spark ignition system and explain their operation.

True or False

1. The contacts of the heating thermostat go closed when temperature falls below the setpoint.
2. The gas valve is basically a solenoid valve with a safety circuit to ensure the pilot light is burning.
3. The oil burner control uses 24 volts to provide the spark to the ignition points.
4. The fan and limit switch provides an operational control to energize the fan and a safety circuit to deenergize the gas valve if the temperature gets too high.
5. The heat anticipator produces heat while the heating thermostat contacts are in the open position.

Multiple Choice

1. The fan switch in the thermostat has _____
    a. two positions (auto and on).
    b. three positions (auto, off, on).
    c. two positions (off and on).
2. The gas valve has a button that must be depressed when the pilot light is being lighted to _____
    a. bypass the high-temperature safety circuit.
    b. bypass the low-temperature safety circuit.
    c. overcome pull-in current of the solenoid in the valve until the thermocouple can produce sufficient current to hold in the solenoid.
3. The cooling anticipator in the cooling thermostat produces

    _____
    a. heat when the contacts of the cooling thermostat are open.
    b. heat when the contacts of the cooling thermostat are closed.
    c. a cooling effect when the contacts of the cooling thermostat are open.
4. The direct-spark ignition system _____
    a. provides a spark to start the pilot flame.
    b. provides a spark to start the main flame.
    c. provides a spark to the thermocouple.
5. The flame rod in the direct-spark ignition system _____
    a. detects the main flame and pilot flame.
    b. detects only the pilot flame.
    c. detects only the main flame.

Problems

1. Sketch the mercury-type switch that is contained in a glass element, and explain how it works when it is attached to a spiral bimetal element.

2. Use the diagram in Fig. 14–6b to identify the terminals that are used in the heating circuit for a heating thermostat.

3. Use the diagram in Fig. 14–8 to identify the terminals that are used in the cooling circuit for a cooling thermostat.

4. Draw a sketch of the fan switch circuit in a heat-cool thermostat and explain its operation in the auto and the on settings.

5. Use the diagram in Fig. 14–21 and draw a sketch of the oil burner control. Explain the operation of each part of the circuit.

6. Use the diagram in Fig. 14–27 to sketch the circuit of a fan and limit switch.

15 Pressure Controls, Timer Controls, and Other Controls

OBJECTIVES:

After reading this chapter, you will be able to:
1. Explain the operation and function of a high-pressure switch.
2. Explain the operation and function of a low-pressure switch.
3. Explain the operation and function of an oil pressure switch.
4. Explain the operation of a defrost timer.

15.0   Overview of Pressure, Timer, and Overcurrent Controls

As a technician, you will need to be able to identify and troubleshoot a variety of pressure controls, timer controls, and overcurrent controls. These controls are used to protect a refrigeration or air-conditioning system against conditions that will cause the system to become damaged. The controls could also be used for normal operation. For example, since pressure and temperature are related, a load pressure switch can be used as a thermostat. When you are troubleshooting a system, you should be able to identify these controls, and locate a diagram of the internal parts of their switches. Next you should be able to troubleshoot the system by activating the pressure timer or overcurrent portion of the switch.

If you find a pressure-type switch, you must be able to test the switch portion and cause the switch to activate by adjusting pressure to the switch. This means a low-pressure switch must be tested by applying pressure to the switch to cause it to activate. The problem with the switch may be on the pressure side, such as a bellows that has a hole in it or a tube that is leaking, or it may be on the electrical side where contacts are corroded. When these devices are in a system, it is important to understand they may open their electrical contacts and shut the system down because they are working correctly. For instance, if a system has a high-pressure switch, it may open when pressure exceeds the normal setpoint. This can occur when the condenser fins become clogged with leaves or other matter, which reduces the air flow across the condenser and increases the condenser pressure to an unsafe level. The high-pressure switch will open and protect the system against the high pressure. If you merely test this electrical portion of the switch, you will find it is open. If you change the switch at this point without testing for the reason of having the high pressure, you will merely change a good part. In many cases you must identify the problem the switch is reporting, and repair the condition that is causing the high pressure.

You will find a wide variety of timer controls on things such as defrost timers for refrigeration units, or defrost timers for heat pump systems. Timers are also used to control circulating pumps and other time control functions. Short-duration timers are used to provide small amounts of time delay when an air-conditioning or refrigeration system is started. This type of time delay is called short-cycle protection. Some of the timers you encounter will be motor driven, and other types will be solid-state electronic time delay.

You will encounter a large variety of overcurrent controls, which will include fuses, circuit breakers, overload devices on compressors, and overload devices on motor starters.

15.1   Pressure Controls

The most common types of pressure controls include high-pressure switches, low-pressure switches, and oil pressure switches. Figure 15–1a shows a picture of a high-pressure switch, and Figure 15–1b shows the electrical symbol for the high-pressure switch. The symbol for the high-pressure switch shows a bellows that activates a contact. The contact will open upward when the high pressure is exceeded. Figure 15–2 shows the internal operation of a high-pressure switch. In this diagram you can see that the high-pressure element is a bellows made of light metal. A spring is used to adjust the amount of pressure placed on the bellows. When refrigerant pressure goes inside the bellows, it will cause the bellows to expand against the spring and activate an electrical switch. The switch can be a single-pole, single-throw switch or a single-pole, double throw-switch.

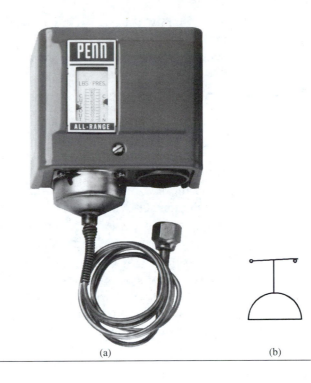

Figure 15–1a A high-pressure switch. **b** Electrical symbol for a high-pressure switch. *(Courtesy of Johnson Controls, Inc.)*

(a)

(b)

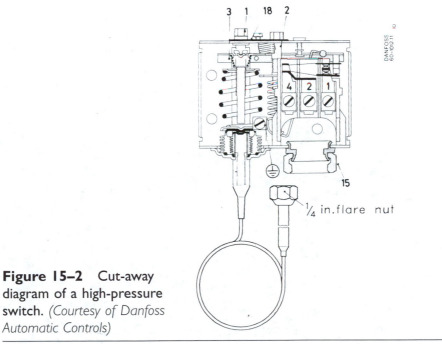

Figure 15–2 Cut-away diagram of a high-pressure switch. *(Courtesy of Danfoss Automatic Controls)*

¼ in. flare nut

The switch will have a screw mechanism provided so that you can adjust spring tension on the bellows. This will set the adjustment for the point at which the switch activates.

15.2   Low-Pressure Switches

Figure 15–3a shows a picture of a low-pressure switch and Figure 15–3b shows the electrical symbol for a low-pressure switch. You should notice that the low-pressure switch looks very similar to the high-pressure switch. The symbol for a low-pressure switch is different than the high-pressure switch because the low-pressure switch opens below the contact. This indicates that the switch contact will open as pressure decreases. The low-pressure switch is constructed with a bellows that is similar to a high-pressure switch, except the low-pressure bellows is more sensitive. Again an adjustment screw and dial are provided to show the pressure setting. When you turn the adjustment screw, you are changing the spring pressure on the bellows, which in turn adjusts the point at which the bellows will move a sufficient distance to activate the switch. The contacts in the low-pressure switch are contacts. When the pressure falls below the setpoint, the switch will activate and return to its open condition. The low-pressure switch is used to detect the loss of refrigerant

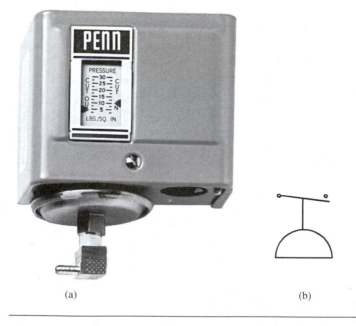

(a)   (b)

Figure 15–3a   A low-pressure switch. **b** The electrical symbol for a low-pressure switch. *(Courtesy of Johnson Controls, Inc.)*

or other conditions such as loss of an evaporator fan, which will cause low refrigerant pressure.

15.3   Fixed Pressure Switches

Figure 15–4 shows examples of fixed pressure switches. These devices are non-adjustable, which means the setpoint must be determined when they are purchased. They are available in a variety of pressures. The basic difference with this type of switch is that it is less expensive than an adjustable switch, and it is simple to install. In many cases the pressure switch can be installed in the system without making any mechanical openings. The fixed pressure switch can be mounted on a valve stem.

15.4   Dual Pressure Switches

Fig. 15–5a shows a picture of a differential pressure switch and Fig. 15–5b shows the electrical diagram of the differential switch. In these figures you can see that the differential pressure switch has two connections with capillary tubes. One connection is for the high-pressure switch, and one is for the low-pressure switch. The differential pressure switch provides a high- and low-pressure switch in one control so that it is easier to install and mount. Each switch inside the dual pressure switch is adjustable just as though it was a single high-pressure switch or

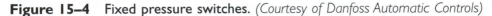

Figure 15–4   Fixed pressure switches. *(Courtesy of Danfoss Automatic Controls)*

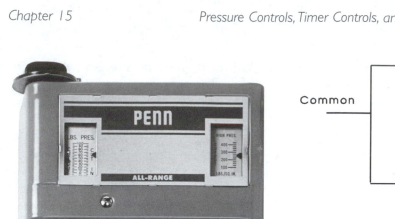

Figure 15–5a   Dual pressure switch. **b** The electrical symbol for a dual pressure switch. *(Courtesy of Johnson Controls, Inc.)*

individual low-pressure switch. Fig. 15–6 shows a diagram of the dual pressure switch.

15.5   Electrical Diagrams for Pressure Switches

When you locate a diagram of a pressure switch as in Fig. 15–2 or Fig. 15–6, you will notice the electrical terminals are numbered, but the diagram does not indicate if the switch is a single-pole, double-throw switch or what type of switch it is. Fig. 15–7 shows a set of diagrams for all of the pressure switches you should encounter in the field. From this diagram you can see that each type of switch is represented. You can test the switch contacts with a continuity test when no voltage is applied to them, and you can adjust the pressure below and above the switch setpoint to ensure the pressure part of the switch is operating correctly. It is important to verify the cut-in and cut-out pressures for each type of switch as well as to test the contacts to ensure that they open and close properly. You must use a set of gauges and an adjustable source of pressure to test the cut-in and cut-out pressure points of each switch.

15.6   Oil Pressure Switches

Fig. 15–8 shows pictures of oil pressure switches. This type of switch is used to protect refrigeration systems against low oil pressure. The switch contains a dif-

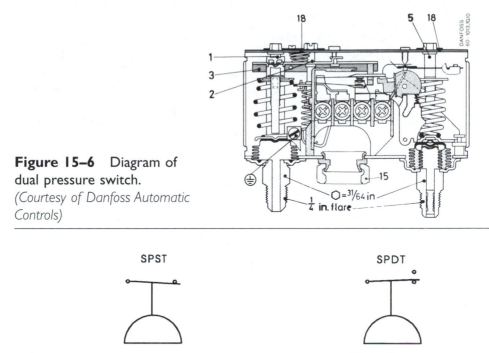

Figure 15–6 Diagram of dual pressure switch.
(Courtesy of Danfoss Automatic Controls)

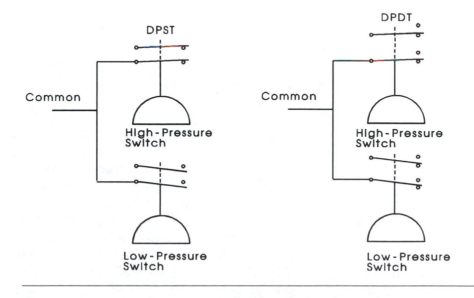

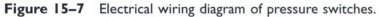

Figure 15–7 Electrical wiring diagram of pressure switches.

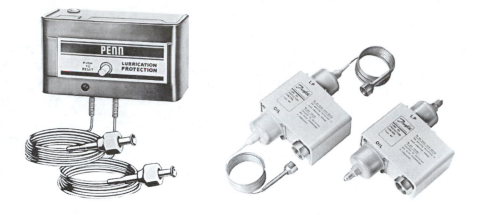

Figure 15–8   Examples of oil pressure switches. *(Courtesy of Johnson Controls, Inc.; Danfoss Automatic Controls)*

ferential pressure mechanism and a time delay. The time-delay mechanism allows a compressor to build oil pressure for approximately 15 to 30 seconds before the switch test for the loss of oil pressure. The differential-type switch is required because the refrigerant in the compressor will provide anywhere from 30 to 60 psi, and the oil pressure must be 30 additional pounds of pressure. This means one side of the switch has a capillary tube connected to the low-pressure side of the refrigerant system. The other side of the switch is connected to a port that can measure oil pressure. If the refrigerant pressure in the system is 30 pounds, the refrigerant side of the switch will measure 30 pounds. The oil pressure measurement may indicate 65 pounds, which means 30 pounds is due to the refrigerant pressure, and 35 pounds is due to oil pressure. The switch has a minimum and a maximum setpoint that causes its contacts to open if oil pressure drops below the lower setpoint. The second setting in the switch is to provide the amount of differential between the oil pressure and the refrigerant pressure. The time-delay circuit is operated by a smaller timer on older models, and by an RC time constant on newer models. The time constant is a resistor and capacitor that are connected to a solid-state switching element such as an SCR, a triac, or transistor. When power is applied to the circuit, voltage will charge the capacitor. The resistor will regulate the amount of time to charge the capacitor. If the oil pressure switch is an older-type switch, it will have an electric heating element and a bimetal strip. A variable resistor is used to help adjust the time delay on the heating strip. When current is applied to the heater element, it will produce heat. When the heat builds to a sufficient level, it will cause the bimetal to open and to warp, which will cause its contacts to open. Both the solid-state and the heating element time delay give the compressor time to build oil pressure before the switch begins to protect it. Fig. 15–9 shows the electrical diagram of the oil pressure switch. Notice that this type of switch needs a time-delay circuit to allow the compressor several seconds to build the oil pressure to a differential above the low side pressure.

15.7   Cut-In and Cut-Out Pressures

The oil pressure has two adjusting screws that allow you to set the cut-out point and the differential. The concept is to protect the compressor against oil pressure failure. If the oil pressure drops below 12 psi, the switch should open. The 12 psi is measured in addition to low-pressure refrigerant pressure. If the oil pressure cuts out, it needs to increase to a point where it will cut back in. Typically it is set to 15 to 20 psi to cut in. You should adjust the cut-in point and differential on an oil pressure switch with regulated oil pressure. This means that you

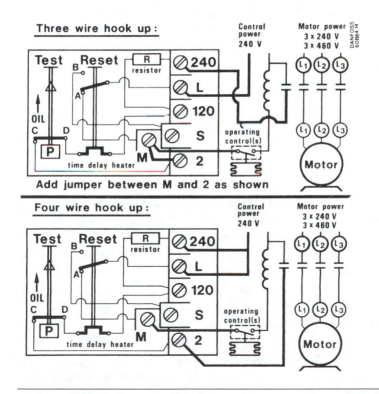

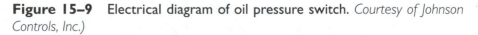

Figure 15–9   Electrical diagram of oil pressure switch. *Courtesy of Johnson Controls, Inc.)*

can set the cut-in point and differential for the oil pressure switch on the bench or when the switch is on the unit.

After you have adjusted the oil pressure switch on the bench, you can install it on the compressor and feel confident that it will protect against loss of oil pressure. When you are troubleshooting a system in the field and the oil pressure switch cuts out the compressor, you must be aware of the loss of oil pressure. Typically the loss of oil pressure will be due to a faulty oil pump in the compressor, or the migration of oil from the compressor into the warmest part of the refrigeration system. This means that many times the oil pressure switch will be open, but it is not broken; it is merely protecting the system against the loss of oil. This means you must correct the reason for loss of oil rather than replace the oil pressure switch.

15.8   Timer Controls

Timers are used in a variety of applications and air-conditioning and refrigeration systems. For example, most refrigerators that have a freezer compartment have some type of defrost clock or defrost timer that cycles the system into a defrost mode as needed when frost develops on the evaporator. In a heat pump system a defrost clock is used to reverse the heat pump cycle when the system is set for heating, anytime that frost may develop on the evaporator. Frost can build up on the evaporator anytime the outdoor temperature is below 36°. The reason the outdoor coil can build up frost when the temperature is above freezing is because the refrigerant temperature on the coil during evaporation is well below freezing. If a defrost cycle is not actuated, frost will build up to a point that it completely covers the coil face so that air does not pass through it.

Commercial refrigeration systems also require a defrost cycle. In many cases a larger defrost timer is used to activate this cycle. Other types of timers that you will find in air-conditioning and refrigeration systems are called short-cycle timers. The short-cycle timer will be activated anytime the thermostat calls for cooling. Instead of starting the compressor, the short-cycle timer will be activated to provide a short time delay of 1 to 3 minutes. This time delay allows time for the refrigeration pressure between the compressor and the high side to equalize to a point that the compressor can start against a lower pressure. A problem occurs if a system is turned off and back on within 1 minute. When the system is turned off and on in a short period of time, high pressure in the high side of the system will prevent the compressor from starting, or cause the compressor to draw large currents. The short-cycle timer can be a motor-driven timer, an electric heating element timer with a bimetal strip, or a solid-state, RC time

Figure 15–10 Typical
defrost timer. *(Courtesy of
Paragon Electric Company, Inc.)*

constant timer. The short-cycle timer may also be included with other compressor protection circuitry.

15.9   Defrost Timer

Fig. 15–10 shows a typical defrost timer and Fig. 15–11 shows the electrical
diagram of a defrost timer. The timer has an electrical motor between terminals
L1 and 3, and a set of single-pole, double-throw contacts between terminals 1,
2, and 4. The normally closed set of contacts is shown between terminals 1 and
4, and the normally open contacts are shown between terminals 1 and 2. The
timer motor can be connected directly across the power supply so that it runs all
the time. This type of timer is called a continuous-run timer. The timer can also
be connected so that it is only energized when the compressor is running. Fig.
15–12 shows a timer connected to a compressor. You can see in this diagram
that when the timer is in the normal run position, contacts 1 to 4 are made so
that power is sent to the compressor. When the timer motor runs to a point that
a defrost cycle is required, a contact switch between 1 and 2 closes a circuit to
the defrost heater. The contact circuit between 1 and 4, which energized the

Figure 15–11 Electrical
diagram of a typical defrost
control. *(Courtesy of Paragon
Electric Company, Inc.)*

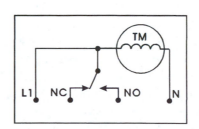

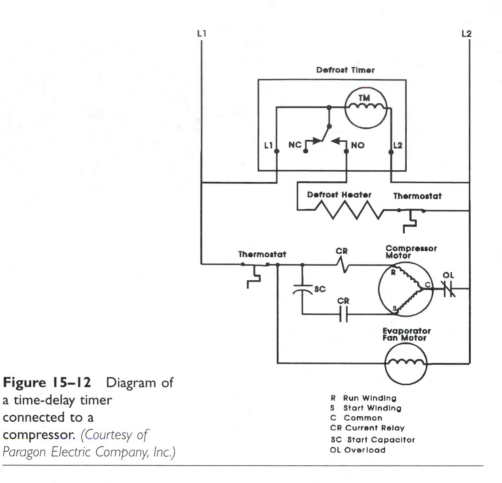

Figure 15–12 Diagram of a time-delay timer connected to a compressor. *(Courtesy of Paragon Electric Company, Inc.)*

R Run Winding
S Start Winding
C Common
CR Current Relay
SC Start Capacitor
OL Overload

compressor, is broken during this period so that the compressor is turned off. When the timer has run for an additional 3 to 5 minutes, it will come out of the defrost cycle and open its contact circuit between 1 and 2, and make the circuit between 1 and 4, which energizes the compressor and evaporator fan again. You can test the timer by using an ohmmeter to measure for continuity in the timer motor. You can also test for continuity between terminals 1 and 4 when the timer is in the normal position and between terminals 1 and 2 when it is in the defrost position. It is possible to advance some timers by manually rotating the cam on the side of the switch with a screwdriver. The defrost timer generally has one of two problems: the timer motor stops running because its winding is open, or the defrost timer contacts tend to stick or weld closed. You can test for these conditions by observing the timer motor to see if it is running and you can check the contacts by manually rotating the cam and testing the contacts.

15.10   Electronic Defrost Controls

Newer types of defrost timers have solid-state electronic controls. These types of timers operate similarly to the motor-driven timers. Fig. 15–13 shows an electronic defrost timer and its diagram.

15.11   Compressor Time-Delay Timers

Timers are available to create a time delay when a compressor is energized. In air-conditioning and refrigeration applications the compressor must remain deenergized for several minutes after it cycles off to allow the refrigeration pressure on its high side to equalize. When the compressor cycles off, the refrigeration pressure may be over 150 psi. If the compressor tries to energize too soon before the pressure is allowed to become lower on its high side, the compressor motor will not be able to start against the high pressure and it will draw locked-rotor amps (LRA), which will cause its fuses to blow. The timer shown in Fig. 15–14a provides a time delay up to 10 minutes to allow the refrigerant pressure to equalize. Fig. 15–14b shows a diagram of the time-delay timer connected in series with the coil of the compressor contactor. When the thermostat closes and sends power to the coil of the compressor contactor, the timer will become energized and provide a time delay up to 10 minutes. After the time-delay period, the timer passes power to the compressor contactor coil and the compressor is allowed to start. This type of timer is especially useful when a power outage occurs during a storm. In the summer, it is typical for lightning storms to knock out the elec-

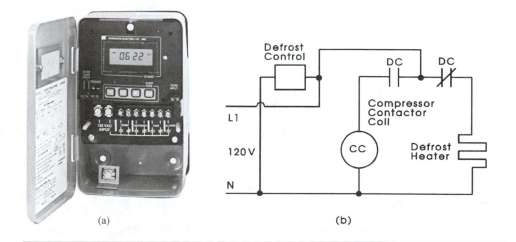

(a)   (b)

Figure 15–13a   An electronic defrost timer. **b** Electrical diagram of the electronic defrost timer. *(Courtesy of Paragon Electric Company, Inc.)*

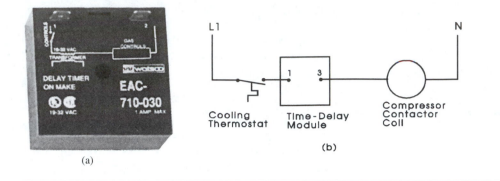

Figure 15–14a A short-cycle protector for a compressor. **b** Electrical diagram of a short-cycle protector connected to a compressor. *(Courtesy of Watsco Components, Inc.)*

trical supply for a few seconds. If a compressor is running during this short power outage, it will stop and then try to start again before the refrigerant pressure has lowered enough to allow the compressor to start. When this occurs, the compressor will draw excessive current and cause its fuses to open. If the fuses are not sized properly, the compressor will be severely damaged by the excess current.

Questions for This Chapter

1. Explain why an oil pressure switch needs a time-delay period.
2. Explain the difference between a fixed pressure switch and an adjustable pressure switch.
3. Identify the basic parts of a high-pressure switch and explain the conditions a high-pressure switch would protect against in an air-conditioning system.
4. Identify the basic parts of a low-pressure switch and explain the conditions a low-pressure switch would protect against in a refrigeration system.
5. Explain the operation of a defrost timer.

True or False

1. The oil pressure switch requires a time delay to allow the oil pressure to build up when the compressor is first turned on.
2. A fixed pressure switch can be adjusted a small amount around a fixed pressure setpoint.

3. The cut-in and cut-out pressure settings on an oil pressure switch will be the same pressure.
4. A defrost timer is normally used to stop the compressor periodically and energize a defrost heating element.
5. A commercial timer can be used to cycle off large commercial air-conditioning systems on weekends and other times when no one is in a building.

Multiple Choice

1. An oil pressure switch compares _____

    a. oil pressure to high pressure.
    b. oil pressure to low pressure.
    c. oil pressure to a setpoint of 15 psi.

2. An oil pressure switch needs _____

    a. a small amount of time delay to allow the compressor to build up oil pressure.
    b. a small amount of time delay to allow the compressor to build up its high pressure.
    c. no time delay because the oil pressure must be detected immediately or the switch will shut the compressor off so that no damage occurs if the oil pressure is low on startup.

3. A low-pressure switch connects the compressor or the coil of the compressor contactor to terminal _____

    a. 4 since it is part of the normally closed contacts.
    b. 4 since it is part of the normally open contacts.
    c. 3 since it is for the time delay.

4. When you are troubleshooting an air-conditioning system and find that it is shut down because its low-pressure switch is open, you should _____

    a. test the switch for continuity a second time and replace it if it is open.
    b. test the switch for continuity a second time and replace it if it is closed.
    c. look for the reason the system has low pressure.

5. If a defrost timer does not cycle the system through a defrost cycle periodically, you should suspect _____

    a. the timer motor is not running continuously or not running at all.
    b. the defrost heater is bad.
    c. the compressor cannot shut off.

Problems

1. Draw a sketch of a high-pressure switch including its electrical diagram and explain its function.
2. Draw the sketch of a low-pressure switch including its electrical diagram and explain its function.
3. Draw a sketch of a dual pressure switch including its electrical diagram and explain its operation.
4. Draw a sketch of an oil pressure switch including its electrical diagram and explain its operation.
5. Draw a sketch of a defrost timer including its electrical diagram and explain its operation.

16 Electronic Devices for HVAC Systems

OBJECTIVES:

After reading this chapter, you will be able to:
1. Explain P-type and N-type material.
2. Identify the terminals of a diode and explain its operation.
3. Explain the operation of PNP and NPN transistors.
4. Identify the terminals of a silicon controlled rectifier (SCR) and explain its operation.
5. Explain the operation of a triac.

16.0 Overview of Electronics Used in HVAC and Refrigeration Systems

Electronic devices such as diodes, transistors, and SCRs have become commonplace in heating, air-conditioning, and refrigeration systems because they provide better control than electromechanical devices and are less expensive to manufacture. When you are troubleshooting, you will find solid-state components in thermostats and in control boards for furnaces and air conditioners. In this chapter you will gain an understanding of P-material and N-material, which are the building blocks of all electronic components. After you get a basic understanding of P-material and N-material, you will be introduced to diodes, transistors, SCRs, and triacs and you will see application circuits of each of these types of components. The theory of operation and troubleshooting techniques for each type of device will also be presented.

16.1   Conductors, Insulators, and Semiconductors

In Chapter 1 you learned that atoms have protons and neutrons in their nucleus and electrons that move around the nucleus in orbits which are also called shells. The number of electrons in the atom is different for each element. For example, you learned that copper has 29 electrons and three of them are located in the outermost shell. The outermost shell is called the valence shell and the electrons in that shell are called valence electrons. The atoms of the most stable material have eight valence electrons, which will be found as four pairs. As an atom for materials that are used as conductors and insulators fills its valence shell with eight electrons, it will open a new orbit with the additional electrons. This means that an atom may have five, six, or seven atoms and it will take less energy to add electrons to get a full shell (eight), or it may have one, two, or three electrons and it will take less energy to give up these electrons to get down to the previous full shell that has eight electrons.

A *conductor* is a material that allow electrons (electrical current) to flow easily through it, and an *insulator* does not allow current to flow through it. An example of a conductor is copper that is used for electrical wiring. An example of an insulator is rubber or plastic. The atomic structure of a conductor makes it easier for electrons to flow through it and the atomic structure of an insulator makes it nearly impossible for any electrons to flow through it.

Fig. 16–1 shows the atomic structure of a conductor. Atoms of conductors can have one, two, or three valence electrons. The atom in this example has one valence electron. Since all atoms try to get eight electrons (four pairs) in their va-

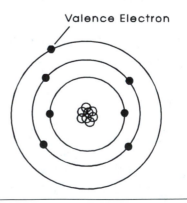

Figure 16–1   Atomic structure of a conductor.

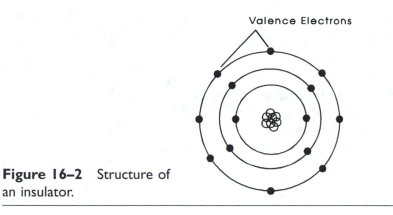

Figure 16–2   Structure of an insulator.

lence shell, it takes less energy for conductors to give up these electrons (one, two, or three) so that the valence shell will become empty. At this point the atom becomes stable because the previous shell becomes the new valence shell and it has eight electrons. The electrons that are given up are free to move as current flow.

Fig. 16–2 shows the atomic structure of an insulator. Insulators will have five, six, or seven valence electrons. In this example you can see that the atom has seven valence atoms. This structure makes it easy for insulators to take on extra electrons to get eight valence electrons. The electrons that are captured to fill the valence shell are electrons that would normally be free to flow as current.

Semiconductors are materials whose atoms have exactly four valence electrons. Since these atoms have exactly four valence electrons, they can take on four new valence electrons like an insulator to get a full valence shell, or they can give up four valence electrons like a conductor to get an empty outer shell. Then the previous shell that has eight electrons becomes the valence shell. Fig. 16–3 shows an example of the atomic structure for semiconductor material.

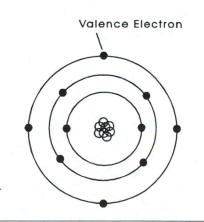

Figure 16–3   Atomic structure for semiconductor material.

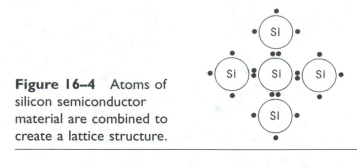

Figure 16–4 Atoms of silicon semiconductor material are combined to create a lattice structure.

16.2   Combining Atoms

When solid-state material or other material is manufactured, large numbers of atoms are placed together. The structure that becomes most stable at this point is called a lattice structure. Fig. 16–4 shows an example of the lattice structure that occurs when atoms are combined. In this diagram you can see that atoms of silicon (Si), which is a semiconductor material with four valence electrons, are combined so that one valence electron from each of the neighbor atoms is shared so that all atoms look and act as if they each have eight valence electrons.

16.3   Combining Arsenic and Silicon to Make N-Type Material

Other types of atoms can be combined with semiconductor atoms to create the special material that is used in solid-state transistors and diodes. Fig. 16–5 shows a diagram of four silicon atoms that are combined with one atom of arsenic. From this figure you can see that arsenic has five valence electrons, and when the silicon atoms are combined with it they create a very stong lattice structure. You can see that each silicon atom donates one of its valence electrons to pair up with each of the valence electrons of the arsenic atom. Since the arsenic atom has five valence electrons, one of the electrons will not be paired up and it will

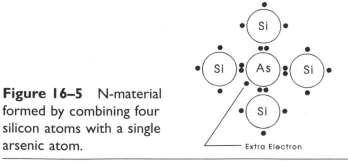

Figure 16–5 N-material formed by combining four silicon atoms with a single arsenic atom.

Extra Electron

become displaced from the atom. This electron is called a *free electron* and it can go into conduction with very little energy. Since this new material has a free electron it is called *N-type material.*

16.4   Combining Aluminum and Silicon to Make P-Type Material

An atom of aluminum can also be combined with semiconductor atoms to create the special material that is called P-type material. Fig. 16–6 shows a diagram of four silicon atoms that are combined with one atom of aluminum. From this figure you can see that aluminum has three valence electrons, and when the silicon atoms are combined with it they create a very stong lattice structure. You can see that each silicon atom donates one of its valence electrons to pair up with each of the valence electrons of the aluminum atom. Since the aluminum atom has three valence electrons, one of the four aluminum electrons will not be paired up and it will have a space where any free electron can move into it to combine with the single electron. This free space is called a *hole* and it is considered to have a positive charge since it is not occupied by a negatively charged electron. Since this new material has an excess hole that has a positive charge, it is called *P-type material.*

16.5   The PN Junction

One piece of P-type material can be combined with one piece of N-type material to make a *PN junction.* Fig. 16–7 shows a typical PN junction. The PN junction forms to make an electronic component called a *diode.* The diode is the simplest electronic device. When DC voltage is applied to the PN junction with the proper polarity it will cause the junction to become a very good conductor. Conversely, if the polarity of the voltage is reversed, the PN junction will become a good insulator.

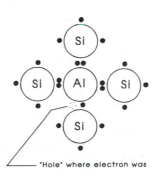

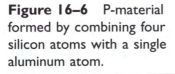

Figure 16–6   P-material formed by combining four silicon atoms with a single aluminum atom.

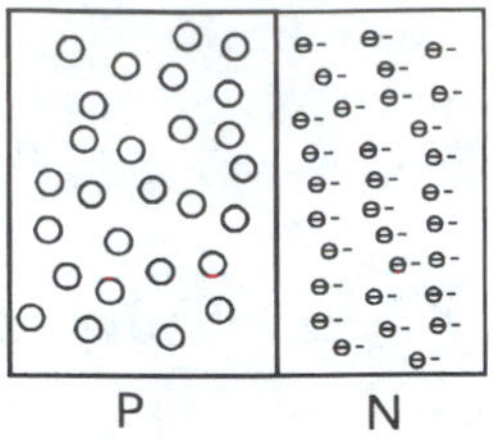

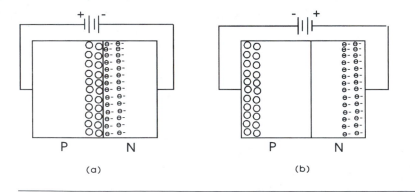

Figure 16–7   An example of a piece of P-material connected to a piece of N-material.

16.6   Forward Biasing the PN Junction

When DC battery voltage is applied to the PN junction so that positive voltage is connected to the P-material, and negative voltage is connected to the N-material, the junction is forward biased. Fig. 16–8a shows a battery connected to the PN junction so that it is forward biased. In this diagram you can see that the positive battery voltage causes the majority of holes in the P-type material to be repelled so that the free holes move toward the junction where they will come into contact with the N-type material. At the same time the negative battery voltage also repels the free electrons in the N-type material toward the junction. Since the holes and free electrons come into contact at the junction, the electrons recombine with the holes to cause a low-resistance junction, which allows current to flow freely through it. When a PN junction has low resistance, it will allow current to pass just as if the junction were a closed switch.

It is important to understand that up to this point in this text, all current flow has been described in terms of *conventional current flow,* which is based on a theory that electrical current flows from a positive source to a negative return terminal. At this point you can see that this theory will not support current flow

Figure 16–8a   A forward biased PN junction. **b** A reverse biased PN junction.

through electronic devices. For this reason *electron current flow theory* must be used when discussing electronic devices. In electron current flow theory, current is the flow of electrons and it flows from the negative terminal to the positive terminal in any electronic circuit.

16.7   Reverse Biasing the PN Junction

Fig. 16–8b shows a battery connected to the PN junction so that it is reverse biased. In this diagram you can see that the positive battery voltage is connected to the N-type material and the negative battery voltage is connected to the P-type material. The negative voltage on the P-type material attracts the majority of holes in the P-type material so that they move away from the junction. Thus, they cannot come into contact with the N-type material. At the same time the positive battery voltage that is connected to the N-type material attracts the free electrons away from the junction. Since the holes and free electrons are both attracted away from the junction, no electrons can recombine with any holes. Thus, a high-resistance junction is formed that will not allow any current flow. When the PN junction has high resistance it will not allow any current to pass just as if the junction were an open switch.

16.8   Using a Diode for Rectification

The electronic diode is a simple PN junction. Fig. 16–9 shows the symbol for a diode. You can see that the symbol for the diode looks like an arrowhead that is pointing against a line. The part of the symbol that is the arrowhead is called the *anode,* and it is also the P-type material of the PN junction. The other terminal of the diode is called the *cathode* and it is the N-type material. Since the anode is made of positive P-type material, it is identified with a + sign. The cathode is identified with a − sign, since it is made of N-type material. Fig. 16–10 shows a diode connected in a circuit that has an AC power source. The AC power source produces a sine wave that has a positive half-cycle followed by a negative half-

Figure 16–9a PN junction for a diode. **b** Electronic symbol of a diode. The anode is the arrowhead part of the symbol, and the cathode is the other terminal.

cycle. The diode converts the AC sine wave to half-wave DC voltage by allowing current to pass when the AC voltage provides a forward bias to the PN junction, and it blocks current when the AC voltage provides a reverse bias to the PN junction. The forward bias condition occurs when the AC voltage sine wave provides a positive voltage to the anode and a negative voltage to the cathode. During this part of the AC cycle, the diode is forward biased and it has very low resistance so current can flow. When the other half of the AC sine wave occurs, the diode becomes reverse biased with negative voltage applied to the anode and positive voltage applied to the cathode. During the time the diode is reverse biased, a high-resistance junction is created, and no current will flow through it.

Rectification is the process of changing AC voltage to DC voltage. One of the main jobs of the diode is to convert AC voltage to DC voltage. Most electronic circuits used in heating, air-conditioning, and refrigeration equipment need some DC voltage to operate. Since the HVAC equipment is connected to AC voltage, a power supply is required to provide regulated DC voltage for the solid-state circuits, and the diode is part of the power supply that rectifies the AC voltage to DC.

16.9   Half-Wave and Full-Wave Rectifiers

When one diode is used in a circuit to convert AC voltage to DC voltage, it is called a *half-wave rectifier* since only the positive half of the AC voltage is allowed to pass through the diode, while the negative half is blocked. The rectifier shown in Fig. 16–10 is a half-wave rectifier. The half-wave rectifier is not very efficient since half of the AC sine wave is wasted.

If four diodes are used in the circuit, they can convert both the positive half-wave and the negative half-wave of the AC sine wave. Fig. 16–11 shows a cir-

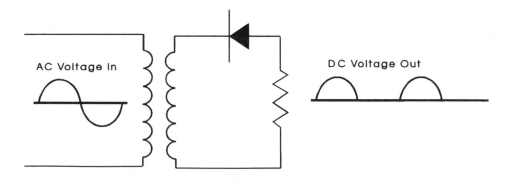

Figure 16–10   A single diode used in a circuit to convert AC voltage to DC voltage.

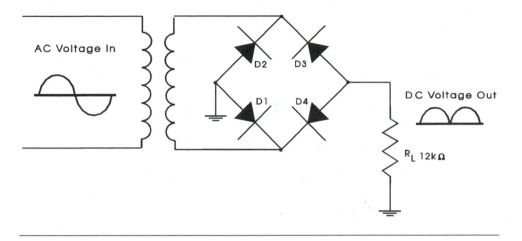

Figure 16–11   Four-diode, full-wave rectifier. This type of rectifier is often called a full-wave bridge rectifier, since the diodes are connected in a bridge circuit.

cuit with four diodes used to rectify AC voltage to DC voltage. Since the four diodes can convert both the positive half and the negative half of the AC sine wave, this type of rectifier is called a *full-wave rectifier.*

16.10   Three-Phase Rectifiers

Larger three-phase AC compressor and fan motors used in air-conditioning, heating, and refrigeration systems can have their speed changed to run more efficiently by changing the frequency of the voltage supplied to them. In these applications, six diodes are used to convert three-phase AC voltage to DC voltage, and then a microprocessor controlled circuit converts the DC voltage back to three-phase AC voltage. The frequency of this voltage can be adjusted to change the speed of the motors. Fig. 16–12 shows a six-diode, three-phase rectifier. Notice that the supply voltage to the diodes is three-phase AC voltage, and the output voltage from the rectifier consists of six positive half-waves.

16.11   Testing Diodes

One of the tasks that you must perform as a technician is to test diodes to see if they are operating correctly. One way to do this task is to apply AC voltage to the input of the diode circuit and test for DC voltage at the output of the diode circuit. If the amount of DC voltage at the power supply is half of what it is rated for in a four-diode, bridge rectifier circuit, you can suspect one of the di-

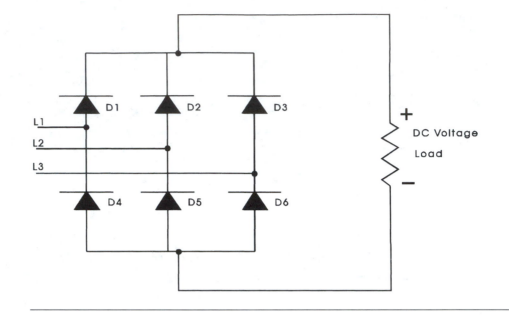

Figure 16–12 A six-diode bridge used to rectify three-phase AC voltage to DC voltage.

ode pairs has one or both diodes opened. If this occurs, you can use an ohmmeter to test each diode to determine which one is faulty.

When you are testing the diodes with an ohmmeter, it is important that all power to the diode circuit is turned off. You should remember from earlier chapters that the ohmmeter uses a battery as a DC voltage source. Since you know the diode can be tested for forward bias and reverse bias with a DC voltage source, you can use the ohmmeter as the voltage source and the meter to test for high resistance and low resistance through the diode junction. Fig. 16–13a shows an example of putting the positive ohmmeter terminal on the anode of the diode, and the negative ohmmeter terminal on the cathode of the diode to cause the diode to go into forward bias. During this test the diode is forward biased and the ohmmeter should measure low resistance. When the ohmmeter leads are reversed as in Fig. 16–13b so that the negative meter lead is connected to the anode of the diode, and the positive meter lead is connected to the anode of the diode, the diode is reverse biased. When the diode is reverse biased, the ohmmeter should measure infinite (∞) resistance. If the diode indicates high and low resistance, it is good. If the diode indicates low resistance during the forward bias test and the reverse bias test, the diode is shorted. If the diode shows high resistance during both tests, it is opened.

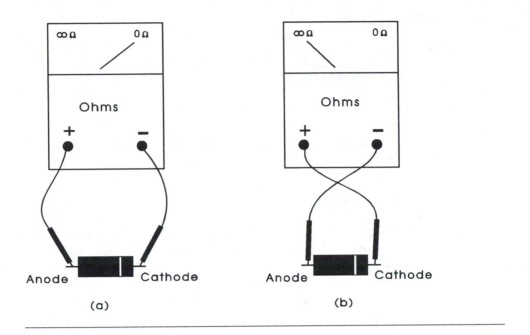

Figure 16–13a   Using the battery in an ohmmeter to forward bias a diode. The diode should have low resistance during this test. **b** Using the battery in the ohmmeter to reverse bias a diode. The diode should have high resistance during this test.

16.12   Identifying Diode Terminals with an Ohmmeter

Since the ohmmeter can be used to determine if a diode is good or faulty, the same test can be used to determine which lead of a diode is the anode and which lead is the cathode. When you use the ohmmeter to test the diode for forward bias and reverse bias, you should notice that the ohmmeter will indicate high resistance when the diode is reverse biased and low resistance when the diode is forward biased. When the meter indicates low resistance, you know the diode is forward biased, so the positive lead is touching the anode and the negative lead is touching the cathode. This method will work when you are testing any diode. If the diode has markings, you can identify the cathode end of the diode because it has a strip around it. Fig. 16–14 shows two types of diodes and the anode and cathode are identified in each.

16.13   Light-Emitting Diodes

A light-emitting diode (LED) is a special diode that is used as an indicator because it gives off light when current flows through it. Fig. 16–15a shows a typi-

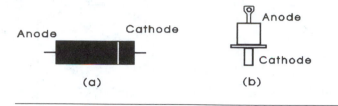

Figure 16–14a Typical diode with anode and cathode identified. **b** Power diode with anode and cathode identified.

cal LED and Fig. 16–15b shows its symbol. From this figure you can see the LED looks like a small indicator lamp. You will likely encounter LEDs on various controls such as thermostats. The major difference between an LED and an incandescent lamp is that the LED does not have a filament so it can provide thousands of hours of operation without failure.

The LED must be connected in a circuit in forward bias. Since the typical LED requires approximately 20 mA to illuminate, it will usually be connected in series with a 600 Ω to 800 Ω resistor which will limit the current. Fig. 16–16 shows a set of seven LEDs that are connected to provide a seven-segment display. The seven-segment display has the capability to display all numbers 0–9. Seven-segment displays are used to display numbers on thermostats and other electronic devices.

LEDs are also used in optoisolation circuits where larger field voltages are isolated from smaller computer signals. The LED is encapsulated with a phototransistor. When the input signal is generated, it will cause current to flow through the LED, and light from the LED will shine on the phototransistor, which will go into conduction and pass the signal on to the computer.

Figure 16–15a A typical light-emitting diode (LED). Notice the LED looks similar to a small indicator lamp. **b** The symbol for an LED.

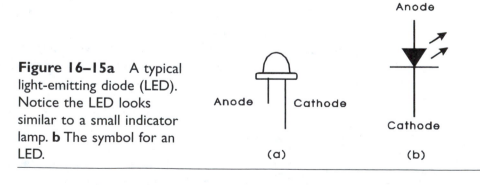

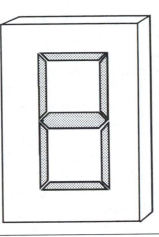

Figure 16–16 LEDs used in seven-segment displays. The seven-segment display can display numbers 0–9.

16.14   PNP and NPN Transistors

Two pieces of N-type material can be joined with a single piece of P-type material to form an NPN transistor. A PNP transistor can be formed by joining two pieces of P-type material with a single piece of N-type material. Fig. 16–17 shows the electronic symbol and the material for both the PNP and the NPN transistors. The terminals of the transistor are identified as the emitter, collector, and base. The base is the middle terminal, and the emitter is the terminal identified by the arrowhead.

16.15   Operation of a Transistor

A transistor can be connected in a circuit to perform a wide variety of functions. The simplest function for a transistor to provide is the function of an electronic

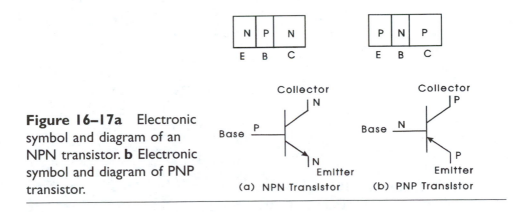

Figure 16–17a Electronic symbol and diagram of an NPN transistor. **b** Electronic symbol and diagram of PNP transistor.

switch. Fig. 16–18 shows a transistor as an electronic switch. In this type of application, the base terminal of the transistor provides a function like the coil of a relay, and the emitter-collector circuit provides a function like the contacts of a relay. When the proper amount and polarity of DC voltage is applied to the base of the transistor, the resistance between the collector and emitter is relatively low, which allows the maximum amount of circuit current to flow through the emitter-collector circuit. The transistor at this time acts like a relay that has its coil energized.

When the polarity of the voltage on the base of the transistor is reversed, the emitter-collector circuit is changed to a high-resistance circuit, which acts like the relay when the coil is deenergized. The major advantage of the transistor is that a very small amount of voltage or current on the base can switch the transistor from high resistance to low resistance. Since the base current is very small and the current flowing through the collector is very large, the transistor is called an *amplifier.* Transistors are used in a variety of applications including thermostats, compressor-protection circuits, and variable-frequency motor drives.

Fig. 16–19 shows a PNP transistor and an NPN transistor as two PN diode circuits. The equivalent diode circuits are shown with each transistor to give you the idea how the two junctions work together inside each transistor. Since each transistor is made from two PN junctions, each junction can be tested just as in the single-junction diode for forward bias (low resistance) and reverse bias (high resistance). If you must work on a number of systems that have electronic circuits, you may purchase a commercial-type transistor tester that allows you to test the transistor while it is in the circuit or when it is out of the circuit.

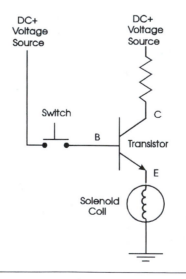

Figure 16–18 A transistor used as an electronic switch.

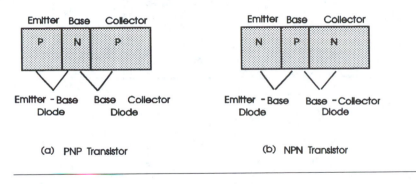

Figure 16–19a   A PNP transistor shown as its equivalent PN junctions. Each PN junction can be tested for forward bias and reverse bias. **b** An NPN transistor shown as its equivalent PN junctions.

16.16   Typical Transistors

You will be able to identify transistors by their shape. Small transistors are used for switching control circuits, and larger transistors will be mounted to heat sinks so that they can easily transfer heat. Fig. 16–20 shows examples of several types of transistors.

16.17   Troubleshooting Transistors

Transistors can be tested by checking each P and N junction from front-to-back resistance. Fig. 16–21 shows these tests. You can see each time the battery in the ohmmeter forward biases a PN junction, the resistance is low, and when the battery reverse biases the junction, the meter indicates high resistance. You can test a transistor in this manner if it has been removed from the circuit. You can also test transistors while they are connected in circuit with a commercial-type tran-

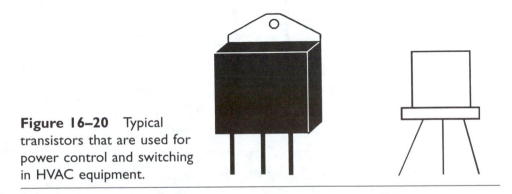

Figure 16–20   Typical transistors that are used for power control and switching in HVAC equipment.

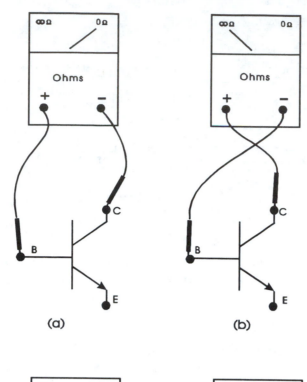

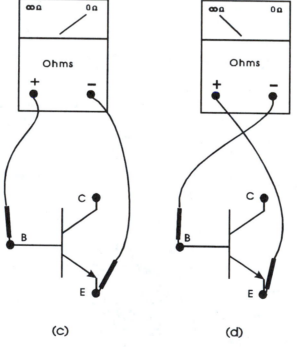

Figure 16–21a Testing the base-collector junction of an NPN transistor for forward bias. **b** Testing the base-collector junction of an NPN transistor for reverse bias. **c** Testing the base-emitter junction of an NPN transistor for forward bias. **d** Testing the base-emitter junction of an NPN transistor for reverse bias.

sistor tester. The transistor tester performs similar front-to-back resistance tests across each junction.

16.18   Unijunction Transistors

There are approximately two dozen different types of electronic devices that are made by combining a number of different sections of P-material and N-material together. Some of these devices are designed specifically for switching larger voltages and currents, and other devices like the unijunction transistor are designed to produce a small pulse of voltage that can be used to turn on or bias other switching devices. The unijunction transistor (UJT) is used to produce a pulse of voltage that is used to turn on silicon controlled rectifiers (SCRs). SCRs are used to control large DC voltages and currents and they will be introduced in the next section of this chapter. As an HVAC technician, you may be more familiar with the current-switching devices such as SCRs or triacs, but you must remember they will not operate without the devices such as UJTs that are used to turn them on and off.

The unijunction transistor is made of a single PN junction. Fig. 16–22 shows a diagram of the PN-material and the electronic symbol for the unijunction transistor. The terminals for the unijunction transistor are called base 1, base 2, and emitter. From the diagram you can see that base 2 and base 1 leads are mounted in the large section of N-type material. This means that if you measured the resistance between base 2 and base 1 terminals, you would measure the same amount of resistance regardless of the polarity of the meter leads. When you measure the resistance between base 2 and the emitter, you would find it is like a diode PN junction, and the polarity of the ohmmeter battery would cause the PN junction to be forward biased one way, and reverse biased when you reversed the leads. The base 1 to emitter junction also acts like a diode junction.

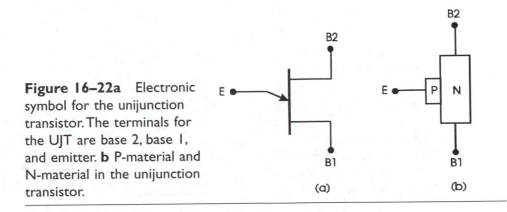

Figure 16–22a Electronic symbol for the unijunction transistor. The terminals for the UJT are base 2, base 1, and emitter. **b** P-material and N-material in the unijunction transistor.

16.19   **Operation of the UJT**

The simplest way to explain the operation of a UJT is to show it in a typical circuit. Fig. 16–23 shows the UJT in a circuit with a resistor and capacitor connected to the emitter terminal, and a resistor connected to each of the base terminals. When voltage is applied to the resistor and capacitor connected to the emitter, the capacitor will begin to charge. The size of the resistor will control the time it takes for the capacitor to charge. This time is referred to as the *time constant* for the circuit.

The same circuit voltage is applied to the resistors that are connected to base 2 and base 1. The resistors connected to base 2 and base 1 create a voltage drop with the internal resistance of the UJT. When the charge in the capacitor grows to a value that is larger than the voltage drop across the UJT and the resistor connected to the base 1 lead, a current path will be developed through the emitter and through the resistor that is connected to base 1. When current flows through this path, it creates a voltage pulse that increases to its maximum voltage very quickly. The waveform of this pulse is shown in the diagram, and you can see that it turns on and increases to its maximum value immediately. The duration of the pulse will depend on the internal resistance of the UJT and the ratio of the size of the base 2 and base 1 resistors.

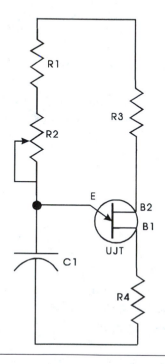

Figure 16–23 A UJT connected to an oscillator circuit to produce sharp output pulses. The size of the resistance and capacitance connected to the emitter controls the frequency of the pulses.

When the UJT is producing the pulse, the capacitor is discharging, and it will begin to charge up immediately and repeat the process as its voltage charge increases to a point where it is larger than the voltage across the UJT. At this point the UJT will produce another pulse. The time between pulses will be determined by the size of the resistance and capacitance of the resistor and capacitor that are connected to the emitter. If the amount of resistance is increased, the time for the capacitor to charge will increase and the time between pulses will increase. If the amount of resistance at the emitter is reduced, the capacitor will charge more quickly and create pulses that are grouped more closely together. This type of circuit is called an *oscillator* and the number of pulses the oscillator produces in 1 second is called the *oscillator frequency*.

16.20   Testing the UJT

The unijunction transistor can be tested like a PN junction. The UJT must be isolated from its circuit and you can test the PN junction between base 2 and emitter, and then test the PN junction between base 1 and emitter. Use an ohmmeter and switch the polarity of its probes so that you forward bias and reverse bias each junction. Remember that when you forward bias a PN junction it should have low resistance, and when you reverse bias a PN junction it should have high resistance.

16.21   The Silicon Controlled Rectifier

The silicon controlled rectifier (SCR) is made by combining four PN sections of material. Fig. 16–24a shows the PN-material for the SCR and Fig. 16–24b shows its electronic symbol. The terminals on the SCR are identified as anode, cathode, and gate. Since the SCR is basically a diode that is controlled by a gate, its symbol uses the arrow from the rectifier diode that you studied at the beginning of

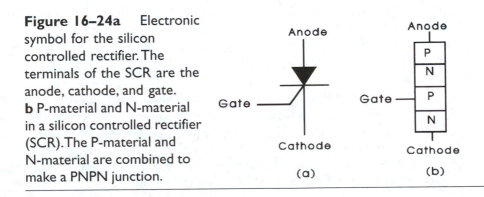

Figure 16–24a   Electronic symbol for the silicon controlled rectifier. The terminals of the SCR are the anode, cathode, and gate. **b** P-material and N-material in a silicon controlled rectifier (SCR). The P-material and N-material are combined to make a PNPN junction.

this chapter. When the SCR is turned on, it can conduct large amounts of DC voltage and current (over 1000 volts and 1000 amps) through its anode and cathode. The major difference between the SCR and the junction diode you learned about in Section 16.5 is that the junction diode is always able to pass current in one direction when the diode is forward biased. The SCR is forward biased by applying positive voltage to its anode and negative voltage to its cathode. At this point the SCR will still have high resistance between its anode-cathode junction. If a positive voltage pulse is applied to the SCR gate, the SCR's anode-cathode junction will have low resistance and the SCR will be in conduction. When the pulse is removed from the gate of the SCR, it will remain in conduction because positive current that comes through the anode will replace the voltage the gate provided. The only way to turn the SCR off is to provide reverse bias voltage to the anode-cathode, or reduce the current flowing through the anode-cathode to zero. You should remember that the AC sine wave has zero voltage right before it provides the negative half of its waveform. This means that if the SCR is powered with AC voltage, the SCR will be turned off when the AC waveform goes through zero volts and then to its negative half-cycle. When the AC voltage waveform goes positive again, a gate pulse can be provided and the SCR can go into conduction again. The gate provides a pulse that is used to cause the SCR to go into conduction.

16.22 Operation of the Silicon Controlled Rectifier

Fig. 16–25 shows an SCR connected in a circuit to control voltage to a DC load. The source voltage for this circuit is AC voltage. The main advantage of the SCR is that it will not go into conduction until it receives a pulse of voltage to its gate. The timing of the pulse can be controlled so that it can be delivered anytime during the halfcycle, which will control the amount of time the SCR will be in conduction. The amount of time the SCR is in conduction will control the amount of current that flows through the SCR to its load. If the SCR is turned on immediately during each half-cycle, it will conduct all of the half-wave DC voltage just like a normal diode rectifier. If the gate delays the point where the SCR turns on and goes into conduction at the 45° point of the half-wave, the amount of voltage and current the SCR conducts will be 50% of the fully applied voltage.

The other important feature of the SCR is that it will only go into conduction when its anode and cathode are forward biased. This means that if the supply voltage is AC, the SCR can only go into conduction during the positive half-cycle of the AC voltage. When the negative half-cycle occurs, the anode and cathode will be reverse biased and no current will flow. This means the SCR will automatically be turned off when the negative half of the AC sine wave oc-

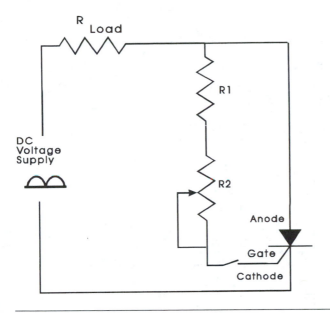

Figure 16–25   A silicon controlled rectifier shown in a circuit controlling DC voltage and current to its load.

curs. Since the positive half of the AC sine wave occurs for 180°, the SCR can only provide control of 0–180° of the total 360° AC sine wave.

It is also important to understand, that since the turn-off point of the SCR is fixed to the point where the sine wave begins to go negative, the SCR can only be controlled by adjusting the point where it is turned on. The point where the SCR is turned on and goes into conduction is called the *firing angle*. If the SCR is turned on at the 10° point in the AC sine wave, its firing angle is 10°. If the SCR is turned on at the 45° point, its firing angle is 45°. The number of degrees the SCR remains in conduction is called the *conduction angle*. If the SCR is turned on at the 10° point, its conduction angle will be 170°, which is the remainder of the 180° of the positive half of the AC sine wave.

16.23   Controlling the SCR

Fig. 16–26 shows an SCR in a circuit with a UJT connected to its gate. The load in this circuit is a DC motor. This type of DC motor is often used as a damper control motor. The circuit is powered by AC voltage, and the variable resistor in the oscillator (capacitor-resistor) circuit sets the timing for the pulse that is used to energize the gate of the SCR. You should notice that a diode rectifier provides pulsing DC voltage for the capacitor, which charges to set the timing for the

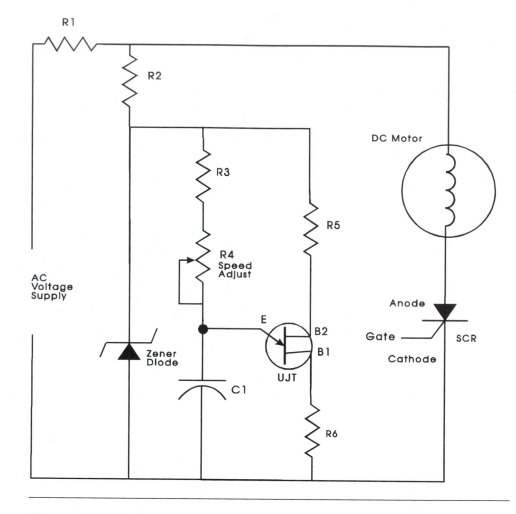

Figure 16–26   An SCR used to control a DC motor. A UJT is connected to the gate of the SCR to control its firing angle.

pulse that comes from the UJT. Since this DC voltage comes from the original AC supply voltage, it will have the same timing relationship of the original sine wave. This means adjusting the pulse from the UJT to turn on the SCR gate at just the right time to control the firing angle of the SCR from 0–180°. In reality the firing angle is usually controlled from 0–90°, which gives sufficient range of control to adjust the output DC voltage that is sent to the DC motor. You should remember that the speed of the DC motor can be controlled by adjusting the voltage sent to the armature and field. This diagram shows the waveform for the voltage at each point in this circuit. The load in this circuit could also be any other DC powered load.

16.24   Testing the SCR

You will need to test the SCR to determine if it is faulty. Since the SCR is made of PN junctions, you can use forward bias and reverse bias tests to determine if it is faulty. In this test you should put the positive probe on the anode and the negative probe on the cathode. At this point the ohmmeter will still indicate the SCR has high resistance. If you use a jumper wire and connect positive voltage from the anode to the gate, you should notice the SCR will go into conduction and have low resistance. The SCR will remain in conduction until the voltage applied to its anode and cathode is reverse biased, or until the voltage applied to the anode and cathode is reduced to zero. This means that you can turn the ohmmeter polarity switch to the opposite setting, or you can remove one of the probes and the SCR will stop conducting. It is important to understand that the amount of current to keep the SCR in conduction is approximately 4 to 6 milliamps. This means that some high-impedance digital volt/ohmmeters will not have enough current when set to the ohms range to keep the SCR in conduction. If this is the case, you may need to test the SCR with an analog ohmmeter. The analog ohmmeter is a type of ohmmeter that has a needle and scale. You should also test the SCR for reverse bias to ensure that it has high resistance. Sometimes an SCR will not go into conduction because it has developed an open in its anode-cathode circuit. Other SCRs may stay in conduction at all times, which means the SCR is shorted.

16.25   Typical SCRs

SCRs are available in a variety of packages and case styles that include three terminals, and larger types that have threads so that they can be mounted directly into heat sink material. The heat sink material will be made from metal and in some cases it will include a fan that moves air over fins that are molded into the material to help remove the large amounts of heat that build up into the devices. Fig. 16–27 shows examples of different types of SCRs.

16.26   The Diac

The unijunction transistor provides a positive pulse for the gate of the SCR that allows the SCR to control large DC voltages and currents. The *diac* is an electronic component that provides a positive and negative pulse used as a trigger

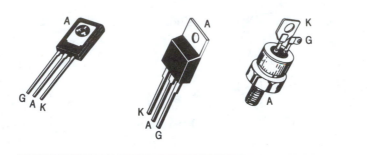

Figure 16–27   Typical SCRs shown in a variety of packages and case styles. *(Copyright of Motorola, used by permission)*

signal for a device called a triac. The triac is similar to the SCR except it can conduct voltage and current in both the positive and negative directions. This means that the triac provides a controlling function for AC voltage and current that is similar to the SCR. As an HVAC technician, you may not notice the diac because it is used as a device to produce a firing pulse for the triac. You will learn in the next section how the triac is used in fan speed controls and other AC voltage control circuits. Fig. 16–28 shows the electronic symbol for the diac. You should notice that the diac symbol consists of two arrows that show voltage pulses can be positive or negative. Since the diac symbol has two arrows, it can be shown in two different styles.

16.27   Operation of the Diac

When voltage is applied to a diac in a circuit, its PN junction will remain in a high-resistance state and block the voltage until the voltage level reaches the breakover level. This means that when AC sine wave voltage is applied to the circuit, the sine wave voltage will increase to its peak level, and return to zero, and then increase to negative peak voltage. Each time the voltage increases toward its peak, and it exceeds the breakover level of the diac, the diac will provide a sharp output pulse of the same polarity as the voltage that is supplied to it. For example, if the breakover voltage level is 18 volts, the diac will block

Figure 16–28   Electronic symbols for the diac. Notice the diac has two arrows, and it can be represented by either symbol.

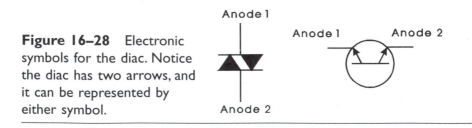

voltage until the AC voltage exceeds the 18 V level. At this point, the diac will produce a pulse that is approximately 18 volts. Fig. 16–29a shows the diac in its circuit and Fig. 16–29b shows the pulse that it creates.

16.28   The Triac

The *triac* is basically two SCRs that have been connected back to back in parallel so that one of the SCRs will conduct the positive part of an AC signal, and the other will conduct the negative part of an AC signal. As you know, the SCR can only control voltage and current in one direction, which means it is limited to DC circuits when it is used by itself. Since the triac acts like two SCRs that are connected inverse parallel, one section of the triac can control the positive half of the AC voltage and the other section of the triac can control the negative half of AC voltage. Fig. 16–30a shows the electronic symbol of the triac and Fig. 16–30b shows the arrangement of its P-type and N-type materials. The terminals of the triac are called main terminal 1 (MT1), main terminal 2 (MT2), and gate. Fig. 16–30c shows examples of typical triac semiconductor devices. Since the triac is basically two SCRs that are connected inverse parallel, MT1 and MT2 do not have any particular polarity.

16.29   Using the Triac as a Switch

The triac can be used in an HVAC circuit as a simple on-off switch. In this type of application, the MT1 and MT2 terminals are connected in series with the AC load. When the gate gets a positive pulse signal, the triac will turn on for the

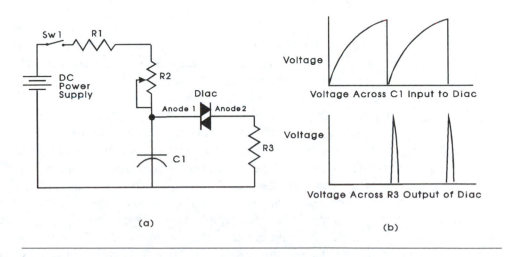

(a)   (b)

Figure 16–29a   Diac in a circuit. **b** The pulse the diac produces.

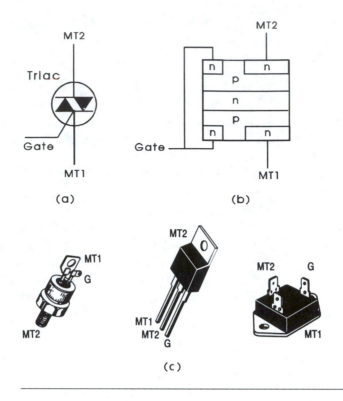

Figure 16–30a Electronic symbol for the triac. **b** P-type and N-type materials for the triac. **c** Typical triac semiconductor devices. *(Copyright of Motorola, used by permission)*

positive half of the AC cycle. When the AC voltage waveform returns from its positive peak to zero volts, the triac will turn off. Next the negative half-cycle of the AC voltage waveform will reach the triac, and it will receive a negative pulse on its gate and go into conduction again.

This means that the triac will look as though it turns on and stays on when AC voltage is applied. The load that is connected to the triac will receive the full AC sine wave just as if it were connected to a simple single-pole switch. The major difference is the triac switch can be used for millions of on-off switching cycles. The other advantage of the triac switch is that the gate pulse can be a very small amount of voltage and current. This allows the triac to be used in temperature control where the temperature-sensing part of a thermostat can be a small solid-state sensing element called a *thermistor.* The sensing element can also be a very narrow strip of mercury in a glass bulb that is very accurate, but can only carry a small amount of voltage or current.

Another useful switching application for a triac is the solid-state relay. Section 16.39 will explain AC and DC solid-state relays. Fig. 16–31 shows a triac in a thermostat connected to a 24 VAC gas valve in a furnace. The temperature-sensing element is connected to the gate terminal of the triac. When the temperature increases, it makes the mercury in the temperature-sensing element expand and make contact between the two metal terminals. The small amount of voltage flowing through the sensing element will be sufficient to cause the triac to go into conduction and provide voltage to the gas valve.

16.30   Using the Triac for Variable Voltage Control

The triac can also be used in variable voltage control circuits, since it can be turned on anytime during the positive or negative half-cycle, which is similar to the way the SCR is controlled for DC circuit applications. In this type of application, a diac is used to provide a positive and negative pulse that can be delayed from 0–180° to control the amount of current flowing through the triac. This type of circuit can be used to control the amount of current and voltage supplied to electric heating elements. This allows the amount of current and voltage to be controlled from zero to maximum, which in turn allows the temperature to be controlled very accurately. Fig. 16–32 shows a diagram of a triac used to control an electric heating element that is powered by an AC voltage source. Notice that a variable resistor is connected with a capacitor to provide an oscillator pulse to the diac. Since the resistor and capacitor are connected to an AC

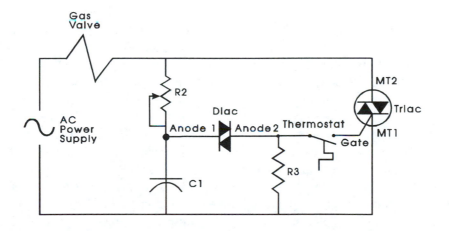

Figure 16–31   A triac uses a switch to turn on voltage to a gas valve for a furnace. The triac receives its gate signal from a small amount of voltage that moves through the temperature-sensing element.

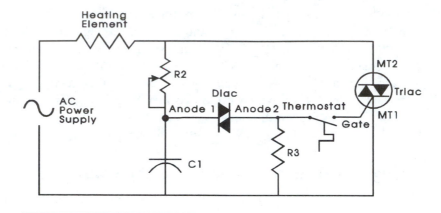

Figure 16–32   A triac used to control variable voltage to an AC electric heating element. Notice a diac is used to provide a positive and negative pulse to the triac gate.

voltage source, the pulse will be both positive and negative as the AC sine wave changes polarity. The triac is connected to the same AC voltage source so that the timing of the pulse from the diac to its gate will always be synchronized with the polarity of the voltage arriving at the main terminals of the triac.

16.31   Testing the Triac

Since the triac is made from P-type material and N-type material, it can be tested like other junction devices. The only point to remember is that since the triac is essentially two SCRs mounted inverse parallel to each other, some of the ohmmeter tests will not be affected by the polarity of the ohmmeter leads. In the first test for a triac an ohmmeter should be used to test the continuity between MT1 and MT2. When no gate pulse is present, the resistance between these terminals should be infinite regardless of which ohmmeter probe is placed on each terminal. Fig. 16–33 shows how the ohmmeter should be connected to the triac. In Fig. 16–33a you can see the positive ohmmeter probe is connected to MT1 and the negative probe is connected to MT2. Since no voltage is applied to the gate, the resistance should be infinite. When voltage from MT2 is jumped to the gate, the triac will go into conduction and the ohmmeter will indicate low resistance.

In Fig. 16–33b you can see that the positive ohmmeter probe is connected to MT2, and the negative probe is connected to MT1. When voltage from MT1 is jumped to the gate, the triac will go into conduction and the ohmmeter will indicate the resistance is low. It is important to remember that the triac will only stay in conduction while the gate signal is applied from the same voltage source

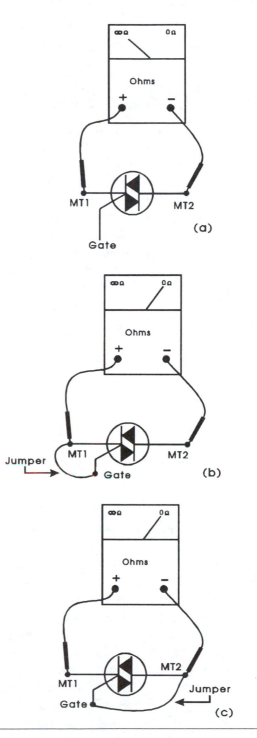

Figure 16–33a Testing the triac by placing the ohmmeter positive probe on MT2 and the negative probe on MT1. When gate voltage is applied from the MT2 probe, the triac goes into conduction. **b** Testing the triac by placing the ohmmeter positive probe on MT1 and the negative probe on MT2. When gate voltage is applied from the MT2 probe, the triac goes into conduction. **c** The jumper is placed between MT2 and gate.

as MT1. As soon as the voltage source is removed the triac will turn off. Fig. 16–33c shows the gate receiving voltage from MT2 and the triac is in conduction.

16.32   Operational Amplifiers

Operational amplifiers have become a common component in heating, air-conditioning, and refrigeration systems because they allow very small-signal voltages from sensors to be amplified to a usable voltage to turn compressors and other loads on and off. The *operational amplifier* is called an *op amp* and it is made of dozens of PN-junction devices such as transistors that are manufactured on a single integrated circuit (IC). As a technician, you will find them in most of the newer solid-state controls such as thermostats, voltage monitors, and other protection and control devices.

The op amp provides two basic functions. First, it can take a very small signal and amplify it. For example, a sensor can send a 100 mV signal to an op amp and the op amp can multiply the signal strength by ten times ($\times 10$) so that the output signal would be 1 volt, or the op amp could multiply the signal by 100 times ($\times 100$) so that the output signal would be 10 volts. The second function the op amp provides is that it can change the polarity of the voltage of the input signal or it can keep it at the same polarity. For example, if a negative sensor voltage is applied to the inverting input terminal of the op amp, the output signal would be positive. If the negative input signal is applied to the noninverting input, the output would remain negative. Each op amp input can also receive a signal that spans from -0.5 volt to $+0.5$ volt and produce an output voltage that spans from -12 volts to $+12$ volts. Fig. 16–34 shows the electronic symbol for an operational amplifier with all of its terminals identified.

In Fig. 16–34a you can see that the op amp has two input terminals identified as terminals 2 and 3, and an output terminal identified as terminal 6. Terminal 2 is identified by a negative sign ($-$), and it indicates this is the *inverting input*. A signal that the op amp receives on the inverting input will have its voltage polarity inverted when the signal reaches the output. Any signal the op amp receives on its noninverting input will remain the same polarity when it reaches the output. The op amp also has two terminals where power supply voltage is applied. It is important to understand that when the op amp amplifies a signal, the output signal cannot be any larger than the supply voltage that is connected to the op amp at its V+ (terminal 7) and V− (terminal 4) terminal. Typical power supply voltage can be 12, 15, or 18 volts DC. You should also notice that the op amp is mounted on an eight-pin integrated circuit, which means the chip will have eight pins. The chip has an identifier mark at the top of the chip to help identify the number 1 terminal.

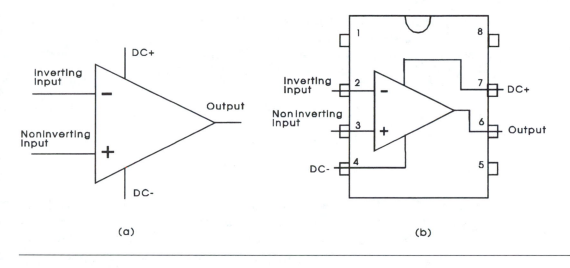

Figure 16–34a   Electronic symbol for an operational amplifier (op amp). **b** Pin outline of integrated circuit (IC) for an op amp.

16.33   The Op Amp as an Amplifier

The op amp has the ability like a transistor to amplify an input signal. The amount of amplification is called *gain*. The major difference between the transistor and the op amp is that the amount of amplification (gain) for the transistor is designed into it when the transistor is manufactured. In contrast, the amount of gain for the op amp is determined by the external resistors connected to it. This means that it is possible to design a circuit where an op amp can have any amount of amplification. The general values of amplification will be between 10 to 100.

Fig. 16–35 shows a typical op amp circuit. The resistor that is connected to the noninverting input is called the input resistor (R_{in}), and the resistor at the top

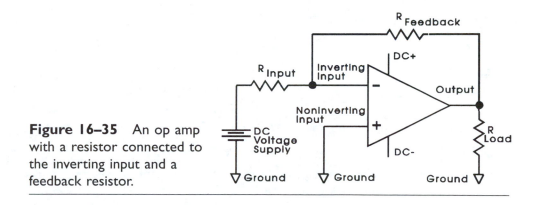

Figure 16–35   An op amp with a resistor connected to the inverting input and a feedback resistor.

of the op amp that is connected between the output terminal and the noninverting input is called the feedback resistor (R_F). When the input signal is connected to the inverting input, the op amp is called an inverting input. The amount of gain for the inverting op amp is determined by the formula:

$$Gain = \frac{R_F}{R_{in}}$$

The formula for the noninverting op amp is:

$$Gain = 1 + \frac{R_F}{R_{in}}$$

In this figure you can see that the resistor connected to the noninverting op amp is 10 kΩ and the feedback resistor is 100 kΩ. This means the gain is 10 for this circuit. If the input resistor is changed to 1 kΩ, the gain for the circuit would be 100. If a 0.1 V signal is sent to the input, the output would produce a 1 V signal if the gain is 10, and the output signal would be 10 volts if the gain is 100.

16.34   Controlling a Relay with an Op Amp

One circuit you will find in air-conditioning and refrigeration systems is a relay that is controlled by an op amp. This type of circuit allows a temperature sensor that produces a millivolt signal the ability to control a 10 hp, three-phase fan motor. Fig. 16–36 shows an example of this circuit. In this circuit the op amp receives a small signal from a temperature sensor. The coil of a small relay is connected to the output of the op amp. The contacts of the small relay control

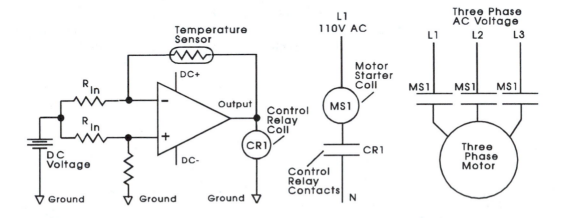

Figure 16–36 A millivolt signal from a temperature sensor is used to control a small relay, and the relay controls a motor starter and 10 hp, three-phase fan motor.

voltage to the coil of a motor starter and the three-phase motor are connected to the contacts of the three-phase motor starter. The gain of the op amp is 100.

When the temperature sensor sends a 0.1 V signal to the input of the op amp, the op amp increases the signal to 10 volts at the output, which sends voltage to the coil of the small relay. When the coil of the small relay is energized, its contacts provide voltage for the coil of the motor starter, and the contacts of the motor starter close and provide 240 VAC to the 10 hp fan motor. When the temperature drops, the voltage from the temperature sensor drops below 0.1 volt, and the small relay coil drops out. When this occurs, the motor starter coil becomes deenergized, and the fan motor turns off.

16.35   A Heating and Cooling Control System

As you know, the op amp has the ability to receive a signal that spans from negative to positive and to produce an output signal that spans from negative to positive. Fig. 16–37 shows a control circuit for a heating and cooling system.

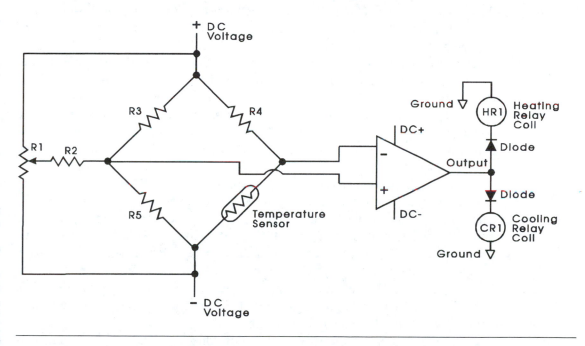

Figure 16–37 An RTD (resistive temperature device) is connected to a bridge circuit. The RTD acts as a temperature-sensing element. The signal from the bridge is sent to an op amp where it is amplified by X100. The coil of a heating relay and the coil of a cooling relay are connected to the output of the op amp. Diodes are used to control when the heating or cooling relay should be energized.

The thermostat for this system is a temperature-sensitive resistor called an RTD. The RTD is connected to three other resistors in a circuit called a bridge. The variable resistor in the bridge is used to set the desired temperature that controls when the air conditioning should turn on and when the heating should turn on. The desired temperature is called the setpoint for the system. For this example circuit, the variable resistor is set to turn on the air conditioning when the temperature is above 72° and turn on the heating when the temperature is below 72°. A dead band of +1° and −1° can be put into this control circuit by adjusting the offset at the op amp.

The bridge circuit allows very small changes in temperature to be converted to changes in voltage. The RTD senses changes in the temperature and it produces a different amount of resistance for every temperature. Since this change of resistance is fairly linear, the system's temperature can be controlled very accurately. When the temperature changes, the resistance in the RTD changes and the bridge will become unbalanced and produce a positive voltage if the temperature is higher than 72°, and a negative voltage if the temperature is lower than 72°. The amount of voltage will increase to a maximum of positive 0.15 volt when the temperature is 5° above the setpoint, and a maximum of −0.15 volt when the temperature is below setpoint.

The second part of the circuit is an op amp. You can see that the signal from the bridge circuit is connected to the inverting input of the op amp. The gain of the op amp is set for ×100 so the output signal will be +15 volts at its maximum and a −15 volts at its minimum. You should notice that two diodes are connected to the output terminal of the op amp. The coil of a cooling relay is connected to the top diode, and the coil of a heating relay is connected to the bottom diode.

When the thermostat senses the temperature is sufficiently above the setpoint, it sends a small positive voltage of 0.1 volt to the inverting input of the op amp. The op amp changes the positive voltage signal to a −10 V signal. Since the voltage is negative, it can only pass through the top diode and energize only the cooling relay coil. When the contacts of the cooling relay close, they energize a compressor motor and the system begins to cool the controlled space. When the temperature in the space cools down, the voltage from the bridge will decrease, and when the input voltage drops below 0.08 volt, the output signal will drop below 8 volts and the relay coil will drop out.

If the temperature in the space gets below the setpoint, a negative signal is sent by the bridge circuit to the input of the op amp. Since the op amp inverts the signal, the output signal will be positive when the temperature is below the setpoint. When the temperature is below the setpoint enough to produce a −0.1 volt signal, the output of the op amp will increase to a −10 volts, which is sufficient to energize the relay coil. Since the signal is positive this time, the current can only pass through the bottom diode to the coil of the heating relay. The contacts of the heating relay close and energize an electric heating coil.

When the controlled space begins to heat up, the signal from the bridge circuit begins to reduce, and the coil to the heating relay will become deenergized when the input signal drops below 0.08 volt. It is important to remember that the amount of positive or negative voltage must be larger than 0.1 volt to energize the coils, and when it drops below 0.08 volt, the coils will drop out.

16.36   Using Proportional Valves Instead of Relays

If the temperature control system must be more sensitive, the control devices can be *proportional valves* that control the flow of hot and cold water through a heat exchanger, rather than relays that turn on and off. A proportional valve is like the dimmer switch that controls a kitchen light, or like the volume control on a radio. When the valve receives a small amount of voltage, it opens slightly, and as the amount of voltage increases, the valve opens wider until it is fully open. The advantage of the proportional valve is that if the thermostat senses the temperature is slightly below the setpoint, the valve controlling the hot water can be opened slightly to allow a small amount of hot water to enter the heat exchanger. Since a fan blows air over the heat exchanger, the amount of Btus that will be moved will be minimal. If the temperature difference at the thermostat becomes larger, the proportional valve can be opened farther to allow more hot water to pass through the heat exchanger.

Fig. 16–37 shows a control circuit that is similar to the previous control circuit that uses an op amp to control relays. In this circuit, the op amps control proportional valves. Fig. 16–38 shows the operation of the valve compared to the voltages it receives. You can see that the valves will be fully open when they receive 10 volts, and they will be half open when they receive 5 volts. The valve will open 10% for every 1 volt it receives.

You should notice in the circuit that diodes are used again to ensure that the correct valve is operating when the signal is positive and when the signal is negative. The RTD and the bridge circuit operate the same as they did in the previous circuit. In this circuit when the temperature changes slightly, the amount of signal will be slight, but it will be amplified and sent to the correct proportional valve so that it can open wider or close down depending on whether the change of temperature is more or less than the previous value. The setpoint is used to determine when the signal will change from positive to negative, and this in turn will determine whether the signal is received by the heating valve or the cooling valve. The amount of dead band can be adjusted by changing the offset resistor on the op amp. Typically the dead band is 1° to 1.5°.

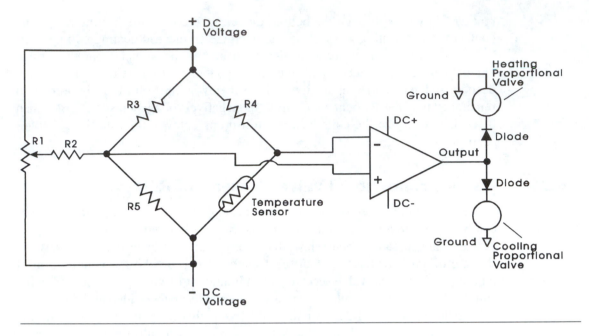

Figure 16–38   An op amp controlled heating and cooling system. This system uses proportional valves instead of relays to control when the system produces heating or cooling.

16.37   Controlling the Speed of a Three-Phase Motor with a Variable-Frequency Drive

The speed of a single-phase or three-phase AC induction motor can be controlled by adjusting the frequency of the voltage that is sent to it. In modern air-conditioning and refrigeration systems, it is important to be able to vary the speed of motors in the system to provide more efficient use of electricity and to provide a system that is able to control temperatures more precisely.

In the 1980s manufacturing technology for transistors and other solid-state devices changed so that transistors could be made that are able to control larger voltages and currents. Up to this time, triacs and SCRs were the devices that were used for controlling large voltages and currents. With the advent of larger transistors, the frequency of larger voltages could be manipulated to a point where the speed of three-phase and single-phase motors could be controlled by changing the frequency of their supply voltage.

Fig. 16–39 shows a typical frequency drive. The drive in the picture costs approximately $450 and it can control compressors or fans up to 1 hp. Larger drives can control motors up to 100 hp. In some air-conditioning systems the frequency

Figure 16–39 A typical
1 hp variable-frequency drive
amplifier. *(Courtesy of
Rockwell Automation's
Allen-Bradley Business)*

control board is built into the control cabinet with the other air-conditioning controls.

16.38   Operation of the Variable-Frequency Drive Circuit

Fig. 16–40 shows the circuit diagram for a typical variable-frequency drive. You should notice that the circuit consists of three sections. The first section is the *rectifier section* where a three-phase diode bridge rectifier changes the three-phase AC voltage to pulsing DC voltage. The second section is the *filter section*, where the pulsing DC voltage is smoothed to pure DC. The third section is the *transistor switching section*, which produces the three-phase AC voltage at the desired frequency.

In the rectifier section you can see that six diodes are connected in a bridge circuit to convert the three-phase 60 Hz voltage to DC voltage. The actual rectifier section is mounted on a single module and can be exchanged by removing the three input voltage wires and the two DC bus connections. Each diode in the module can still be tested for front-to-back resistance ratio with an ohmmeter just as if they were individual diodes. The output of the rectifier section is 12 half-wave pulses that are 60° apart. The output of the bridge rectifier is connected to two large copper conductors that are called the *DC bus*.

The filter section of the circuit consists of several capacitors that are connected in parallel with the DC bus and a large inductor that is connected in series with the DC bus. The capacitors charge and discharge in synchronization with the input voltage. This causes the half-wave signal to be converted to pure DC. The

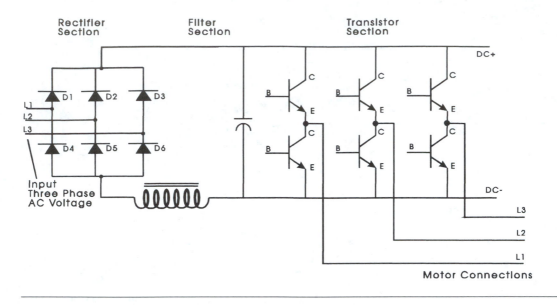

Figure 16–40　Circuit diagram of a variable-frequency drive. *(Courtesy of Rockwell Automation's Allen-Bradley Business)*

capacitors are used to filter the voltage part of the waveform, and the inductor is used to filter the current part of the waveform.

The transistor section consists of six transistors, two for each output phase. You should notice that one transistor of each phase is connected to the positive DC bus and the second transistor is connected to the negative DC bus. Each transistor is controlled by a base firing circuit that is controlled by a microprocessor chip (small computer). At the appropriate time each transistor is turned on in six distinct steps. Fig. 16–41 shows an example of the waveform that will result from turning the transistor on in six steps. You can see that the transistor that is connected to the positive DC bus will conduct and produce the positive part of the output AC waveform, and the transistor that is connected to the negative DC bus will produce the negative part of the output AC waveform. The first set of transistors produces the A phase for the three-phase output waveform, and the second set produces the B phase. The third set of transistors produces the C phase of the AC output waveform. You can see that the output frequency can be adjusted by changing the timing of the base firing circuit for each set of transistors. Typical control for a variable-frequency drive will be from 0 Hz to 120 Hz, which allows the motor to be controlled between 0 and 200% of its rated rpm. In reality the motor is generally adjusted from 60% rpm to 130% rpm, which provides sufficient control for the system that is being adjusted.

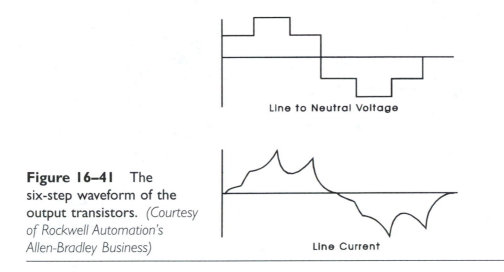

Line to Neutral Voltage

Figure 16–41 The six-step waveform of the output transistors. *(Courtesy of Rockwell Automation's Allen-Bradley Business)*

Line Current

You can see how important it would be to be able to adjust the speed of a very large fan to provide minimal air flow when the temperature conditions are near setpoint, and larger air flows when the system conditions warrant. The same is true of controlling the speed of a compressor motor. When the temperature is slightly above the setpoint, the compressor can turn more slowly and pump less refrigerant, and when the temperature is far above setpoint, the frequency of the voltage that is sent to the compressor motor can be increased to cause the motor to rotate more quickly and pump more refrigerant.

16.39   Solid-State Relays

Another solid-state device that you will see frequently used in HVAC systems is called a *solid-state relay (SSR)*. The solid-state relay uses DC voltages for the control part of the circuit that acts like the coil of a regular relay, and it can switch AC voltages through the part of the circuit that acts like the contacts. The solid-state relay uses SCRs, transistors, or triacs to control various loads. It operates similarly to an electromechanical relay in that it is energized by a small voltage which controls a larger voltage and current.

Fig. 16–42 shows a picture of a typical solid-state relay and it provides a block diagram of the operation of the relay. Fig. 16–43 shows the electrical diagram for the solid-state relay that uses a light-emitting diode (LED) and a phototransistor to accept the DC input, and a triac is used to control the AC voltage and load in the output. The LED and phototransistor are called an *optoisolation device* or an *optocoupler*. When a DC input signal is received at the SSR, it is converted to light through an LED. The LED and phototransistor are manufac-

Figure 16–42 A solid-state relay rated for DC input and 120 VAC output. *(Courtesy of Johnson Controls, Inc.)*

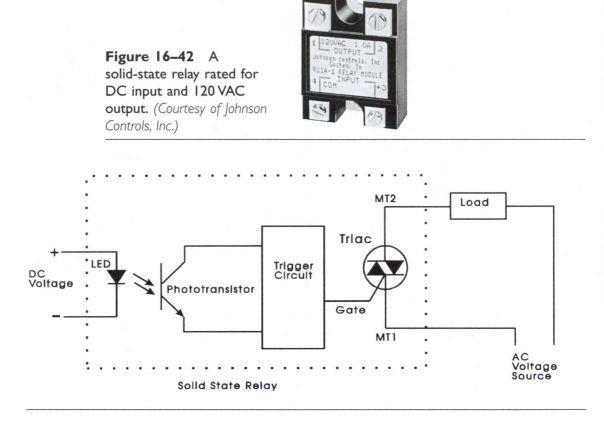

Figure 16–43 An electronic diagram of a solid-state relay that has a DC input signal and an AC output signal.

tured in an integrated circuit (IC) so that its light shines from the LED directly on a phototransistor. When the light strikes the phototransistor, its collector-emitter circuit goes to low resistance and the collector current flows through it to the base of the triac. The AC load is controlled directly by the triac. Notice that the power source for the load must be isolated from the input signal.

Questions for This Chapter

1. Explain P-type and N-type materials.
2. Explain the operation of a diode (PN) junction and show the input AC waveform and the output DC waveform.
3. Identify the terminals of a transistor and explain its operation.

4. Explain the operation of the SCR and the type of circuit where you would find one.
5. Explain the function of the unijunction transistor when it is used with an SCR.

True or False

1. The function of the triac is to provide switching similar to an SCR except the triac operates in an AC circuit.
2. The light-emitting diode (LED) is similar to a DC light bulb in that it has a very tiny filament.
3. The main function of diodes and SCRs is to convert AC voltage to DC voltage.
4. The variable-frequency drive has the ability to change the frequency of voltage sent to a motor so that it controls the motor's speed.
5. The UJT and diac are important solid-state devices that provide pulse signals to other components to use as a firing signal.

Multiple Choice

1. A PN junction is forward biased when _____
   a. positive voltage is applied to the N-material and negative voltage is applied to the P-material.
   b. negative voltage is applied to the N-material and negative voltage is applied to the P-material.
   c. its junction has high resistance.
2. The op amp is an important solid-state device because it _____
   a. can amplify very small signals from thermocouples and RTDs to larger voltages and currents.
   b. can provide a negative firing pulse to the gate of an SCR.
   c. can be used as a seven-segment display for numbers.
3. A circuit that has one SCR in it can _____
   a. control AC voltage.
   b. control DC voltage.
   c. control both AC and DC voltage.
4. The transistor operates like a relay in that _____
   a. its emitter is like the coil and its base and collector are like a set of contacts.
   b. its collector is like the coil and its base and emitter are like a set of contacts.

    c.   its base is like a coil and its emitter and collector are like a set of contacts.

5.   The solid-state relay _____

    a.   uses an LED to receive a signal and a transistor to switch current.
    b.   uses a capacitor to receive a signal and an IC to switch current.
    c.   uses a transistor to receive a signal and an IC to switch current.

Problems

1.   Draw the symbol for a diode and identify the anode and cathode.
2.   Draw the symbol for an NPN and a PNP transistor and identify the emitter, collector, and base.
3.   Draw the symbol for a silicon controlled rectifier (SCR) and identify the anode, cathode, and gate.
4.   Draw the electrical symbol for a unijunction transistor and identify base 1, base 2, and emitter.
5.   Draw the electrical symbol for a triac and identify MT1, MT2, and gate.
6.   Draw the electrical symbol for an op amp and identify the inverting input, the noninverting input, the output, and the feedback resistor.

17 Electrical Control of Heating Systems

OBJECTIVES:

After reading this chapter, you will be able to:
1. Identify the major electrical parts of an electrical furnace and explain their function.
2. Identify the major electrical parts of a gas heating system and explain their function.
3. Identify the major electrical parts of an oil heating system and explain their function.
4. Troubleshoot the electrical system of a gas, electric, and oil furnace.

17.0 Overview of Heating Systems

You will find different types of heating systems as you move to different parts of the country. For example, you may find electric heating coils, gas furnaces, or oil furnaces. Each of these heating systems uses similar controls to provide heat. This chapter will explain the operation of each type of heating system and provide diagrams that will help you to locate faults within each system.

All heating systems will have a means to provide heat, and controls to ensure the amount of heat does not exceed safe temperatures. All heating systems will

409

have a means to move the heat to areas where it is needed, such as a fan to pump air or a pump motor to pump water that is heated. If water is used as the transfer medium, the heating system will be called a boiler. If a fuel such as gas or oil is used to provide the heat, the furnace will need a control valve that allows the fuel to enter the area where it is burned. If the furnace uses oil, it will need a pump to move the fuel to the burner. As you learn about each system, you will see that the major parts of each system will be similar, and a few new components are added to each system to make their type of control somewhat unique.

17.1   Electric Heating Systems

Perhaps the simplest type of heating system is the electric heating system. Electric heating coils can act as the only source of heat for a system, or they can be an add-on for heat pumps or solar heating systems. Fig. 17–1 shows a picture and cut-away diagram of a typical electric furnace. From this diagram you can see that the furnace has a blower that moves air over two separate sets of electric heating coils. You will be able to identify all of the parts of the system and the electrical controls from the wiring diagram and ladder diagram. Fig. 17–2 shows the wiring diagram and Fig. 17–3 shows the ladder diagram for the furnace. The wiring diagram will be used to help you to locate the parts of the furnace and to identify wire colors and terminal points. The ladder diagram will show the system in the sequence of operation. Both diagrams will provide a legend to show the symbol and the abbreviation used to identify each of the parts.

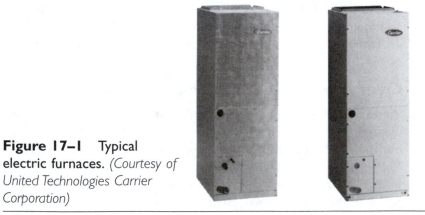

Figure 17–1   Typical **electric furnaces.** *(Courtesy of United Technologies Carrier Corporation)*

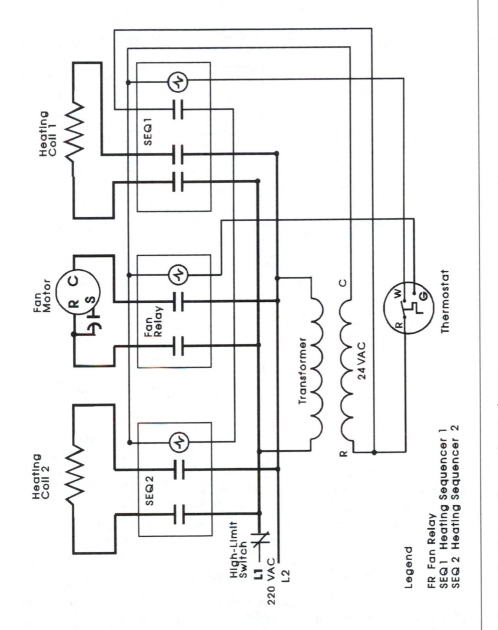

Figure 17–2 Wiring diagram for an electric furnace.

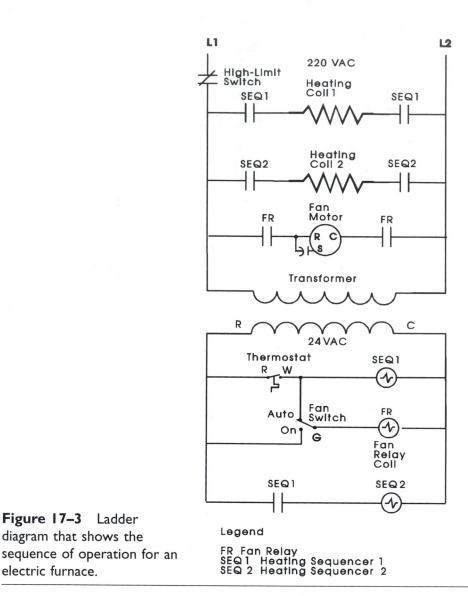

Figure 17–3 Ladder diagram that shows the sequence of operation for an electric furnace.

Legend

FR Fan Relay
SEQ 1 Heating Sequencer 1
SEQ 2 Heating Sequencer 2

17.2   The Main Parts of the System

In the electrical diagram you can see that the main parts of the electric heating system are the electric heating coils, the high-temperature limit control, and the fan control. Two heating contactors are used with sequencers to provide a means to stage the heating coils by turning one heating element on at a time. A thermostat is used to control the point where the heat is turned on or off during the first stage of heat, and the sequencer energizes the second stage on time delay.

17.3   Normal Operation

In normal operation the thermostat is mounted on an inside wall in the space that is to be heated, and it is set to a temperature such as 70°. When the temperature inside the space drops below 70°, the thermostat closes the circuit between the R and W contacts, which provides 24 volts to the coil of the first heating contactor HC1. When the coil of HC1 is energized, its contacts close and provide 230 VAC single-phase voltage to the first heating coil.

The current flowing through the heating coil produces a large amount of heat, and this heat causes the fan switch to close its contacts and send voltage to the furnace fan. The furnace fan in this diagram is rated for 230 VAC, but in some electric furnaces the fan can be powered with 120 VAC. When the furnace fan is energized, it turns a squirrel-cage blower that moves air through the duct work into the controlled space. When the temperature in the controlled space increases to the setpoint and exceeds it, the thermostat opens its contacts and the coil of the heating contactor is deenergized.

If the controlled space continued to get colder and its temperature dropped 2° below the setpoint, the thermostat would close a set of second-stage contacts, which would energize heating contactor HC2. The contacts of HC2 would close and provide 230 VAC to a second set of heating coils, which would double the amount of heat provided for the room. As the temperature in the controlled space increases back to setpoint, the second-stage heating would drop out when the temperature difference at the thermostat became less than 2°, and when the temperature reached the setpoint both of the heating contactors would be deenergized. In some cases, both heating contactors are controlled by the first stage of the thermostat, which makes the two heating coils act as one larger coil. If the temperature in the furnace became too hot, the high-limit control would open its contacts and cause the voltage to the heating contactors to be interrupted. In most cases, the only way the temperature can get too hot is if the furnace fan fails to come on and remove the heat from the furnace. When this occurs, the high limit provides protection against overheating.

17.4   Troubleshooting the Electric Heating Furnace

If the electric heating system fails to provide heat, you will need to troubleshoot it to locate the source of the fault. You will see that in this simple system either the wiring diagram or the ladder diagram can be used to trace wires and locate faults. As systems become more complex, you will find that the wiring diagram will be more difficult to follow for troubleshooting purposes, and that it will be used primarily to locate parts and to identify wire colors and terminals. The ladder diagram will always provide the sequence of operation.

When the system does not work properly, you will need to conduct several voltage tests to determine if the fault exists in the wiring, controls, or the major load of the circuit. For example, if a wire between the thermostat and the coil of heating contactor HC1 is broken, voltage cannot get to the coil and energize the HC1 coil. Another type of problem could occur if the thermostat is faulty so that voltage could not pass through its contacts to the coil of HC1. Yet another problem could be caused if the coil of HC1 is open, or if an open develops in the main heating coils themselves. In each of these conditions, you can see the result is the same in that the furnace would not produce any heat, and you will need to trace the circuit and test the wires and components to locate the fault.

The first step you should perform when you are asked to troubleshoot a system is to try and make the system run. You must remember that the person setting the thermostat or controlling the furnace may not understand how to set the thermostat and may not understand that the circuit breaker for the furnace must be turned on for the system to operate properly. By trying to make the system run, you may find that no fault actually exists, and that the problem is in the way the furnace was being operated. If the furnace does not operate when you try to make it operate, you will need to use a voltmeter to take several measurements. The check lists in the following sections will provide a detailed picture of how far voltage is moving through the system and where the fault exists. You will find by making these voltage measurements that the fault will be in the wires and controls or that the motor or heating coils are faulty.

17.5   Troubleshooting the Heating Coils

Use the following tests to determine if voltage is reaching the main heating coils. You can also use a clamp-on ammeter to take a current test to determine the amount of current flowing through the main coils.

1. Test at L1 and L2 for 230 VAC _____
2. Test across the primary terminals of transformer _____
3. Test for 24 volts at R to C (secondary of transformer) _____
4. Test from W to C for 24 volts _____
5. Test across coil terminals of HC1 for 24 volts _____
6. Test from T1 on HC1 to L2 for 230 volts _____
7. Test from T2 on HC1 to L1 for 230 volts _____
8. Test directly across heating coil for 230 volts _____

The reason you should write the amount of voltage down for each test point is that you can look at the list and basically determine the location of the fault by determining the last point that has voltage and the next point where voltage is lost, and this will be the area where the fault exists.

If the voltage at test 1 is not 230 volts, it indicates that the supply voltage to the system has a fault. Locate the source of voltage for the system and check to see that the circuit breakers are not tripped. If you have voltage at test point 1 but not at test point 2, it means that you lost voltage in the wires between the primary terminal of the transformer and the point where voltage comes into the system at L1 and L2. You should notice in the wiring diagram that after voltage is provided to the incoming terminal lugs, it is sent immediately to the L1 and L2 terminals of HC1.

If you have voltage at the first two test points and do not at test point 3, you have determined that the transformer is faulty. If no voltage is present at the secondary terminals of the transformer, you can turn power off to the system and remove the transformer and test it for continuity. You can also refer to earlier chapters in this text that explain the operation of the transformer in more detail. If voltage is present at test point 3, and you do not have voltage at test point 4, you should suspect the thermostat is not set correctly, or that it is faulty. You can test the thermostat in depth using the information provided in Chapter 20. If voltage is present at test point 4, and you do not have voltage at test point 5, you should suspect the wiring between the W and C terminals on the terminal board and the point where the wires connect to the coil of HC1.

Voltage tests at test points 5, 6, and 7 should help you to determine if the coil of HC1 is energized and if voltage is passing through the contacts. If voltage is present at the coil and the contacts do not close, you should suspect the coil is faulty. You can turn power off to the system and remove HC1 and test the coil for continuity. If the coil causes the contacts to close, but the contacts do not provide voltage at test point 6, or test point 7, you should suspect faulty contacts in the heating contactor. You can visually inspect these contacts, or turn off power to the system and isolate the contacts so that you can test them for continuity. If you have 230 VAC at test point 8 and the heating coils do not produce any heat, you should suspect the heating elements are open, and you can turn off all power to the system and test them for continuity. As you locate any faulty parts, you will need to change them out and continue testing the system until it operates correctly.

17.6   Testing the Fan Motor

If the heating element produces heat but the fan motor does not turn to move the air, you must test the motor circuit for faults to determine if the problem is in the wiring, controls, or the fan motor itself. If the fan does not operate, use the following test points to determine the location of the fault.

1. Test L1 to L2 for 230 volts _____
2. Test T1 on fan switch to L2 _____

Note: Temperature must be high enough to cause contacts of fan switch to close.

3. Test T2 on fan switch to L2 _____
4. Test across the fan motor _____

You can see that if you have voltage at all four steps, the circuit wires are good and the fan control is working properly, and you should suspect the problem is in the fan motor. If you have lost voltage at any point along the test, you can suspect the point between the last place voltage is measured and where it is lost. Be sure to turn off voltage to the furnace and isolate any components that you test for continuity. It is also important to understand that some electric furnaces use a relay to control the fan in addition to the fan limit. The relay allows smaller voltage and current to be controlled by the fan switch, and the 230 volts for the fan is controlled by the relay contacts. You should also remember to test the fan for bearing failure, which may cause the fan motor to draw excessive current.

17.7   The Gas Heating System

The main difference between the electric heating system and a gas heating system is that a gas heating system uses a gas valve that is controlled by the thermostat to control the flow of gas to the burners. The gas valve has an ignition system that is interlocked with it to provide the source of ignition for the fuel. The ignition system has controls to lock out the gas valve if a problem occurs and the fuel does not ignite correctly. You can review Chapter 14 if you need more information. The gas heating system has a fan and limit control that is similar to the electric furnace. Fig. 17–4a shows a picture and Fig. 17–4b shows a cut-away diagram of the gas furnace. This diagram allows you to identify the main parts of the system. The furnace in this system is an up-flow furnace which means warm air is moved out of the top of the furnace. If the furnace is a counterflow furnace, the warm air is moved out of the bottom of the furnace. An up-flow furnace is typically located in a basement, and a counterflow furnace is typically located on the first floor or main floor of a building.

It should be pointed out at this time that the gas furnace was a fairly simple system before controls were added to make the system more efficient. The example in this chapter is a high-efficiency furnace that has a combustion blower and a more complex ignition system.

17.8   The Main Parts of the Gas Furnace

The main parts of the gas furnace are the heat exchanger (combustion chamber) where the gas fuel is burned, the gas valve and ignition controls, the combustion blower that provides combustion air to and removes exhaust gases from the com-

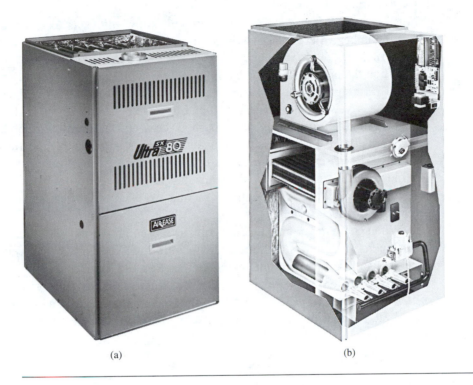

(a)   (b)

Figure 17–4a   A gas furnace. **b** Cut-away view of a gas furnace. *(Courtesy of Armstrong Air Conditioning, Inc.)*

bustion chamber, the furnace fan, and fan controls. The furnace fan is also used as the evaporator fan if air conditioning is added to the furnace. The furnace also has an air pressure switch that ensures the combustion blower is operating, and a high-temperature limit switch.

17.9   Normal Operation of the Gas Furnace

In normal operation, the thermostat is set for a specific temperature and it will provide voltage to the furnace when the temperature drops below the setpoint. When the furnace receives the signal voltage from the thermostat at terminals W and C, it causes the ignition control to begin the ignition process. The ignition process begins with the combustion blower beginning to provide air to the combustion chamber and to begin removing air from the combustion chamber. When air flow is established, it will cause the air pressure differential switch to close and allow fuel to flow to the pilot portion of the system. At the same time a hot surface igniter (also called a glow plug) receives current so that it can heat up to a point where it glows red hot. After the igniter is allowed to become completely hot for several seconds, the gas valve will allow pilot gas to enter the combustion chamber where it comes into contact with the igniter plug and ignites the pilot fuel.

When the pilot light ignites, a flame sensor will detect the presence of a flame and send a signal back to the integrated furnace control, which will send a signal to the gas valve that allows it to open completely and allows fuel to enter the combustion chamber where it is burned, creating large amounts of heat. When the temperature in the heat exchanger increases sufficiently, the furnace fan is energized to circulate room air across the heat exchanger. In most furnaces the furnace fan is energized by a time delay rather than by temperature. After the fan is energized, it will continue to operate and move heat from the furnace through the duct system.

When the temperature in the room increases above the setpoint, the thermostat opens its contacts and deenergizes the burner controls, which causes the gas valve to close and extinguish the flame, and the furnace begins a cool-down process. The furnace fan continues to run on a time-delay circuit for several minutes to move the remaining heat from the heat exchanger. When the time delay has expired, the control turns off the furnace fan and the cycle is ready to start again.

17.10   Troubleshooting the Gas Furnace

Fig. 17–5a shows a wiring diagram and Fig. 17–5b shows a ladder diagram of a simple standing pilot gas furnace. Fig. 17–6 shows a wiring diagram and ladder diagram of a more complex high-efficiency gas furnace. The wiring diagrams show the location of all of the terminal boards in the furnace and the color of wires connected to each terminal. You should also notice that each terminal is numbered, and all of the components are identified. You should also notice that it will be rather difficult to troubleshoot this circuit using the wiring diagram by itself.

The ladder diagrams show the high-voltage circuit (110 VAC) with the main loads (furnace fan and combustion blower) at the top of the diagram and the control circuit with the ignition, fan control, and gas valve at the bottom of the diagram in the 24 V circuit. The high-temperature limit switch and air pressure switch are also shown in the control circuit wired in series with the gas valve so that if they open for a fault condition, the gas valve will be deenergized immediately.

17.11   Troubleshooting the Control Circuit of a Gas Furnace

When you are called to troubleshoot a gas furnace, the first thing you should try to do is to set the thermostat at a high temperature that is well above room temperature and try to operate the furnace. When the thermostat closes its heating

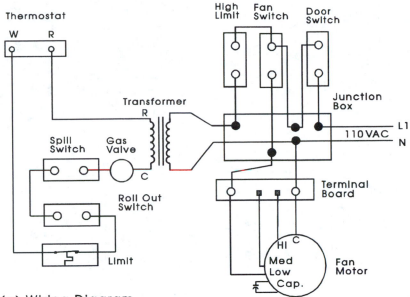

(a) Wiring Diagram

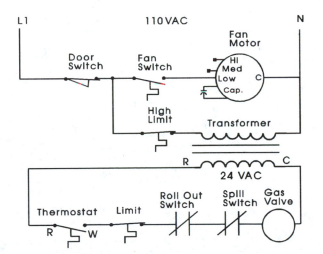

(b) Ladder Diagram

Figure 17–5a   A wiring diagram for a simple standing pilot gas furnace. **b** A ladder diagram for a standing pilot gas furnace.

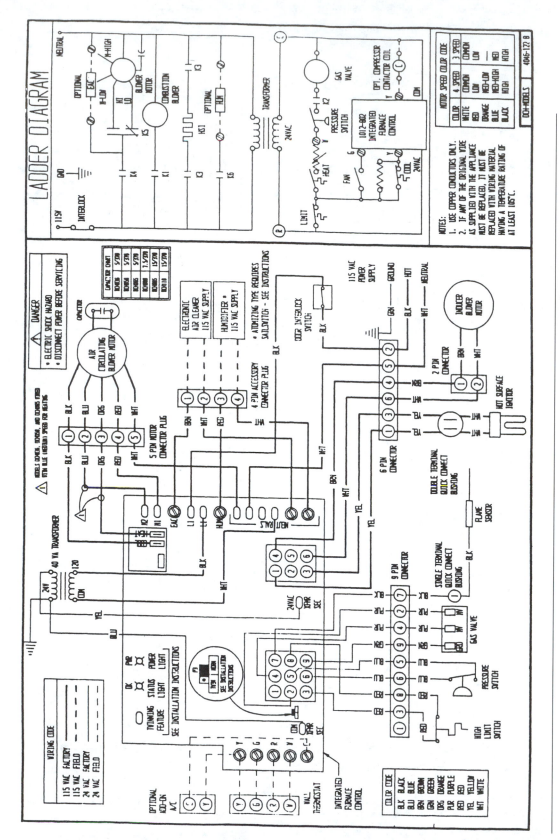

Figure 17–6 A wiring diagram and ladder diagram of a high-efficiency gas furnace. (*Courtesy of Bard Manufacturing Co.*)

circuit (R-W), the furnace should begin its ignition process and you should hear the combustion blower begin to operate. If the furnace does not operate correctly, you can break the problems down into two basic problems: problems with the fuel system, and problems with the furnace fan.

If the pilot did not ignite and the main gas valve did not open, you can begin the troubleshooting process by taking the following voltage measurements. Refer to the diagram in Fig. 17–6.

1. Terminals R to C at the terminal board of integrated furnace control for 24 VAC _____
2. Terminal C to terminal 8 of terminal board for integrated furnace control for 24 VAC to test to see if high-limit switch closed _____
3. Terminals W to C at the terminal board of the integrated furnace control for 24 VAC to test if contacts of the thermostat are closed _____
4. Terminal C to terminal 5 at the terminal board of the integrated furnace control for 24 VAC to see if the air pressure switch is closed
5. Terminal C to terminal K2b on the K2 relay for 24 VAC _____
6. Terminals 1 to 3 on the six-pin connector for 110 VAC going to the hot surface igniter (HSI), which is the glow plug _____
7. Terminals 4 to 2 on the terminal board of the integrated furnace control for the pilot signal to the gas valve for 24 VAC _____
8. After the pilot has been established, terminals 4 to 2 on the terminal board of the integrated furnace control for 24 VAC to main gas valve

   _____
9. Across the two terminals of the gas valve identified as ground and MV

   _____

If voltage is not present at values indicated at each point, you can begin to determine the problem by identifying the last test that has the proper amount of voltage and the next test point where voltage is lost. For example, if you do not have 24 at the first test, it indicates the transformer is not providing the 24 VAC used as the supply of control voltage. This means that you should test for 110 VAC at the black and white wires connecting the transformer primary terminals at L1 and neutral on the integrated furnace control board. If 110 VAC is not present at these terminals, you will need to test for L1 voltage through the door switch. You can do this by connecting one voltmeter terminal to the neutral terminal on the integrated furnace control board and the other terminal to T2 on the six-pin connector. If you do not measure 110 VAC, either the door is not closed correctly, or the door switch is faulty. Another possibility is that the main 110 VAC power to the furnace at terminals 5 and 2 on the six-pin connector do not have power and you will need to check the supply voltage to the furnace to see why it is lost. (For example, the circuit breaker is open, or a fuse is blown.)

If you have voltage at test point 1, and you do not have voltage at test point 2, you should suspect the high-limit switch is faulty and you can turn off power and isolate the switch by removing its wire from terminal 8 and testing it for continuity. If the high limit is open, be sure to check if a high temperature has occurred, or if the switch has failed. If a high temperature has occurred, be sure to fix the cause. If the switch is faulty, change the switch. It is important to understand that if the high limit is open because of an overtemperature condition, you will not solve the problem by changing the switch. You must fix the problem that is causing the overtemperature condition.

If you have voltage at test point 2, but you do not have voltage at test point 3, you should suspect the thermostat. Be sure it is set to heating, and that the setpoint is significantly above the room temperature. If you still do not have voltage at test point 3, you should remove the thermostat and test the R-W circuit for continuity. You can also visually inspect the thermostat to see if the heat anticipator resistor is burnt open. If you cannot get continuity through the thermostat, you should change it and try to start the furnace again.

If you do not have voltage at test point 4, you should suspect the air flow switch. It is important to understand that the air flow switch is measuring the difference of air pressure at two different points in the combustion air circuit. If the combustion fan is not moving sufficient air, the air switch contacts will be open. You must correct any problems that prohibit the proper amount of air flow to move. It is also important to understand that the small plastic tubing that connects the switch to the two test points in the combustion air circuit may become clogged so that the switch cannot sample the differential pressure correctly. You should also visually inspect the intake air pipe and the exhaust air pipe to see that they are clear. It is possible for them to be partially clogged with snow or a small animal or debris may obstruct the pipes.

If you do not have voltage at test point 5, you should suspect the K2 relay is not closed. The K2 relay will close when a pilot flame is detected. Notice that the coil of the K2 relay is not shown in either the wiring diagram or ladder diagram because it is part of the flame detection safety circuit. You can only test for voltage through its contacts. If they are not closed, you should suspect the flame detector. Be sure to look into the combustion chamber and visually confirm the pilot flame has been established.

If you do not have a pilot flame, you can test for the 110 VAC at the hot surface igniter (glow plug) at test point 6, and you should also visually confirm that the glow plug is glowing red. If it has voltage and is not glowing, you should remove voltage from the igniter and test it for continuity. If it is open, you will need to change it. If you do not have voltage at test point 7 or test point 8, and you have a pilot flame established and voltage at all of the other six test points, you can suspect the integrated furnace control and remove and replace it with a new one. If you have 24 VAC at both coils of the gas valve and it will not open, you can suspect the gas valve is faulty. You should return to Chapter 14 and review the operation and troubleshooting of the gas valve. Remove and replace the gas valve if needed and try to operate the furnace again.

17.12   Troubleshooting the Combustion Blower and Furnace Fan

The combustion blower and the furnace fan are both powered by 110 VAC. Each motor is controlled by the contacts of a relay. The furnace fan is a PSC motor and it is controlled by the fan relay K4 and the speed relay K5 determines if the fan runs at high speed or low speed. The coil of the fan relay is energized by the R-G circuit in the thermostat and by voltage from terminals G and C on the terminal board of the integrated furnace control. The combustion blower is also called the induced blower motor in the wiring diagram. It receives voltage from the six-pin connector. When you find that the furnace fan will not turn on, you can uses steps 1 to 4 to troubleshoot the fault. If you find the combustion blower will not turn on, you can use steps 1 to 3 of the second set of measurements to troubleshoot it.

Test points for the furnace fan:

1. Terminal L1 to N for 110 VAC _____
2. Terminal K4b to N for 110 VAC _____
3. Terminal 1 (black wire) to N for cooling for 110 VAC _____
4. Terminal 3 (orange wire) to N for heating for 110 VAC _____

If you do not have 110 VAC at L1 and N, you should test the incoming voltage at the fuse box or circuit breaker. You should also test to see that the door switch is operating correctly. If you have voltage at test point 1 but not at test point 2, you should test to see that the fan relay coil is energized by testing for 24 VAC at terminals G-C. If the coil voltage is present, you should suspect the fan relay contacts or coil is faulty and replace the fan relay. If you have voltage at test point 2 but not at test point 3 when you are in the cooling mode, or at test point 4 when you are in the heating mode, you should be sure to test for voltage directly across the red and common terminal for low speed, or across orange and common for medium speed, or across black and common for high speed. If you have voltage at any of these points, the motor should run. If you do not have voltage at any of these points, you have lost voltage between the fan relay contacts and the motor itself. This means one of the wires or one of the contacts on the relay is faulty.

Test points for combustion blower:

1. Terminal L1 to N for 110 VAC _____
2. Terminal K1b to N for 110 VAC _____
3. Terminal 1 to terminal 2 (brown and white wire) on blower for 110 VAC _____

If you do not have voltage at test point 1, use the same inspections as for step 1 for the furnace fan. If you have voltage at test point 1 but not at test point 2, you should suspect the combustion blower relay is not pulling in or the contacts are faulty. Be sure that the thermostat is calling the furnace for heat by sending a 24 VAC signal to terminals W-C on the furnace terminal board. If voltage is not present at test point 3, it indicates that wires between the combustion blower relay and the fan terminals are faulty, causing the loss of voltage. If voltage is present at the motor terminals and the motor is not running, you should suspect the motor. Remove power and isolate the motor leads so that it can be tested for continuity. Replace the motor if it is found to be faulty. After you have found the problem and corrected it, you should operate the furnace through several cycles to ensure that you have fixed the problem.

17.13   The Oil Heating System

The main difference between the oil heating system and the gas heating system is that in the oil heating system oil must be pumped from a storage tank to the nozzle in the end of the oil burner under high pressure to ensure the oil is atomized as it leaves the nozzle. When the oil atomizes, it becomes easy to ignite and burn. Fig. 17–7a shows a picture and Fig. 17–7b shows a cut-away diagram of a typical oil furnace. From this diagram you can see the main parts of the oil furnace are the furnace fan, the primary ignition control, the oil pump motor, the transformer, the fan and limit switch, and the high limit.

17.14   Normal Operation of the Oil Furnace

Fig. 17–8 shows a wiring diagram of a typical oil furnace and Fig. 17–9 shows a ladder diagram of the oil furnace. When the oil furnace is operating correctly, it will receive a 24 VAC signal from the thermostat on terminals W-C. This signal will cause the primary control for the oil burner to become energized and it will cause the pump to pressurize the oil to the nozzle and cause a continuous high-voltage spark for ignition at the tip of the burner. When the spark ignites the atomized oil, it will create a high-temperature flame in the heat exchanger. The increase in heat causes the fan switch to close and energize the furnace fan. When the thermostat is satisfied, it will open its contacts and deenergize the furnace

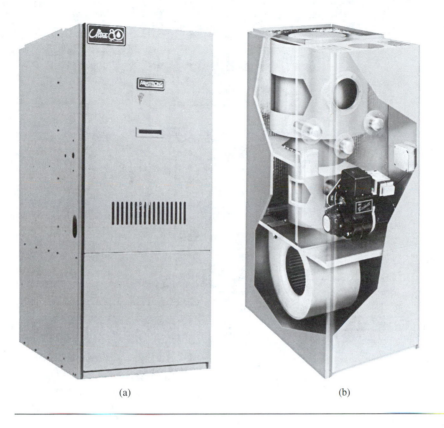

(a)   (b)

Figure 17–7a   A picture of an oil furnace. **b** Cut-away diagram of an oil furnace.   *(Courtesy of Armstrong Air Conditioning Inc.)*

primary control. If the furnace develops an overtemperature problem, it will cause the high limit to open and deenergize the primary control.

17.15   Troubleshooting the Combustion Control of an Oil Furnace

If the oil furnace does not operate correctly, you will need to make several voltage measurements to determine the location of the fault. Make the following voltage measurements and record the amount of voltage.

1. L1 to N for 110 VAC _____
2. T1 (black wire side) of limit switch inside fan/limit control to N for 110 VAC _____
3. T1 (yellow wire side) of upper limit control to N for 110 VAC _____
4. T1 (black wire side) of control transformer to N for 110 VAC _____

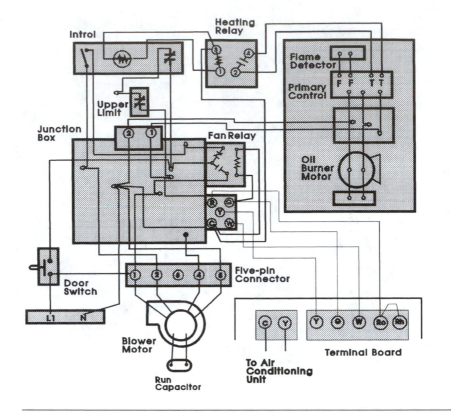

Figure 17–8   Wiring diagram of an oil furnace.

5.   T1 (black wire side) of burner assembly to N for 110 VAC _____
6.   Terminals R to C on secondary of transformer for 24 VAC _____
7.   Terminals W-C furnace terminal board for 24 VAC _____
8.   Terminals 1–3 on the heating relay for 24 VAC _____
9.   Black wire to white wire on oil burner pump for 110 VAC _____

If you do not have voltage at L1 to N at test point 1, you do not have incoming voltage to the oil furnace and you need to check the power source for any fuse or circuit breaker that may have tripped or be disconnected. If you have voltage at test point 1 but not at test point 2, you should suspect the door switch. Turn power off and test for continuity in the door switch and the two wires connected to it. If you have voltage at test point 2 but not at test point 3, the limit switch part of the fan/limit control or the two wires connected to it are open. Turn power off and isolate the switch and wires and test them for continuity. If the limit is open, it may be functioning correctly to protect the furnace against

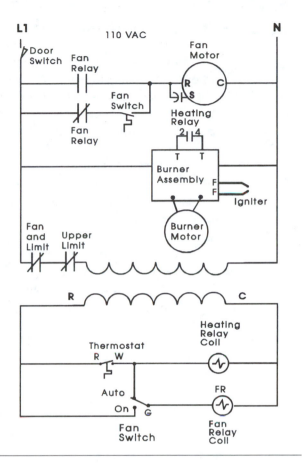

Figure 17–9 Ladder diagram of an oil furnace.

an overtemperature condition. If the furnace is overheated, you must find the source of the problem and correct it. For example, the furnace fan may not be operational and it is causing the furnace to overheat. If the furnace is cool and the limit is open, it may be defective and you should change it. If you have voltage at test point 3 but not at test point 4, you should suspect the upper limit control. Again, if the furnace is overheated, this limit switch will be open as well as the limit switch in the fan/limit control. If the furnace is cool, you should suspect the upper limit or the wires connected to it of being open. Turn off power and test them for continuity.

If you have voltage at test point 4 but not at test point 5, you should suspect the black and white wire that connects the burner assembly to the main furnace controls. Since these wires are connected to the other wires in the furnace controls with wire nuts, it is possible that you have a loose connection. You can also turn off power and test the wires for continuity. If you do not have voltage at terminal 6, you should suspect the transformer of being faulty. If the voltage at test point 4 indicates you have voltage at the primary side of the transformer, you should also have 24 VAC on the secondary side or the transformer may have an open circuit. Turn off power and test the transformer for continuity.

If you have 24 VAC at test point 6, but you do not have 24 VAC at test point 7, you should suspect the room thermostat is set incorrectly or it has an open in the heating circuit or the two wires connected to it. If you have voltage at test point 7 but not at test point 8, you should suspect the low-voltage wires between the 24 V terminal board on the furnace and terminals of the heating relay. Turn off power and test these two wires for continuity. If you have voltage at test point 8, but not at test point 9, you should suspect the coil or contacts of the heating relay. You can turn off power and test the coil and contacts of this relay for continuity.

If you have voltage at test 8 but not at test point 9, you should suspect the wires connecting to the heating relay or the contacts of the heating relay. If you have voltage at test point 9 and the pump motor and ignition probes do not provide a spark, you will need to troubleshoot the burner assembly itself. You may need to replace the pump motor or high-voltage transformer that provides the spark. If it is running and the fuel does not ignite, you may have problems with getting fuel to the nozzle and in getting a good spray pattern to ignite properly.

17.16   Troubleshooting the Oil Furnace Fan

If the burner control operates but the furnace fan does not turn on, you will need to troubleshoot the furnace fan. Use the following test points to help you locate the problem.

1. L1 to N at the voltage terminal board for 110 VAC _____
2. Terminals G-C on the low-voltage terminal board for 24 VAC _____
3. Terminal T2 (brown wire) at the fan relay to N for 110 VAC _____
4. Terminal T2 (red wire) at fan switch to N for 110 VAC _____
5. Terminals 2–5 on the furnace fan terminal board for 110 VAC _____
6. Terminal T3 (red wire) at fan relay to N for 110 VAC _____
7. Terminals 1–5 on the furnace fan terminal board for 110 VAC _____

If you do not have voltage at test point 1, use the same troubleshooting procedure as in the first set of tests since this indicates you have lost incoming voltage. If you have voltage at test point 1, you can continue to test point 2. If you have voltage at test point 2, it indicates the thermostat is sending a signal to the furnace to operate the furnace fan at high speed. If you do not have voltage at test point 2, the heating relay will allow voltage to pass through its normally closed contacts to the fan switch part of the fan/limit switch. This is considered the heating mode for the fan and it will run at low speed when the fan switch closes as the furnace heats up. You can continue to follow the voltage that makes the fan run at low speed.

If voltage is available at test point 3, it means voltage is available to the fan switch. If voltage is not available at test point 4, you must determine if the temperature in the heat exchanger is warm enough to cause the fan switch to close.

If voltage is available at test point 4 but not at test point 5, you should suspect the fan switch or the wires connected to it. If you have voltage at test point 5 and the motor does not run, you should suspect the motor has an open and you can turn off power and test the motor for continuity. You should recognize the motor as a PSC motor.

If the thermostat turns on the furnace fan by sending a 24 VAC signal to terminals G-C on the furnace control board, the fan relay should energize and turn the fan on for high speed. This would be used if the furnace fan is used as the evaporator fan when an air-conditioning system is added to the furnace. If you do not have voltage at test point 2, and the furnace fan switch is set to the *on* position, you should suspect the transformer secondary voltage (24 VAC) or the thermostat fan switch or the wires between the thermostat and fan relay (R wire for hot and G wire for fan). If you want the fan to run in high speed, you can omit test points 3 to 6 because these are part of the fan's heating circuit that cause the motor to run at low speed. When the fan switch is on in the thermostat, the furnace fan motor should run at high speed.

If you have voltage at test point 2 but not at test point 7, you should suspect the normally open contacts or the coil of the fan relay. Turn off power and test the coil and the contacts for continuity. You should also test the wires connected to these terminals for continuity. If you have voltage at test point 7, you should suspect the fan motor itself, and use the tests for the PSC-type motor that are found in Chapter 10. If you do not have voltage at test point 7, you should suspect the 110 VAC wire (red wire) between the fan relay and the motor terminal board terminal T1. When you have found the problem, you can change the parts or wires as needed and then you should cycle the furnace several times to make sure it is running correctly.

Questions for This Chapter

1. Identify the major parts of an electric furnace and explain their operation.
2. Explain what must occur for the heating elements to be energized in an electric furnace.
3. Explain what causes the furnace fan to turn on in an electric furnace.

4. How much voltage do the electric heating elements require in the electric furnace?

5. Identify the major parts of a gas furnace and explain their operation.

True or False

1. The pilot light must be established before the thermostat can energize the gas valve in the gas furnace.

2. The differential air pressure switch in the high-efficiency gas furnace will close when the exhaust fan is moving a sufficient amount of air.

3. The combustion blower in a high-efficiency furnace will turn on after the gas valve opens and the main burner flame is established.

4. The spark for the ignition system on the oil furnace turns off after the flame is established.

5. After the flame is established in the gas furnace, the temperature increases until the fan switch is enabled and energizes the furnace fan.

Multiple Choice

1. If the main heating coils of an electric furnace are not heating up, you should suspect _____

   a. the coil of the fan relay is open or the fan relay contacts are faulty.

   b. the fan switch contacts are not closed or the fan switch is faulty.

   c. the system has an open fuse, the contacts of the first sequencer are not closed, or the main heating element is open.

2. If the main gas valve of a standing pilot gas furnace is not energized and allowing gas to flow to the main burner you should suspect _____

   a. the heat anticipator of the thermostat is set incorrectly.

   b. the pilot flame is not established, the thermocouple is faulty, or the coil in the gas valve is open.

   c. the fan switch is open, the fan is faulty, or the filter may be dirty.

3. The sequence of operation of the high-efficiency furnace is _____

   a. the thermostat calls for heat, the combustion fan turns on, the pilot is established, the main burner comes on.

   b. the thermostat calls for heat, the pilot is established, the combustion fan turns on, the main burner comes on.

   c. the thermostat calls for heat, the pilot is established, the main burner comes on, the combustion fan turns on.

4. The sequence of operation for the oil burner is _____
    a. the thermostat calls for heat, the ignition spark is established, the furnace fan comes on, the main flame is established.
    b. the thermostat calls for heat, the ignition spark is established, the main flame is established, the furnace fan comes on.
    c. the thermostat calls for heat, the furnace fan comes on, the ignition spark is established, the main flame is established.
5. The sequence of operation of the standing pilot gas furnace is

    _____

    a. the thermostat calls for heat, the pilot is ignited, the main gas valve is energized, the furnace fan comes on.
    b. the thermostat calls for heat, the furnace fan comes on, the pilot light is ignited, the main gas valve comes on.
    c. the pilot light is ignited and remains on continually, the thermostat calls for heat, the main gas valve is energized, the furnace fan comes on.

Problems

1. Draw a sketch of the electrical diagram of the control circuit that includes the thermostat for a standing pilot gas furnace.
2. Draw a sketch of the electrical diagram of the control and load circuit for an oil furnace.
3. Draw a sketch of the electrical diagram of the control and load circuit for an electric furnace.
4. Draw a sketch of the electrical diagram of the control and load circuit for a high-efficiency gas furnace.
5. Draw a sketch of the electrical diagram of fan and limit control that controls the furnace fan.

18 Electrical Control of Air-Conditioning Systems

OBJECTIVES:

After reading this chapter, you will be able to:

1. Identify the basic electrical components of an air-conditioning system and explain their operation.
2. Explain why a window-type air conditioner has one fan motor for both the evaporator fan and the condenser fan.
3. Draw a sketch of a split-type air-conditioning system and explain which parts are in the outdoor unit, and which parts are in the indoor unit.
4. Explain the difference between a split-type air-conditioning system and a package-type air-conditioning system.

18.0   Overview of Air-Conditioning Systems

As a technician you will be required to work on a variety of air-conditioning systems. These systems will include rather small window-type air conditioners, split-system residential air conditioners that are part of the heating system, and larger commercial package systems. This chapter will help you to identify all of the components in each system and develop a troubleshooting strategy to help you to determine the location of faulty parts quickly. The main point you should remember in this chapter is that all of the parts in the different types of air-conditioning systems are very similar to each other. In many cases, the components such as compressors and fans are just different sizes in the different units, and they may be powered with single-phase voltage in the smaller units and with three-phase voltage in the larger units. You should also remember that you have learned how all of the components in the system operate individually, and now you will see how they will operate together as a system. You will also get experience reading both the ladder diagram and wiring diagram for each system. Remember, if you need help in moving from a ladder diagram to the wiring diagram, you can refer to Appendix A. You should remember that sometimes only one of the two types of diagrams will be provided, and you may need to sketch the other type of diagram to get sufficient information to understand the system.

18.1   Room Air Conditioners (Window Air Conditioners)

The simplest type of system to troubleshoot is a window air conditioner, which is also called a room air conditioner. Fig. 18–1 shows a cut-away picture of a typical window air conditioner. Fig. 18–2 shows a wiring diagram of a window air conditioner and Fig. 18–3 shows a ladder diagram of a window air conditioner. You should remember that the wiring diagram shows the relative location of each of the components in the system. We will use this diagram to identify all of the components in the system. You should also see that the wiring diagram provides a legend that lists the abbreviations of each part of the system and identifies the types and colors of wires for each part of the system. A table is also provided to identify the contact closures for the main operating switch for the system.

Figure 18–1   A typical window air conditioner. *(Courtesy of Comfort-Aire Heat Controller Inc.)*

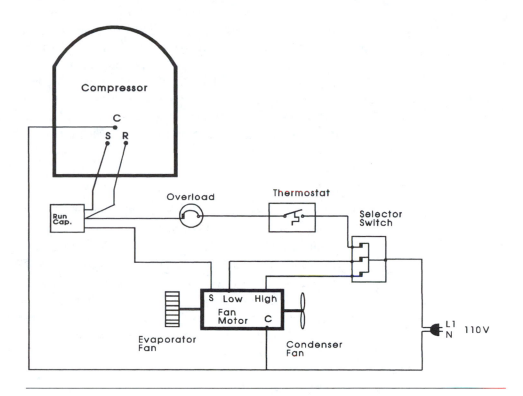

Figure 18–2   Wiring diagram of a window air conditioner.

18.2   The Main Components of the System

You can see in the electrical diagrams that the window air conditioner is powered by 110 VAC and it has a single-phase compressor that is wired as a PSC
motor. It also has a single fan with a double shaft that has the evaporator squirrel-

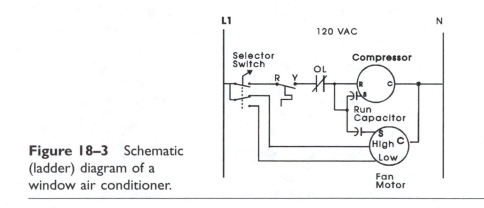

Figure 18–3   Schematic
(ladder) diagram of a
window air conditioner.

cage blower mounted on one end and a blade fan mounted on the other end for the condenser fan. This fan is a three-speed PSC motor. You can review the troubleshooting operation of these motors in previous chapters.

The controls for this system include the fan switch, which controls the speed of the fan and whether the compressor runs with the system or if the fan runs alone. The compressor is turned on and off with the thermostat that is connected in series with it. The other components in this system include the external overload and run capacitor for the compressor.

18.3   Normal Operation

In normal operation, the compressor will run and pump refrigerant through the system anytime the contacts in the thermostat close and send it power. Since the evaporator fan and condenser fan are the same motor, it will be called the *fan motor* and it will run anytime the compressor is energized. The fan motor can run with the compressor turned off. This means that the system is fairly simple to understand since the controls only have to supply power to the compressor and fan motor to get them to run. If voltage cannot get to either motor, or if a motor is faulty, the system will not run correctly. When this occurs, you can use either the wiring diagram or the ladder diagram to troubleshoot the circuit.

You can see that it is more difficult to trace the route of voltage to each component in the wiring diagram than in the ladder diagram, so we will use the ladder diagram to trace voltage to each system. In the ladder diagram you can see that all of L1 voltage comes from the 110 VAC plug and directly goes into the main control switch. When this switch is set to the fan only position, the fan receives power and the compressor does not. This allows the fan to circulate air in the room for a moderate cooling effect. When the switch is adjusted to the low cooling setting, the fan will receive voltage to the low speed, and voltage will be sent to the thermostat. If the room is warm enough, the thermostat contacts will close and send voltage to the C terminal on the compressor. The S and R terminals are connected directly to the other neutral potential through the run capacitor, so that when voltage comes from the thermostat, the compressor will start and run. When the room has cooled sufficiently, the thermostat will open and stop the voltage supply to the C terminal of the compressor. It is important to note that some manufacturers connect the C terminal of the compressor to the neutral potential in the circuit and the R and S terminals are connected to the L1 side of the circuit.

18.4   Troubleshooting the System

At times you will encounter a window air conditioner that does not run correctly when power is applied. You will need to determine the exact nature of the fault in order to correct the problem and get the system running again. You will be able to use some of the same troubleshooting tests you used previously to locate faults. The first test that you should always make when you start to work on a system is to try and make the system run. This sounds rather simple, but you must remember that the customer may not be familiar with the terminology on the operating switches on the front of the unit and may not know that the system must be plugged into a wall outlet, the switch must be set to cooling, and the thermostat must call for cooling for the compressor to run.

The next test you should do is to determine if any motors run or try to start or if power is interrupted completely. If either the compressor or fan runs without the other, you at least know that the system has some voltage and the likely problem will be in the wiring or components inside the window air conditioner. If both the compressor or fan fail to run when power is applied, the next test should be for supply voltage. You must use a voltmeter and measure the exact voltage available at the wall outlet. This is also a good time to feel the power cord and wall outlet to determine if they are warm or hot, which could cause a fire. You must remember that some older homes are not wired for the larger current required by the window air conditioner. If either the power cord or the wall outlet is hot, you must ensure that it is at room temperature when you have completed your repairs. If it continues to be warm, you will need to instruct the homeowner of the problem. Remember that if the unit is trying to start and draws large amounts of current, the wall outlet may be warm until you repair the problem.

18.5   Testing the Fan Motor

One problem that can occur with the system is if the fan motor or if the circuit that supplies voltage to it is faulty. The simplest way to test for a faulty motor or the loss of voltage is to place one terminal of the voltmeter at the locations identified here and record the voltage at each point. If you have voltage at the motor, you must suspect the motor is faulty, and if you do not have voltage, you must suspect an open in the circuit that supplies voltage to the motor.

1. Terminal 1 of fan switch to N _____
2. Terminal Low of fan switch to N _____
3. Terminal Low of the fan to N _____
4. Terminal Low to N directly on fan _____

If you have 110 VAC at each of these tests, the circuit is complete and the controls that supply voltage to the motor are good, and the motor should run. Be sure that the fan blades can rotate freely. If the blade does not turn freely, you may need to lubricate the fan bearing. If the motor does not run, you can change the motor at this time or use additional troubleshooting tests found in Chapter 10 and determine what part of the motor is faulty, but you will still need to replace the motor.

If voltage is present at the supply but is not present at the motor at reading 4, voltage has been interrupted at the switch or in the wiring and the measurements you have for tests 1, 2, and 3 will locate the fault. If you have full voltage at test 1 but not at test 2, the switch is faulty. If you have full voltage at test 1 and test 2, but not at test 3, the wire between the switch and the motor or between the motor and C is faulty. Make repairs as needed and try to start the system.

If the fan runs on one speed but not on others, you can use a similar test to determine if the problem is in the switch, the motor, or the wire between the switch and motor. Repeat the test with the fan switch set for high or medium speed and place the voltmeter probes on the high- and low-speed terminals instead of the low-speed terminals.

18.6   Testing the Compressor

You can use these troubleshooting tests or the ones found in Chapter 11 to locate faults. When you test the compressor you must break the test into two parts: Does the compressor try to start and make a noise or is it completely quiet? If the compressor tries to start, it indicates that voltage is making its way to the compressor, and the fault is in the starting winding or the capacitor, and you can use the troubleshooting procedure provided in Chapter 10 for permanent split-capacitor (PSC) compressors.

If the compressor is not trying to start and it does not make any noise, it means that voltage to the compressor has been interrupted by the controls or an open in the motor itself. Use the following test to identify the location of the open circuit.

1. L1 in switch to N _____
2. C in switch to N _____
3. Incoming terminal of thermostat to N _____
4. Outgoing terminal of thermostat to N _____
5. Incoming terminal of overload to N _____
6. Outgoing terminal of overload to N _____
7. C terminal of compressor to N _____
8. Terminal A of capacitor to C of compressor _____
9. Terminal B of capacitor to C of compressor _____
10. Terminal R of compressor to C of compressor_____
11. Terminal S of compressor to C of compressor _____

If the circuit is providing voltage to the compressor correctly, you should have voltage at each of the 11 points listed. If you lose the voltage at any of the points, you can back up one test and the open circuit in between the last place you have voltage and the point where voltage is lost. This is a very simple testing procedure that will help you to determine where the open in the circuit has occurred. If the open occurs in a component such as the thermostat, you may need to return to the chapter that explains the operation and troubleshooting of the individual component. If you have voltage at test points 9, 10, and 11, the circuit is operating correctly and the problem is in the compressor motor itself. Be sure to consult Chapter 10 about the PSC compressor motor for more troubleshooting details.

18.7   Split-System Residential Air-Conditioning Systems

The split-system residential air-conditioning system adds a condenser, which includes the compressor and condenser fan to the existing furnace. Fig. 18–4 shows

(a)   (b)   (c)

Figure 18–4a   The condenser unit is part of a split-system residential air conditioner and it is outside the house. **b** The furnace is the part of the split-system residential air conditioner inside the house. **c** A cut-away picture of the condenser unit. *(Courtesy of Armstrong Air Conditioning Inc., United Technologies Carrier Corporation)*

a picture of the components for a typical split-system air conditioner. Fig. 18–4a shows the condenser that is outside the house and contains the compressor, condenser fan, and controls. Fig. 18–4b shows the furnace part of the system that is inside the house and has the evaporator fan and the transformer to provide control voltage to the coil of both the evaporator fan relay and compressor contactor. Fig. 18–4c shows a cut-away picture of the condenser. The thermostat that controls the system is mounted on an interior wall in the house.

18.8   The Main Components of the System

Fig. 18–5 shows a wiring diagram of the split-system air conditioner and Fig. 18–6 shows a ladder diagram of the system. You can see in the electrical dia-

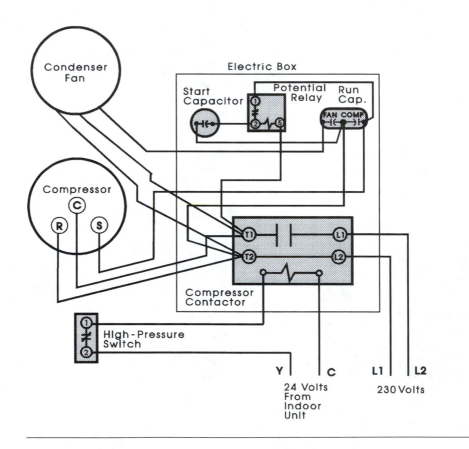

Figure 18–5 A wiring diagram for a split-system air-conditioning unit with the evaporator fan in the furnace, and the compressor and condenser fan in the condensing unit that is outside the house.

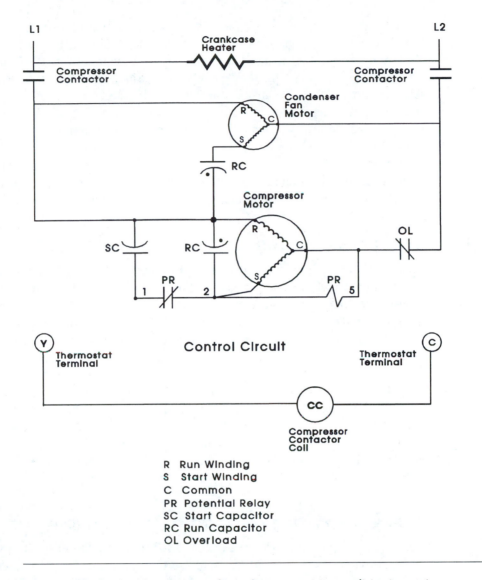

Figure 18–6   Ladder diagram for split-system air conditioning unit.

grams that the condensing unit is powered by 230 VAC and the control circuit for the system is powered by 24 VAC. The top part of the ladder diagram shows the components that are mounted in the condensing unit. The system uses a single-phase compressor that is wired as a PSC motor. You can also see that a current relay can be used with a start capacitor if the compressor requires additional starting torque. (Note: The diagram shows where the start capacitor and relay would go if they are used, otherwise the compressor is connected as a PSC motor). The condenser fan motor is also powered with 230 volts and it is also wired as a PSC motor. You can review the troubleshooting operation of PSC open motors and compressor motors in previous chapters. Since the compressor for this unit sits outdoors, it has a crankcase heater to keep the compressor pump warm so that oil does not migrate into the evaporator coil inside the house.

The main control for this system is the compressor contactor whose contacts close to provide voltage to the compressor. The 24 V coil of the compressor contactor is connected to the low-voltage control circuit where it is energized by the thermostat that is connected in series with it. Other components in this system include the external overload and run capacitor for the compressor.

18.9   Normal Operation

In normal operation the temperature in the room will increase above the setting on the thermostat and the contacts in the thermostat will close and energize the coil of the compressor contactor and cause its contacts to close. When the compressor contacts close, the compressor motor will run and pump refrigerant through the system. The condenser fan is also connected to the contactor and it will run anytime the compressor runs. The evaporator fan is mounted in the furnace and it is controlled by a fan relay. As you know, the thermostat can be set to run the fan continuously or in the auto position to run only when the compressor is energized. This system like the window air conditioner is fairly simple to understand since the controls only have to supply power to each of the motors to get them to run. If voltage cannot get to a motor or if a motor is faulty, the system will not operate correctly. When this occurs, you can use either the wiring diagram or the ladder diagram to troubleshoot the circuit.

18.10   Using the Wiring Diagram to Troubleshoot

The wiring diagram is provided in Fig. 18–5 and you can basically use it to locate the parts of the system. You can see that since the diagram of this system has several additional components and more wires than the simple window air conditioner, the wiring diagram is more difficult to use to trace the circuit. The wiring diagram is better used to locate the components, identify terminals, and locate the points where actual wire connections are made. For example, you can see that the supply voltage for the system is actually connected at the L1 and L2 terminals of the compressor contactor. You can also see that a wire is connected directly to the C terminal on the compressor and the other end of that wire is connected to the T1 terminal on the compressor contactor. The wiring diagram also shows the run capacitor for the compressor motor and the fan motor is located in the top right part of the diagram and its terminals will be identified as common (COMM), fan, and hermetic (HERM).

18.11   Using the Ladder Diagram to Troubleshoot

You can see that since it is more difficult to trace the route of voltage to each component in the wiring diagram than in the ladder diagram, we will use the ladder diagram to trace voltage to each part of the system. In the ladder diagram you can see that all of the supply voltage comes from L1 and L2, which are the terminals for the 230 V supply. As you know, this voltage must come from a disconnect switch or a circuit breaker.

When the disconnect or circuit breaker is turned on, 240 volts will be supplied directly to terminals L1 and L2 on the compressor contactor. Since the sump heater is also connected to these same two terminals, it will always draw current anytime power is applied to the system. The ladder diagram clearly shows that both the compressor and condenser fan motors are controlled by the compressor contactor that is located between terminals L1 and T1 on the compressor contactor.

18.12   Troubleshooting the System

At times you will encounter a split-system air conditioner that does not run correctly when power is applied. You will need to determine the exact nature of the fault in order to correct the problem and get the system running. Again, the first test that you should always make when you start to work on a system is to try to make the system run. Remember, the customer may not be familiar with the terminology on the thermostat and may not know that the disconnect or circuit breaker for the system must be turned on, that the thermostat must be set to cooling, and the temperature in the room must be above the setting on the thermostat for the compressor to run.

The next test you should do is to determine if any motors run or try to start, or if power is interrupted completely. If the compressor and either fan try to start or run, you at least know that the system has some voltage and the likely problem will be in the wiring or components. If the compressor and both fans fail to run when power is applied, the next test should be to test for supply voltage. You must use a voltmeter and measure the exact voltage available at the L1 and L2 terminals at the compressor contactor.

18.13   Testing the Condenser Fan Motor

One problem that can occur with the system is if the condenser fan motor or if the circuit that supplies voltage to it is faulty. The simplest way to test for a faulty motor or the loss of voltage is to place one terminal of the voltmeter at the following locations identified here and record the voltage at each point. If the motor has voltage supplied to it, you should suspect the motor of being faulty. If you do not have voltage to the motor, you must suspect an open in the wiring that supplies voltage to the motor. Remember that you must always use the terminals shown for each reading because they provide a reference for each measurement to opposite voltage potential.

1. Terminal T1 of compressor contactor to L2 _____
2. Terminal C of motor (black wire) to L2 _____
3. Red terminal on capacitor to L1 _____
4. Yellow wire on fan to L1 _____
5. Red wire on fan to L1 _____

If you have 230 VAC at each of these tests, the circuit is complete, the controls that supply voltage to the motor are good, and the motor should run. If the fan does not run, it may have an open or it may have a mechanical problem. You can remove the wires to the fan and test the fan windings for continuity to see if one of the wires or windings has an open circuit. Since the problem may also be mechanical you should try to turn the fan blades to see if they can rotate freely. If the blade does not turn freely, you may need to lubricate the fan bearing. If the motor does not run, you can change the motor at this time or use additional troubleshooting tests found in Chapter 10 and determine what part of the motor is faulty, but you will still need to replace the motor.

If voltage is present at the supply but is not present at one of the other points, it has been interrupted by an open in the wiring. The measurement that indicates zero volts also indicates the point that has the problem. You could also use a clamp-on ammeter to test for current in the wire for the start winding, run winding, and common. If the wire connected to the start or run winding does not draw current, it is an indication of an open circuit in the wire that supplies voltage to that winding, or the winding itself has an open.

18.14   Testing the Compressor

The test for the compressor in the split system is very similar to the tests for the compressor in the window unit. The main difference is that the compressor in the split system is usually powered with 240 VAC rather than 115 VAC. The compressor motor may also be connected as a PSC motor or CSCR motor depending on its horsepower. You can use the same two tests as before with the other compressor when you determine if the compressor tries to start and make a noise or if it completely quiet. If the compressor tries to start, it indicates that voltage is making its way to the compressor, and the fault is in the starting winding or the capacitor, and you can use the troubleshooting procedure provided in Chapter 10 for permanent split-capacitor (PSC) compressors.

If the compressor is not trying to start and it does not make any noise, it means that voltage to the compressor has been interrupted by the controls or an open in the motor itself. Use the following test to identify the location of the open circuit.

1. L1 on compressor contactor to L2 _____
2. T1 on compressor contactor to L2 _____
3. Terminal C on compressor to L2 _____
4. Terminal R on compressor to L2 _____
5. Capacitor RED terminal to L1 _____
6. Terminal S on compressor to L1 _____

If the circuit is providing voltage to the compressor correctly, you should have voltage at each of the six points listed. If you lose the voltage at any of the points, you can back up one test and the open circuit is between the last place you have voltage and the point where voltage is lost. This is a very simple testing procedure that will help you to determine where the open in the circuit has occurred. If you have voltage at test points 3, 4, 5, and 6, you have determined that the circuit wires are doing their job and they are supplying voltage directly to the motor and you can suspect the motor has an open.

If the open occurs between terminals L1 and T1 on the compressor contact, it indicates the contacts are not closed or they are faulty and are not passing voltage. Be sure to test the two terminals of the coil of the compressor contactor for 24 volts. If 24 volts is present, you must test the coil and the contacts for continuity. Be sure to turn off all power to the system before you make any continuity tests.

If 24 volts AC is not present at the coil terminals, you must test the low-voltage control circuit for voltage. Use the following tests (remember you should measure 24 VAC at each of these test points):

1. R and C terminals on the transformer in the furnace _____
2. Terminal Y to C on the terminal board in the furnace _____
3. Terminal Y to C in condensing unit _____
4. Incoming terminal on contactor coil to C _____
5. Incoming terminal high-pressure switch to Y _____
6. Outgoing terminal of high-pressure switch to Y _____
7. Across both terminals of compressor contractor coil _____

Some technicians find it is easier to start at step 7 and work their way back to the voltage source at the transformer. You are trying to locate the point where voltage is lost. If voltage is not present at the R and C terminals of the furnace transformer, be sure that power to the furnace is turned on. If voltage is present on the 110 V primary side of the transformer and not at the secondary, you can disconnect the transformer and test it for continuity.

If you have lost voltage at the thermostat, make sure that it is set so that the room is warmer than the setting. Remember that the R to Y circuit of the thermostat must have continuity for it to pass voltage. If you have voltage inside the house on the furnace terminal board but lose it when it comes outside to the condensing unit terminals, you should suspect broken wires between the furnace and condenser. You should find this system is a fairly easy way to locate the source of the open circuit.

18.15   Commercial Packaged Air-Conditioning Systems

The commercial packaged air-conditioning system is very similar to the split-system air conditioner and the window air conditioner in the fact that it has a compressor, condenser fan motors, and evaporator fan motor. It also has a room thermostat that controls when the compressor and evaporator fan should run. The packaged air-conditioning system usually is mounted on the roof of a commercial building and the conditioned air and return air are both ducted through the roof into the attic where ducts distribute the air to each part of the building through ceiling registers. Since this type of unit is mounted on a commercial building, it will most likely be powered with three-phase voltage and it may have an air economizer circuit.

Fig. 18–7a and Fig. 18–7b shows a picture of a typical package-type system. Fig. 18–8 shows a wiring diagram and Fig. 18–9 shows a ladder diagram of the

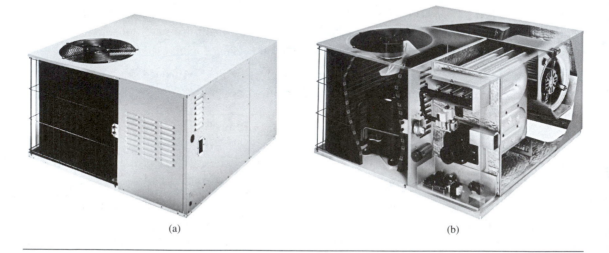

(a)   (b)

Figure 18–7a   A typical packaged air-conditioning system. This type of system may also be called a rooftop system. **b** A cut-away picture of a packaged air-conditioning system. *(Courtesy of Armstrong Air Conditioning Inc.)*

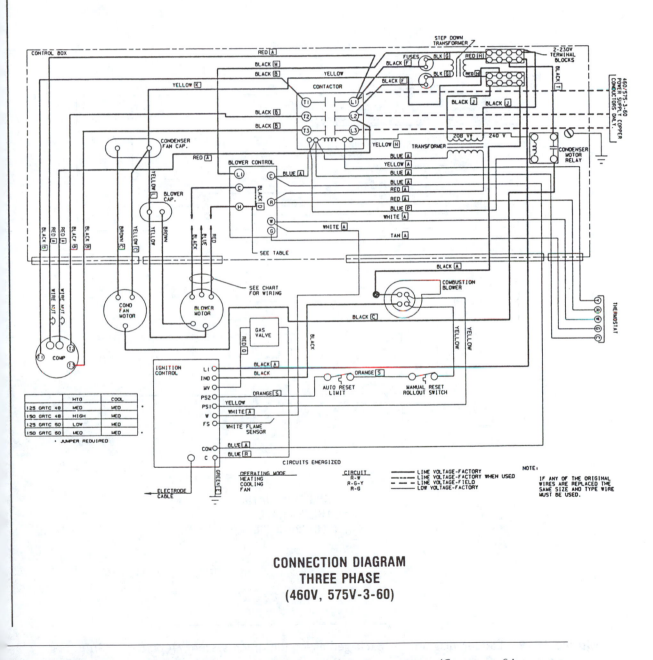

CONNECTION DIAGRAM
THREE PHASE
(460V, 575V-3-60)

Figure 18–8 Wiring diagram for a packaged air-conditioning system. (*Courtesy of Armstrong Air Conditioning Inc.*)

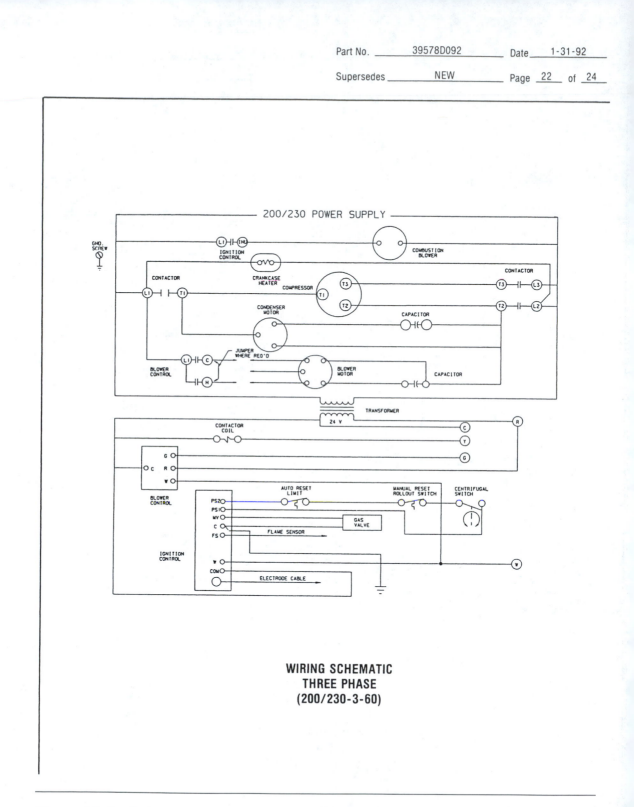

Figure 18–9 Ladder diagram for packaged air-conditioning system. *(Courtesy of Armstrong Air Conditioning Inc.)*

electrical system. In the electrical diagram you can see that this type of unit has similar components as the previous air-conditioning systems. The major difference in this system is that the unit is powered with 460/575 three-phase VAC. The compressor motor and crankcase heater are connected to this three-phase voltage. The primary winding of a step-down transformer is connected across L1 and L2 of the three-phase voltage and it drops the voltage to single-phase 230 volts. The 230 volts is used to power the condenser fan and evaporator fan. Since the condenser is powered with different voltage than the compressor, a condenser fan relay is used to energize and deenergize the condenser fan.

A primary winding of a second transformer is connected across terminals S1 and S2 of the first step-down transformer. This transformer converts 230 volts to 24 volts AC for the control voltage that is used to power the coil of the contactor relay, the condenser fan relay, and the evaporator fan blower control. The thermostat is used to control the voltage that energizes and deenergizes the coils of these relays. You can use the procedures provided earlier in this chapter to trouble-shoot this type of system.

Questions for This Chapter

1. Explain the function of the electrical components of a window air conditioner.
2. Identify the electrical components of a split-system air conditioner and explain their function.
3. Explain why a window air conditioner has only one fan motor.
4. Identify the electrical components of a packaged-type air-conditioning system and explain their function.
5. Explain which electrical components are in the indoor unit and which electrical components are in the outdoor unit of a split-system air-conditioning system.

True or False

1. The Y terminal on the thermostat is used to control the air-conditioning outdoor system.
2. In the split-system air conditioner the condenser fan can run without the compressor running.
3. The evaporator fan can run without the compressor running.
4. The evaporator fan is located in the outdoor unit in a split-system air-conditioner.
5. The G terminal on the thermostat is used to control the evaporator fan relay.

Multiple Choice

1. The evaporator fan for a split-system air conditioner is a _____
   a. squirrel-cage fan.
   b. blade fan.
   c. shaded-pole fan.

2. If the compressor tries to start but does not make any noise, you should suspect _____
   a. the run winding is good and drawing current but the start winding is open.
   b. the main fuse is blown or the compressor is not receiving voltage.
   c. the start winding is good and drawing current but the run winding is open.

3. If the condenser fan in a split system is running and the compressor will not run and it is humming, you should suspect _____
   a. the main fuse is blown or the main power to the outdoor section is not receiving power.
   b. the indoor temperature is at the thermostat setpoint so the compressor does not need to run.
   c. either the start or run winding is open or the components connected to the start windings of the condenser fan are malfunctioning.

4. If the evaporator fan in a window unit is operating correctly and little or no air is moving across the condenser fins, you should suspect _____
   a. the condenser fan is faulty or not running at all.
   b. the condenser fan blade may be loose or the condenser fins may be covered with dirt since the condenser and evaporator use the same fan motor.
   c. the fuse to the condenser fan is probably open.

5. If the condenser fan in a packaged-type air-conditioning system is not running and the compressor and evaporator fan are running, you should suspect _____
   a. the condenser fan has an open winding or a wire to the fan is open.
   b. one of the main fuses to the packaged unit is open.
   c. there is a short in the compressor motor which is causing low voltage in the condenser motor.

Problems

1. Draw a sketch of the wiring diagram or ladder diagram that includes all of the components in a window air conditioner and explain what controls the voltage to each of the motors.

2. Draw a sketch of the wiring diagram or ladder diagram that includes all of the components in a split-system air conditioner and explain what controls the voltage to each of the motors.

3. Draw a sketch of the wiring diagram or ladder diagram that includes all of the components in a packaged unit air conditioner and explain what controls the voltage to each of the motors.

4. Identify the steps you would use to troubleshoot the compressor of a split-system air conditioner if it does not try to run when power is applied.

5. Identify the steps you would use to troubleshoot the condenser fan of a packaged air conditioner if it does not try to run when power is applied.

19 Electrical Control of Heat Pump Systems

OBJECTIVES:

After reading this chapter, you will be able to:

1. Identify the basic parts of a heat pump system when it is in the cooling mode.
2. Explain how the reversing valve changes the system from cooling to heating.
3. Identify the basic parts of a heat pump system when it is in the heating mode.
4. Explain the defrost cycle of a heat pump.
5. Troubleshoot a heat pump system and locate problems in the heating and cooling cycles.

19.0   Overview of Heat Pump Systems

A heat pump operates like a normal air-conditioning system in the summer to provide cool air to the conditioned space, and in the winter a reversing valve is energized to reverse the refrigerant flow so that the indoor coil becomes the condenser. In the heating cycle, the indoor fan moves air across the indoor coil where it moves heat from the coil to the indoor space like a regular furnace. The heat

453

pump gets its name from the way it takes heat from outside air that may be as cool as 10°F and pumps it into the indoor coil in the winter where it is converted to temperatures of 80° F by increasing refrigerant pressures. And in the summer it pumps heat from inside the conditioned space to the outdoor coil.

It is sometimes difficult to understand how the heat pump gets heat out of very cold air in the winter until you remember that the heat pump system is a basic refrigeration cycle and it is possible to get the refrigerant to evaporate in the outside coil even at very low temperatures. When this occurs, the heat that causes the refrigerant to evaporate is absorbed into the refrigerant and the compressor pumps the refrigerant in vapor form and compresses it into a hot gas. The hot gas is pumped to the indoor coil where air from the conditioned space is moved over it to condense the refrigerant. The condensing process allows the refrigerant to give up heat it picked up in the outdoor coil to the indoor air as air is moved over the coil by the furnace fan. The amount of heat that is available at the indoor coil is sufficient to heat the indoor space, but the temperature of the air will not be over 100° as it moves over the indoor coil, so sometimes people inside the conditioned space will complain that it feels like cool air is coming out of the air registers. Basically the heat pump operates like a regular air conditioner in the summer, and in the winter it reverses its cycle and provides warm air like a furnace. Fig. 19–1a shows a picture of the indoor unit for a split-system heat pump. Fig. 19–1b shows a picture of the outdoor unit for a split-system heat pump and Fig. 19–2 shows a cut-away picture of the outdoor unit for a split-system heat pump.

This means that the indoor coil acts as the evaporator when the heat pump operates like an air conditioner, and it acts like the condenser for the system in the winter when the refrigerant flow is reversed. A valve called a *reversing valve* makes the refrigerant change its flow when the heat pump is switched from cooling mode to heating mode. The heat pump requires a two-circuit thermostat; one of the circuits is for the heating cycle, and the other circuit is for the cooling cycle. The remainder of this chapter will explain several other modes, including the defrost mode, second-stage heating mode, and emergency heating mode, which the heat pump must utilize to operate efficiently. You will need to determine if the heat pump is trying to operate in its heating cycle or in its cooling cycle when it is in these modes so that you will know what part of the system to troubleshoot when it is faulty.

19.1   The Main Parts of a Split-System Heat Pump

The main parts of the heat pump are discussed in regards to the indoor part of the system and the outdoor part of the system. The main parts of the indoor system include the furnace fan and the indoor coil. The indoor section also will

(a)   (b)

Figure 19–1a   The indoor unit for a split-system heat pump. **b** The outdoor unit for a split-system heat pump.   *(Courtesy of Bard Manufacturing Co.)*

have some type of alternative heating source such as an electric heating coil, or a gas valve and heat exchanger, or an oil burner and heat exchanger. The alternative heating source can be used as a second stage of heating if the first stage does not provide enough heat, and it will also be used to temper the indoor air when the heat pump is in its defrost cycle. The main parts of the outdoor system include the outdoor coil, the outdoor fan, the compressor, the reversing valve, defrost timer, and overpressure controls.

19.2   Normal Operation in Cooling

When the heat pump is in the cooling mode, the indoor coil will act as the evaporator coil and the outdoor coil will act as the condenser coil. The furnace fan will act as the evaporator fan and the outdoor fan will act as the condenser fan. The compressor will pump the refrigerant as a hot vapor to the outside coil where it will be condensed and flow as a liquid to the indoor coil. The furnace fan will circulate inside air across the indoor coil and cause the refrigerant to evaporate and cool the air. Fig. 19–3 shows a diagram of the heat pump system. You can see in each diagram that the control circuit is shown in relationship to the heat

Figure 19–2 A cut-away picture of the outdoor unit for a split-system heat pump. *(Courtesy of Armstrong Air Conditioning Inc.)*

pump thermostat. This will make it easier to understand the operation based on the setting of the thermostat.

You can see that the wiring diagram looks very complex for the heat pump system, and that the ladder diagram simplifies the sequence of operation somewhat. The problem with the ladder diagram from this manufacturer is that it tries to combine the wiring diagram of the heat pump control module with the ladder diagram. Anytime that you try to show part of the diagram as a wiring diagram and part of the same diagram as a ladder diagram, it will lead to extreme confusion. Fig. 19–4 shows the manufacturer's ladder diagram in a more simplified form so that you can see the sequence of operation. It is important to understand that the ladder diagram in this figure does not try to show the relationship between the location of terminals for the heat pump control module, rather it shows where each switch in the heat pump control module is located in the ladder diagram. You should also notice that the components in the furnace (indoor unit) and the control circuit are shown separately from the high-voltage circuit at the top of the ladder diagram.

In normal cooling operation the thermostat circuit R-Y is closed, which sends voltage to terminal Y of the heat pump control module and energizes it. When the heat pump control module is energized, it closes its contacts LRI-B, which provides voltage to energize the coil of the compressor contactor (CR2) in the first line of the ladder diagram. When the contacts of CR1 close, it energizes the

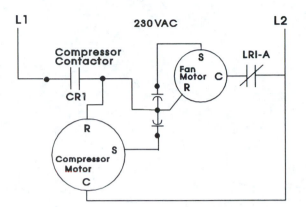

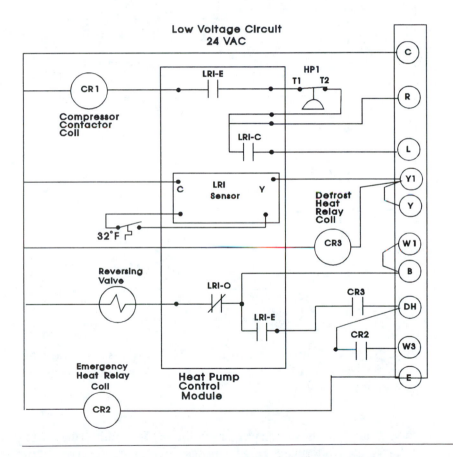

Figure 19–3   A ladder diagram of a split-system heat pump.

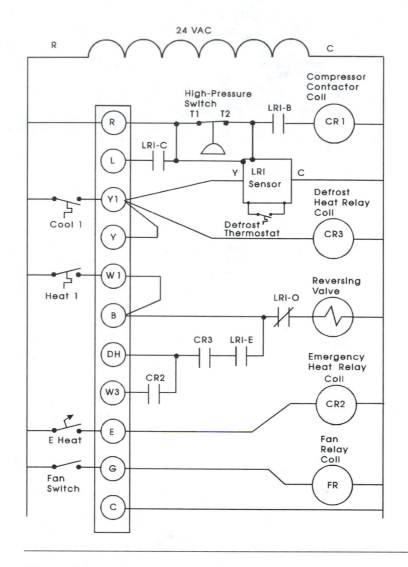

Figure 19–4 Simplified ladder diagram of a split-system heat pump. This diagram does not try to show the relationship between the terminals on the heat pump control module.

compressor and the condenser fan in the outdoor unit. The thermostat also provides voltage to terminal G, which causes the fan relay to be energized. The contacts of the fan relay close and provide voltage to the furnace fan. Voltage from terminal Y of the thermostat also provides voltage to the coil of the defrost relay

(CR3), but at this point it does not affect anything until the heating cycle. The remainder of the heat pump circuits are utilized for heating and defrost circuits so they do not provide any function at this time.

19.3   Normal Operation in the Heating Cycle

When the heat pump is in normal operation in the heating cycle, the compressor will run and the reversing valve will be energized so that the hot gas from the compressor will move to the indoor coil. The furnace fan will be energized to blow the indoor air over the indoor coil to heat the air. The outdoor fan will also be energized with the compressor to move sufficient outdoor air over the outdoor coil so that heat can be removed from it and moved to the indoor coil. The heating part of the thermostat uses two or three stages to provide sufficient heat. When the conditioned space is 1 to 2 degrees below the setpoint of the thermostat, the stage-one circuit of the thermostat is closed and 24 VAC is sent to the W1 terminal, which is jumpered to the B terminal. This causes the reversing valve to become energized, and the heat pump module to become energized. The heat pump module causes the CR2 (compressor contactor) to energize. When the contacts of the CR2 contactor close, they provide 230 VAC to the compressor and outdoor fan. During the heating cycle the outdoor coil and outdoor fan will act like the evaporator part of the refrigeration cycle. An outdoor thermostat is connected across terminals Y and Y1. In normal operation it provides a circuit until the outdoor temperature drops below 10°, which is the lowest temperature that the heat pump can efficiently remove heat from the outdoor air. If the temperature drops below this point, it will cost more to run the compressor than it would to get the heat from the electric heating coil.

The heat pump control module also has an outdoor thermostat that is connected to two terminals identified as *sensor.* This sensor is mounted close to the outdoor coil and it is used during the defrost cycle to bring the system out of the defrost cycle when the temperature reaches above 57°. If the temperature inside the conditioned space drops more than 2° below the setpoint, a second stage of heating is energized. In this system the second stage of heating thermostat energizes the W3 terminal, which closes and sends 24 VAC to the CR4 coil. The CR4 contacts close and provide 230 VAC to the electric heating coil. The electric heating coil can provide additional heat, which will also cause the temperature of the indoor air to increase several degrees and make the air coming through the warm air vents feel warmer. This will also cause the conditioned space to warm up quickly. When the indoor temperature increases to within 2° of the setpoint, the second-stage contacts of the thermostat open and deenergize the CR4 contacts.

19.4   **Normal Operation During the Defrost Cycle**

If the heat pump runs long enough during the heating cycle, the outdoor coil will begin to form a frost over it. This will cause the outdoor coil to become less efficient as the frost blocks the air passages through the coil. At a point where the frost completely covers the coil, the temperature will drop sufficiently to bring in the defrost circuit. During the defrost cycle, the heat pump is switched back to air-conditioning mode where the outdoor coil becomes the condenser, and the compressor pumps hot gas to it. During the defrost cycle the outdoor fan is deenergized so that the outdoor coils can receive maximum heat from the hot gas flowing through the outdoor coil. This heat causes frost to melt from the outdoor coils. When a sufficient amount of frost has melted from the coil, it will begin to warm up quickly from the hot gas flowing through it because the outdoor fan is not running. The defrost thermostat will sense the increase in temperature and switch the system out of the defrost cycle and back to the heating cycle. Some heat pump systems use a defrost timer or a differential air pressure switch to bring the system in and out of the defrost cycle. If a defrost timer is used, it will automatically send the heat pump into defrost at timed intervals. You should understand that the heat pump must be set up differently for each area of the country. For example, in the southern part of the country, the outdoor temperature routinely gets above 45° every day or two. This temperature is warm enough to help the outdoor coil defrost without using the defrost cycle. In the northern part of the country, the temperature can stay below 32° for several weeks at a time. When this occurs, the outdoor coil would continue to add frost and a defrost cycle must be used to keep the outdoor coil free of frost so the heat pump can run efficiently. If the pressure differential switch is used to determine the need for a defrost cycle, it samples the pressure differential that is created as air tries to move across the coil as it becomes blocked by frost. As more frost builds up, the pressure differential of the air moving across the coil becomes larger.

When the heat pump switches to its defrost cycle, the reversing valve is deenergized so that the system runs in its air-conditioning mode and liquid refrigerant is pumped to the inside coil where it removes heat from the inside air. The heat is pumped to the outdoor coil where it melts the frost, but it also causes the system to blow cold air from the indoor vents during the time the unit is in defrost. Since this is not desirable, the second stage of heating is energized during the defrost cycle to provide additional heat to the indoor air. In the circuit, you can see that a jumper is provided between the DH terminal to the CR2 contacts to ensure that the second-stage heating relay coil (CR4) is energized during defrost.

19.5   The Heat Pump Thermostat

The heat pump thermostat is slightly more complex than a normal heating and cooling thermostat. Fig. 19–5 shows a diagram of a typical heat pump thermostat. In this diagram you can see that the heat pump has an R terminal where all of the 24 VAC power source originates. The R lead of the transformer is connected directly to the R lead of the thermostat. The system switch determines if the thermostat is set to heating mode or to cooling mode. When the thermostat is set to cooling mode, the reversing valve is energized directly from the R-B terminal. The B terminal is identified as the O terminal in some heat pumps. When the thermostat system switch is set to the heating mode, the reversing valve is deenergized.

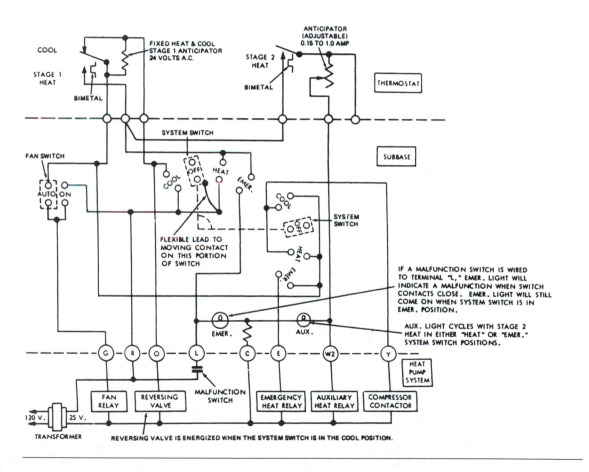

Figure 19–5   Heat pump thermostat. *(Courtesy of White Rogers)*

You can see that this thermostat has two stages of heating: W1 and W3. The W1 circuit is the circuit that powers the heat pump during the first stage of heating. It is connected to the Y1 circuit through internal circuitry in the thermostat so that anytime the circuit between R and W1 is made, 24 VAC will also be sent to the Y1 terminal to ensure that the compressor contactor is energized.

The second stage of heating circuit (W3) is used to energize the auxiliary heat. In most cases the auxiliary heat is an electric heating coil and it is energized by a heating contactor. Some heat pump thermostats provide three stages, so you will find a W1, W2, and W3. The second and third stages of heating will each be a bank of electric heat. In some parts of the country where natural gas or fuel oil is available, the second stage of heating could be a gas furnace or an oil furnace.

The E terminal in the thermostat provides a signal to the emergency heat circuit. The emergency heat circuit would be utilized if a problem occurred in the outdoor unit and it needed to be shut off. The emergency heat circuit activates the coil of a control relay whose contacts energize the auxiliary heating system directly, and deenergizes the compressor and outdoor fan by opening the circuit to the compressor contactor. When the thermostat is switched to emergency heat, an indicator will be illuminated on the thermostat to remind the person operating the thermostat that the outdoor unit requires service.

The cooling circuit in the thermostat is controlled through the Y1 and Y terminals. When the thermostat is switched to automatic or cooling, power will be switched from the R terminal to the Y1 terminal. Power from the Y1 terminal is used to energize the heat pump control module, which closes a set of contacts that provides voltage to energize the compressor contactor coil. The contacts of the compressor contactor provide high voltage to the compressor and outdoor fan.

The Y terminal is provided to allow an outside thermostat to deenergize the heat pump cycle if the outdoor temperature drops below a specified temperature called the *balance point*. The balance point is determined by calculating the cost of electricity versus the cost of the auxiliary heating energy. The outdoor thermostat is set at the temperature where the efficiency of the heat pump becomes too low to make the heat pump cost-effective. Anytime the outdoor temperature is below this temperature, the outdoor thermostat deenergizes the heat pump cycle and the auxiliary heat is used to provide all of the heat for the system. When the outdoor temperature returns above this temperature, the heat pump will become the primary heating source for the system and the auxillary heat only comes on during defrost or when the indoor temperature drops more than 2° below the thermostat setpoint.

The thermostat also has a set of contacts called DH, which receives power from the outdoor unit when it goes into the defrost cycle. If the heat pump is operating when the outdoor temperature drops below 40°, the surface of the outdoor coil will be below 32°. This allows any moisture in the air to freeze on the surface of the coil and begin a process of building up and blocking the air flow across the coil. The defrost cycle can be energized by time and temperature or by sensing the pressure of the refrigerant in the outdoor coil, which can be used to calculate the temperature on the surface of the coil. Some brands of heat pump also use various methods of measuring the pressure drop of air flowing across the coil to help determine when the coil needs to be defrosted.

The G terminal on the thermostat provides voltage to the coil of the indoor fan relay. The G terminal receives power from the R terminal when the fan switch is on. When the fan switch is set to auto, the G terminal receives power from the Y terminal when the thermostat is calling for cooling or heating. The fan relay energizes the fan to run at high speeds because the heat pump needs to move a lot of air, since the temperature of the air is not as warm as the air coming from a gas or oil furnace.

The thermostat also has a C terminal to provide the return potential for the 24 VAC source. The C terminal on the thermostat is connected to the C terminal on the transformer. This provides a connection to the common side of the 24 V circuit for any indicators on the thermostat such as the emergency heat indicator, and the second-stage heating indicator if one is used. When you suspect the thermostat is not operating correctly, you can disconnect power and test each of the circuits for continuity with respect to the R terminal. Be sure the thermostat is set correctly and that the temperature is above or below the setpoint as needed to activate the heating or cooling circuits. This means that during the troubleshooting period, you may need to set the thermostat substantially higher or lower than normal.

19.6   Troubleshooting the Heat Pump

When you troubleshoot a heat pump system, you must be sure which mode you are trying to make it run in. For example, if it is summer and you are trying to make the system run in the cooling mode, you would only need to make the compressor, outdoor fan, and indoor fan run. In winter if you were trying to make the heat pump run, you would also need to check to ensure that the second stage of heating operated correctly, and that the defrost cycle operated correctly. This means that the customer complaint may be with the second stage of heating and not the heat pump, or the heat pump may work correctly to provide heat, but it may not be able to go into defrost mode. In all of these cases, you will need to refer to the diagram and make the correct tests.

19.7   Troubleshooting the Heat Pump in the Cooling Mode

When you are requested to troubleshoot the heat pump in the cooling mode, you should set the thermostat for a temperature that is lower than the room temperature and try to make the system run. If the system does not run, you should make the following voltage tests: Use Fig. 19–6 for these tests.

1. L1-L2 for 230/208 volts _____
2. R to C on the thermostat or transformer (24 VAC) _____
3. Y to C on heat pump terminal board (24 VAC) _____
4. T1 of high-pressure switch to C (24 VAC) _____
5. T2 of high-pressure switch to C (24 VAC) _____
6. Coil of CR2 to C (24 VAC) _____
7. L1 terminal of CR2 contacts to L2 (230 VAC) _____
8. T1 terminal of CR2 contacts to L2 (230 VAC) _____
9. R-C on compressor (230 VAC) _____
10. S-C on compressor (230 VAC) _____
11. R-C on condenser fan (230 VAC) _____
12. S-C on condenser fan (230 VAC) _____
13. Across the terminals of reversing valve (24 VAC) _____

If you do not have 230 VAC at L1-L2, you have lost the source voltage for the system, and you should suspect the circuit breaker is tripped, the fuses are blown, or the disconnect is not on. Locate the source of the problem and correct it. If you have 230 volts at L1-L2, but you do not have 24 VAC at terminals R and C inside the unit, you should suspect the transformer or the wire supplying voltage to it is not operating correctly. You can make a voltage test directly across the terminals of the primary side of the transformer for 230 VAC if it is connected to L1-L2, or 110 VAC if it is connected to L1-N. If you do not have voltage directly across the primary terminals, but you have supply voltage to the unit, you should suspect that one of the wires connecting the transformer to the power supply is open.

If the primary side of the transformer has the proper amount of voltage, but the secondary does not, you should suspect that one of the windings in the transformer is open. Turn power off and test the primary and secondary sides of the transformer for continuity.

If you have voltage at the R-C terminals, but do not have voltage at Y-C on the heat pump terminal, you should suspect the thermostat is not supplying voltage to the heat pump. Check the thermostat to ensure it is set for cooling and that it is set well below the room temperature. If the thermostat is set correctly and you still do not have voltage at Y-C, you should suspect the thermostat, or the wire connecting the thermostat to the heat pump is faulty. You can place a

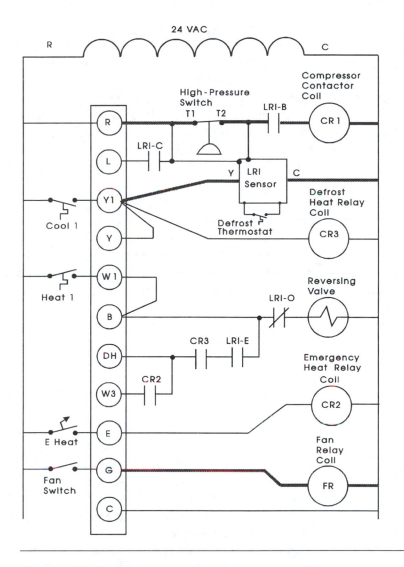

Figure 19–6 Diagram of heat pump with cooling mode circuit in bold.

jumper across the Y to R terminals in the thermostat to see if the problem is in the thermostat, or you can remove power from the system, and test the thermostat cooling circuit for continuity at the R-Y wires at the furnace. Be sure to turn off power and remove the wires from the terminal board so that they are isolated. If you get voltage at terminals Y-C in the heat pump when the jumper is placed on terminals R-Y in the thermostat, you can suspect the thermostat is faulty and you should replace it. If you still do not have voltage when the jumper is in place, you should suspect the wires between the thermostat and the furnace, or the wires between the furnace and the heat pump. Remember that the wires run from the thermostat and the furnace, and then from the furnace to the outdoor unit. Visually inspect the terminals in the thermostat, furnace, and the outdoor unit to ensure the wires are connected securely. You may have to test each wire involved for continuity until you locate the faulty wire and replace it.

If you have 24 VAC at test point 2 but do not have voltage at test point 4 (terminal T1 of the high-pressure switch and C), you should suspect the wire between the R terminal in the heat pump terminal board, and the high-pressure switch. Turn off power and test this wire for continuity. If the wire is faulty, replace it, and try the system again.

If you have voltage at test point 4 but not at test point 5, you should suspect the high-pressure switch is open. You must determine if the high-pressure switch is open because the pressure is too high, or because the switch is broken. This means that you must put gauges on the system to determine the pressure and compare it to the pressure chart for the refrigerant that is used and the ambient temperature at the coil. If the pressure is actually too high, you should suspect the outdoor unit is not moving air across its coil correctly and you should check the outdoor fan to see if it is operating correctly. If the fan is running, the outdoor coil may be plugged with debris such as grass or leaves so sufficient air cannot pass through it. If the pressure is high, you must correct the problem that is causing it. If the pressure is normal, you should suspect the switch is faulty, and you can test it and replace it if necessary.

If you have voltage at test point 5 but not at test point 6 you may have a faulty heat pump control module or the wire between the LRI-B contacts and the coil may be open. The heat pump control module causes contacts LRI-B to close when a 24 V signal is received at terminal Y1 on the module. You determine if the contacts in the module are closed by testing for 24 VAC from the common terminal (C) to the terminal where voltage comes in and then test a second time for voltage on the terminal where the wire is connected to the coil. If you have voltage at the input terminal and not at the output terminal, you can suspect the contacts are not closed and the problem is in the module or the signal coming from the Y1 terminal. If you have 24 VAC at both the incoming terminal and the outgoing terminal, the contacts are closed and passing power correctly.

If you do not have voltage at the coil, you can suspect the wire between the LRI-B contacts and the coil is open. Turn off power and test this wire for continuity and replace the wire if necessary. If you have voltage (24 VAC) across the coil of the compressor contactor (CR2), its contacts should be closed. If its contact do not close, you should suspect the contactor is faulty and you can turn off all power and test the coil for continuity. If the contacts close, you should move to test points 7 and 8 to determine if 230 VAC is passing through the contacts. You can also notice if the outdoor fan is running. If the outdoor fan is running, you have determined that the contacts are passing voltage, and you can move on to the next test. If you do not have 230 VAC at test point 7 (L1 to L2), you should suspect a problem with the fuse disconnect or the wires between the disconnect and the terminal box in the outdoor unit. You really should have noticed this loss of voltage with test 1 of this procedure. If you have 230 VAC at test point 7 but not at test point 8, it indicates that the compressor contacts are not closed or they are malfunctioning. Turn the power off and test the contacts for continuity by physically activating (pressing the contacts closed) the contactor. If you do not have low resistance, remove and replace the contactor.

If you have voltage at test point 8 but not at test point 9, you should suspect the wire between the compressor contactor and the R terminal on the compressor. It is important to note that the compressor is wired as a permanent split-capacitor (PSC) motor and the terminal connection for this wire may be on the run capacitor where voltage will be supplied for both the compressor and the outdoor fan. You can check this wire with a continuity test or by using a jumper wire.

If you have voltage at test point 9, you should hear the compressor trying to start, since voltage at the R terminal should cause current flow through the compressor run winding to terminal C. You could also use a clamp-on ammeter to measure the exact amount of current in this circuit. If the compressor does not make a noise, or if no current flow is measured, you should suspect the compressor has an open in its run winding, its overload is open, or the wire from terminal C on the compressor to L2 may be open. Turn the power off and test each of these parts of the circuit for continuity. If the compressor is "humming" or trying to start, you have determined that there is a complete path from either the R terminal to L2 or the S terminal to L2. This means that the overload is operating correctly and that you have voltage.

If you have voltage at test point 9 but not at test point 10, you should suspect the capacitor is faulty since the voltage for test point 10 must come through the run capacitor. You should notice that the capacitor in this diagram is a two-circuit capacitor, which means it has one common terminal. It also has a terminal for the capacitor for the compressor, and a second capacitor on the outdoor fan. You can test the capacitor or you can try a different capacitor to make the system run. If you have voltage at test point 10, you should suspect the start winding is open. At this point you can confirm if the problem is in the start winding by measuring the current in the start winding circuit. If this winding does have current flow, you should suspect the start winding is open, and you can turn off power and test the start winding for continuity.

If the compressor is running and the outdoor fan is not running, you should test the circuit of the outdoor fan. Remember that the outdoor fan is connected as a PSC motor and it receives voltage from the same source as the compressor. This means you do not need to test the voltage source if the compressor is running and the fan is not running. If the fan is not running, you should suspect the fan itself, its run capacitor, or the normally closed LRI-1 contacts. These contacts are opened when the system is placed into the defrost mode. If they are open, you will have voltage from R-L2 but not at R-C on the motor. You should suspect the LRI-1 contacts are open and you can check the heat pump control module for proper operation. If you have voltage at the R-C terminals (test point 11) and at S-C terminals (test point 12), you should suspect the motor is faulty. Turn off all power and test the motor windings for continuity.

The final test for voltage across the terminals of the reversing valve would only need to be made if the system was producing warm air at the indoor coil and cool air at the outdoor coil. This condition would indicate the reversing valve is not energized. Test for 24 VAC, and if it is present and the valve is in the wrong position, you will need to remove power and test the heat pump reversing valve solenoid coil for continuity. If the voltage is not present, you will need to make several other voltage tests to locate the loss of voltage. Remember, the voltage should come directly from the R terminal to the B terminal when the system switch on the thermostat is set to cooling. When you have completed all tests, and replaced the appropriate parts as necessary, you should cycle the system several times through its cooling cycle to ensure that it is operating correctly.

19.8   Troubleshooting the Heat Pump in the Heating Mode

When the heat pump is placed in the heating mode, the signal to energize its cycle comes from the W1 terminal of the thermostat. You should notice in the diagram that the W1 circuit is jumpered to the B terminal, and the B terminal controls voltage to the reversing valve. Voltage is also sent to the heat pump control module to energize the CR2 coil, which ensures that the compressor and outdoor fan are powered. Voltage is also sent to the FR1 fan relay coil to ensure that the indoor fan is also powered. Fig. 19–7 shows the diagram of the heat pump with the heating circuit highlighted. If you must troubleshoot a heat pump system that will not operate in the heating mode, you can use the previous set of tests to ensure that the compressor, outdoor fan, and indoor fan are running, and you would use the following tests to determine the location of the faults if the reversing valve is not deenergized and if the second stage of heating is not energized.

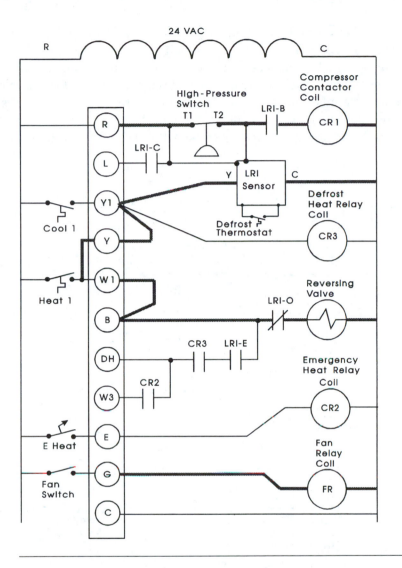

Figure 19–7   Heating circuit for heat pump is shown in bold.

19.9   Troubleshooting the Heat Pump in the Defrost Mode

When the heat pump is in its heating cycle and it goes into defrost mode, it is basically changing back to the cooling mode, and the second stage of heating is turned on. When this occurs, the reversing valve is powered again to cause the refrigeration cycle to switch. The types of problems you will get with the heat pump defrost cycle are that it will not go into defrost, it will not come out of the cycle correctly, or it will not provide back-up heat during the defrost cycle. If the system does not go into defrost often enough, you can suspect the defrost timer is set incorrectly. Remember the defrost timer uses the clock-type timer to set the amount of time between defrost cycles. This is activated by the outdoor thermostat that indicates it's cold enough to cause frost to form. If the system does not go into defrost, you should test for voltage through the circuit that has the defrost relay coil. If the defrost cycle does not come out of the defrost cycle correctly, you can

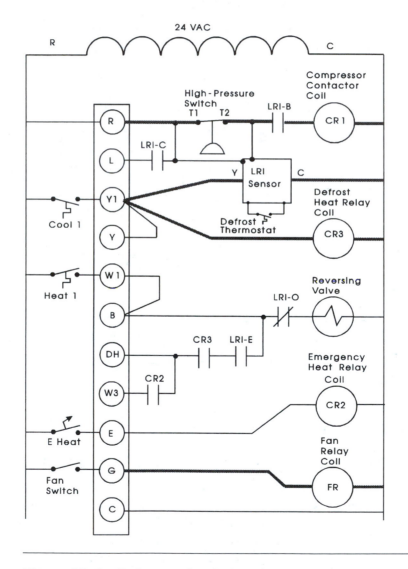

Figure 19–8   Defrost cycle of a heat pump is shown in bold.

suspect the defrost timer is faulty if the cycle is terminated by time, or the thermostat for the defrost cycle is faulty if the cycle is terminated by temperature.

If the second stage of heating does not energize when the system is in defrost, you should test for voltage across the second-stage heating coil for 24 VAC. If the coil has 24 VAC, and the second stage of heating is not energized, you can use the test found in the previous section. If the coil of the heating relay does not have 24 VAC, you should suspect the contacts of the defrost relay. You can test for a voltage loss in this circuit and test for continuity when you find the suspect parts or wire. Fig. 19–8 shows an electrical diagram of the heat pump in the defrost mode.

19.10   Package-Type Heat Pumps

Fig. 19–9 shows an example of a package-type heat pump, which is similar to the split-system heat pump except in the package system the indoor fan is in the same cabinet as the outdoor equipment. In the package system the wiring between the indoor and outdoor sections is all completed through a single terminal board, since both sections are in the same cabinet. The other main difference is that all of the fan motors in the package system are powered with 230 VAC, whereas the indoor fan in a split system is usually powered with 110 VAC. You can use the troubleshooting tests for cooling and heating cycles found in the previous sections to find similar problems in a package system. Fig. 19–10 shows the electrical diagram of a package system heat pump.

Figure 19–9   A package-type heat pump. *(Courtesy of Bard Manufacturing Co.)*

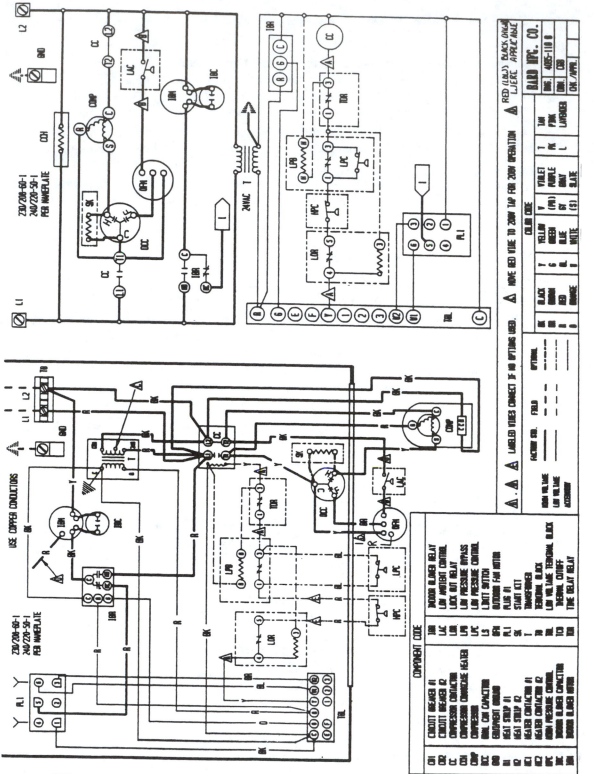

Figure 19-10 Electrical diagram of a package system heat pump. *(Courtesy of Bard Manufacturing Co.)*

472

Questions for This Chapter

1. List the parts of a heat pump when it is in the cooling cycle.
2. List the parts of a heat pump when it is in the heating cycle.
3. Explain how the defrost cycle starts and ends in the heat pump. Be sure to include the condition that must occur to get the heat pump to go into the defrost cycle.
4. Identify the ways a heat pump can provide second stage heating.
5. Explain the difference between a package-type heat pump and a split-system heat pump.

True or False

1. When the heat pump is in the cooling cycle, the indoor coil will be cool.
2. When the heat pump is in the defrost cycle, the outdoor coil will be cool.
3. When the heat pump is in the defrost cycle, the auxiliary heat will be on.
4. If the reversing valve has an open in its coil and it becomes inoperative, the heat pump will remain in the heating cycle.
5. When the heat pump is in the defrost cycle, the outdoor fan will be deenergized, and the indoor fan will be running.

Multiple Choice

1. When the heat pump is in the cooling cycle, _____ will be warm.

   a. the indoor coil
   b. the outdoor coil
   c. both coils

2. When the heat pump is in the defrost cycle, _____ will be warm.

   a. the indoor coil
   b. the outdoor coil
   c. both coils

3. When the heat pump is in the emergency heat cycle, the _____

   a. compressor will be turned off and the auxiliary heat will be turned on.
   b. outdoor section will be energized, and the indoor section will be turned off.
   c. reversing valve must be in the cooling mode.

4. The auxiliary heat will be energized _____
    a. only when the heat pump is in the defrost mode.
    b. only when the heat pump is in the heating mode and second-stage heating is needed.
    c. when the heat pump is in the defrost mode or when it needs additional second-stage heating in the heating mode.
5. The balance point is _____

    a. the outdoor temperature where it is no longer economical to operate the heat pump in the heating cycle and the auxiliary heat is used when the temperature is below this point.
    b. the indoor temperature where it is no longer economical to operate the heat pump in the heating cycle and the auxiliary heat is used when the temperature is below this point.
    c. the comparison of the indoor temperature and the outdoor temperature where the heat pump is the most efficient.

Problems

1. Use the diagrams found in this chapter to sketch an electrical diagram of the components that are energized when the heat pump is in the cooling mode.
2. Use the diagrams found in this chapter to sketch an electrical diagram of the components that are energized when the heat pump is in the heating mode.
3. Use the diagrams found in this chapter to sketch an electrical diagram of the components that are energized when the heat pump is in the defrost mode.
4. Use the diagrams found in this chapter to sketch an electrical diagram of the heat pump thermostat and explain what part is energized during the heating cycle, cooling cycle, defrost cycle, and emergency heat cycle.
5. Use the diagrams found in this chapter to sketch an electrical diagram of the heat pump reversing valve and indicate when it is energized or deenergized during the heating cycle, cooling cycle, defrost cycle, and emergency heat cycle.

20 Electrical Control of Refrigeration Systems

OBJECTIVES:

After reading this chapter, you will be able to:
1. Identify the main parts of a reach-in cooler and explain their function.
2. Identify the main parts of a reach-in freezer and explain their function.
3. Identify the main parts of a package system ice maker and explain their function.
4. Identify the main parts of a split-system ice maker and explain their function.
5. Troubleshoot the electrical circuit of a reach-in cooler, a reach-in freezer, a package-type ice maker, and a split-system ice maker.

20.0 Overview of Refrigeration Systems

You will find different types of refrigeration systems in use as you make service calls. This chapter will discuss reach-in and walk-in freezers and coolers as well as commercial ice makers. You will find the basic parts of each of these systems are similar, and that troubleshooting them will be very similar. All commercial refrigeration systems will have a compressor and condenser fan motor. Some systems will have a plate-type evaporator so that no fan is used, and other systems use an evaporator fan to help aid in moving the cool air. The thermostat for this system can be a temperature-type control or a pressure-type control, since the low pressure can be used to determine the temperature in a system. Most refrigeration systems also utilize a defrost system to remove unwanted frost from the evaporator when it builds up.

This chapter will present a reach-in cooler, a reach-in freezer, and two examples of commercial ice makers. One of the ice makers has a self-contained condenser, and the other has a split condenser that is placed outside the building. The chapter will also explain the functions of a variety of switches such as low-pressure switches, high-pressure switches, and oil pressure switches that are used to protect the system against all types of high- and low-pressure problems or the loss of oil pressure. You will learn the theory and sequence of operation and the relationship between the parts of each system. You will also learn the proper sequence of voltage and current tests to help you isolate problems when you troubleshoot the system.

20.1   Reach-In Coolers and Refrigeration Cases

The reach-in cooler is commonly used in grocery stores, convenience stores, and floral shops. Fig. 20–1 shows a typical reach-in cooler. Some coolers have the fin-type evaporator mounted on a wall or in the ceiling of the cooler and a fan is built into the evaporator housing to move air over the fins. The condenser fan and compressor can be mounted in the bottom of the case, on top of the case, or it can be located remotely, which means it will be placed outside the building.

Figure 20–1 A reach-in cooler used in grocery stores or convenience stores. *(Courtesy of Tyler Refrigeration Corporation)*

Fig. 20–2 shows an electrical diagram of a reach-in cooler. From this diagram you can see the main electrical parts of the reach-in cooler are two compressors, two condenser fans, two evaporator fans, three door heaters, and a cabinet light. The compressors in this system are capacitor-start, induction-run compressors and they each have a current relay, starting relay and a thermal overload. The condenser fans and the evaporator fans are shaded-pole motors. The compressors and condenser fans are energized through the contacts of a relay with a thermo-

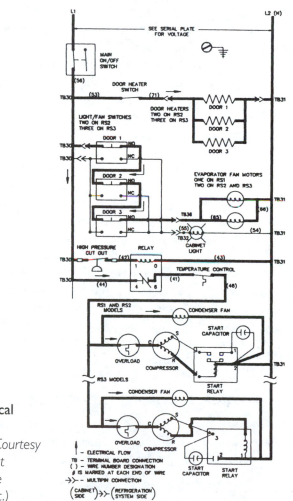

**REFRIGERATOR MODELS
RS1, RS2 AND RS3**

Figure 20–2 Electrical diagram of a reach-in refrigeration cooler. *(Courtesy of Manitowoc Equipment Company, Division of The Manitowoc Company Inc.)*

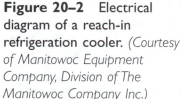

stat switch that controls the temperature in the cooler. A high-pressure cut-out switch is used to protect the compressor and condenser fan. The evaporator fans run anytime all of the doors are closed. The system can be powered with 110 VAC (L1-N) or with 230 VAC (L1-L2).

20.2   Normal Operation for the Reach-In Cooler

During normal operation the evaporator fans will run anytime the disconnect switch is closed for the unit. The heaters in the door seals are also powered anytime the disconnect is closed. These heaters keep moisture from building on the door seals. The normally closed contacts of the high-pressure switch provide voltage to the coil of the compressor relay. Voltage for the compressors and condenser fans is supplied through the contacts of the compressor relay and through the contacts of the thermostat when the temperature in the case is above the setpoint of the thermostat.

When voltage is first supplied to the compressors, it will cause current to flow through the run winding to the C terminal. Since the coil of the current relay is connected in series with the run winding, this large current will cause the coil of the current relay to become energized and pull in its contacts. The current relay contacts are connected in series with the start capacitor to supply voltage to start winding. When the current relay contacts close, it allows voltage to flow to the start winding, which causes a current to flow through the compressor start winding. When the compressor begins to run, its counter EMF causes the current in the start winding to return to lower levels, which will allow the current relay coil to return its normally open contacts to the open position. The compressor continues to run with current only flowing through its run winding. The shaded-pole condenser fan will begin to run immediately when power is applied to its winding.

When the temperature in the case cools down and reaches the setpoint, the contacts in the thermostat open and deenergize the compressor and the condenser fan. The evaporator fan runs continuously to keep air circulating in the case. Anytime the temperature increases above the setpoint, the thermostat contacts close and the compressor and condenser fan are energized again and the temperature in the cooler begins to lower.

20.3   Troubleshooting the Refrigeration System

When you must troubleshoot the refrigeration case, you should make the following tests to determine where voltage loss has occurred. Be sure to start at the source of voltage for the case.

1. L1-L2 for 230 VAC or L1-N for 110 VAC _____
2. T4 on compressor relay to L2 for 230 VAC _____

3. T6 on compressor relay to L2 for 230 VAC _____
4. T1 thermostat to L2 for 230 VAC _____
5. T2 thermostat to L2 for 230 VAC _____
6. C on compressor to L2 for 230 VAC _____
7. 2 on potential relay to L1 for 230 VAC _____
8. R on compressor to L1 for 230 VAC _____
9. Across condenser fan motor terminals for 230 VAC _____
10. Across evaporator fan motor terminals for 230 VAC _____
11. Across door heaters for 230 VAC _____

Fig. 20–3 shows the part of the electrical diagram where current flows for test points 1–6. From this diagram you can see that all voltage measurements must be made in reference to the L2 terminal. If you do not have 230 VAC at L1-L2 for a system connected for high voltage or 110 VAC at L1-N for a system that is connected for low voltage for the first test point, you do not have supply voltage and you should see if the disconnect or circuit breaker is set to the on position. If you have voltage at test point 1, and you do not at test point 2 (terminal T4 on the compressor relay), you should suspect the wire between L1 and terminal T4 on the relay. Turn power off and test the wire for continuity.

If you have voltage at test point 2, and you do not at test point 3 on the compressor relay, you should suspect the high pressure switch is open, the coil of the relay is bad, or the contacts of the relay is bad. You can test across the coil of the relay for 230 VAC. If you have voltage, you should suspect the coil or contacts of the relay is bad. You can turn off power and test the coil and contacts for continuity. If you do not have voltage across the coil, you should suspect the high-pressure switch is open. When the pressure switch is open, you must verify the system refrigeration pressure to see if it is too high. Many times, the high-pressure switch is open because of high pressure, which can be caused by a dirty

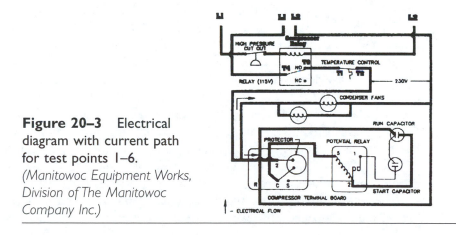

Figure 20–3 Electrical diagram with current path for test points 1–6. *(Manitowoc Equipment Works, Division of The Manitowoc Company Inc.)*

condenser coil, faulty condenser fans, or blocked air supply to the condenser. If refrigeration pressure is normal and the switch is open, you will need to replace the switch.

If you have voltage at test point 3 but do not at test point 4, you should suspect the wire between terminal 6 on the relay and terminal 1 on the thermostat. Turn power off and test the wire for continuity. If you have voltage at terminal point 4 but not at terminal point 5, you should suspect the thermostat is open. You will need to verify if the temperature in the cooler is above the setpoint of the thermostat. You should remember that if the temperature in the cooler is below the setpoint of the thermostat, the contacts of the thermostat are supposed to be open and it is operating correctly. You can temporarily set the temperature of the thermostat to the lowest setting while you troubleshoot the system and then set it back to its normal setting when you are finished with your tests. If the contacts of the thermostat are still open after you have it set at its lowest temperature, you should turn off power and test the thermostat for continuity. Change the thermostat if it is faulty. If you have voltage at test point 5 but not at test point 6, you should suspect the thermal overload is open. You should determine if the compressor has been trying to start. If it has an overtemperature condition, you must wait for the compressor to cool down. If the compressor is cool and the overload does not return to its normally closed condition, you should replace the thermal overload.

If you have voltage at test point 6 but not at test point 7, it indicates that you have lost voltage between the contacts of the start relay and the S terminal of the compressor. You should suspect either the capacitor or the wire between the capacitor and the S terminal on the compressor. Turn off power and test the capacitor for proper operation, and test the wire for continuity.

If you have voltage at test point 7 but not at test point 8, you should suspect the wire between the start relay and the R terminal on the capacitor or the coil of the start relay. You can turn off power and test each of these for continuity.

If you have voltage at test point 9 across the terminal of the condenser fan motor and the condenser fan is not running, you should suspect the condenser fan is faulty. You can turn off power and test the condenser fan for continuity. Since this is a shaded-pole motor, it should run if the coil has continuity. Be sure and test to see if the fan shaft can turn freely, or if it needs lubrication.

If you have voltage at test point 10 across the evaporator fan motor terminals, the fan motors should be running. If they do not run, you can turn off power and test them for continuity. The evaporator fan motors are also shaded-pole motors, which means they should turn anytime power is applied to the windings. If you do not have voltage at test point 10, you should suspect that one of the door switches is open. Notice that the door switches are connected in series so that if one of them is opened power will be lost to the evaporator fan.

The last test is for voltage at the door heaters. You should suspect the door heaters are not functioning properly if you notice moisture around the door seals. The job of the door heater is to provide sufficient heating on the door seal to prevent moisture from forming on them. It is important to understand that the three door heaters are wired in parallel with each other, which means that one of them could have a problem, and the other two are operating correctly. The one that has a problem will be the one that has moisture on it. You can also test each door heater for the amount of current it is drawing.

20.4   Freezer Cases

Freezer cases are used in grocery stores and convenience stores to display frozen foods. The main difference between a freezer case and a refrigeration case is that the evaporator on the freezer case will reach temperatures below freezing so a method must be provided for a defrost cycle. The defrost cycle is controlled by a defrost timer control. The remainder of the components of a freezer case are very similar to the refrigeration case. The compressor for the freezer case may be mounted directly below the case or it may be part of a remote condensing unit. Fig. 20–4a and Fig. 20–4b shows two typical freezer cases and Fig. 20–5 shows an electrical diagram of a freezer case. In this diagram you can see that the freezer case has a compressor, condenser fan, evaporator fan, door heaters, and a defrost heater. A defrost timer is also used to make the system go into defrost.

20.4.1   Normal Operation of a Freezer Case

In normal operation, the freezer case energizes its compressor with a thermostat. When the temperature in the freezer case rises above the setpoint of the thermostat, the thermostat closes its contacts, which energizes the compressor and condenser fan. The evaporator fan in the freezer case runs continuously just as in the refrigeration case. The operation of the freezer case is exactly like the operation of the refrigeration case, except for the defrost cycle. When the temperature in the freezer case drops below the setpoint, the thermostat contacts open and the compressor and the condenser fan are turned off.

20.4.2   Operation of the Freezer Case During Defrost Cycle

When the defrost control mechanism senses frost on the evaporator coil, or when time is expired on the defrost cycle time clock, the contacts in the defrost cycle timer will close and cause the system to go into a defrost cycle. During the defrost cycle, the compressor and the condenser are deenergized and the defrost heater is energized. The defrost heater provides heat directly on the evaporator coil, which causes the frost to melt. The defrost cycle will last for the duration

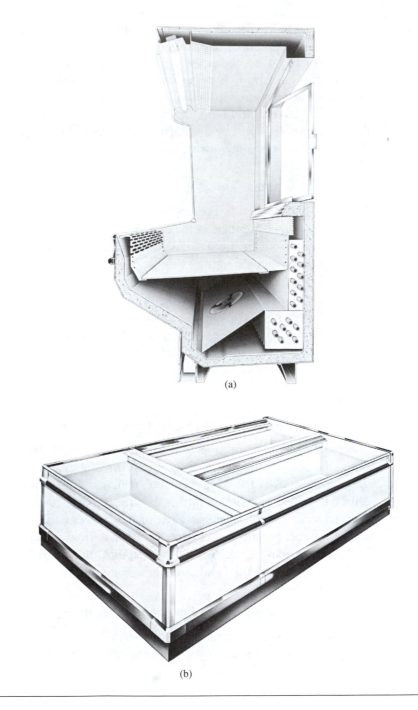

Figure 20–4a   A display-type freezer case. **b** A deep-type freezer case.
(Courtesy of Tyler Refrigeration Corporation)

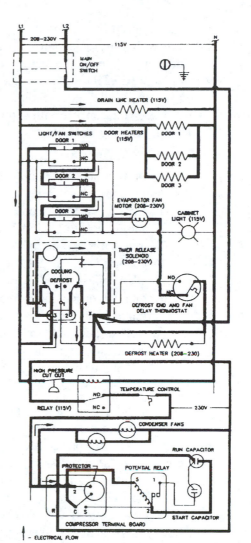

Figure 20–5 Electrical
diagram for a freezer case.
*(Courtesy of Manitowoc
Equipment Works)*

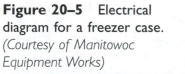

of the timer, or until the system determines the frost has been eliminated. In some
systems hot gas from the condenser is used to melt the frost in much the same
way as a heat pump cycle is reversed. This type of defrost is called a hot pass
gas defrost. The diagram for this freezer shows an electric heater defrost.

20.5   Troubleshooting the Freezer Case

When you must troubleshoot a freezer case, you can use the following tests to
help you determine the location of the problem. These tests are very similar to

the tests for the refrigeration case, except the freezer case has a defrost cycle. It is also important to understand that the compressor for this system is larger than the ones in smaller refrigeration cases and it is wired as a capacitor-start, capacitor-run (CSCR) motor. Notice that the CSCR compressor uses a potential relay to energize the start winding. Fig. 20–6 shows the ladder diagram for the freezer case. You can use either the wiring diagram or the ladder diagram for this troubleshooting exercise.

1. L1-L2 for 230 VAC _____
2. T4 on compressor relay to L2 for 230 VAC _____
3. T6 on compressor relay to L2 for 230 VAC _____
4. T1 thermostat to L2 for 230 VAC _____
5. T2 thermostat to L2 for 230 VAC _____
6. Terminal 2 to L2 on compressor TB(230VAC) _____
7. C on compressor to L2 for 230 VAC _____
8. R on compressor to L1 for 230 VAC _____
9. Terminal 1 on potential relay to L1 for (230 VAC) _____
10. Terminal 2 on potential relay to L1 for 230 VAC _____
11. Across condenser fan motor terminals for 230 VAC _____
12. Across evaporator fan motor terminals for 230 VAC _____
13. Across door heaters for 230 VAC _____
14. Across defrost heater for 230 VAC _____

If you don't have voltage at test point 1, you have lost the supply voltage. Check for the main disconnect to be closed and to be sure that voltage is coming through the fuses. If you have voltage at test point 1 but not at test point 2, you should check the high-pressure cut-out. Be sure that the refrigeration pressure is normal. Remember if the refrigeration pressure is high, the high-pressure compressor cut-out will be open. You can turn off power and test the high-pressure cut-out for continuity. If you have voltage at test point 2 but not at test point 3, you should suspect the compressor relay has not been energized. You can test the coil for continuity or the contacts for continuity. If you have voltage across the compressor relay coil, its contacts should be closed. If you voltage at test point 3 but not at test point 4, you should suspect the wire between the compressor relay and the temperature control. Turn off power and test this wire for continuity.

If you have voltage at test point 4 but not at test point 5, you should suspect the temperature control. Set the temperature control to the lowest possible temperature and try to make the system run. If the compressor still does not try to start, you can turn off power and test the temperature control for continuity. If you have voltage at test point 5 but not at test point 6, you should suspect the wire between the temperature control and the compressor terminal 2. Turn off power

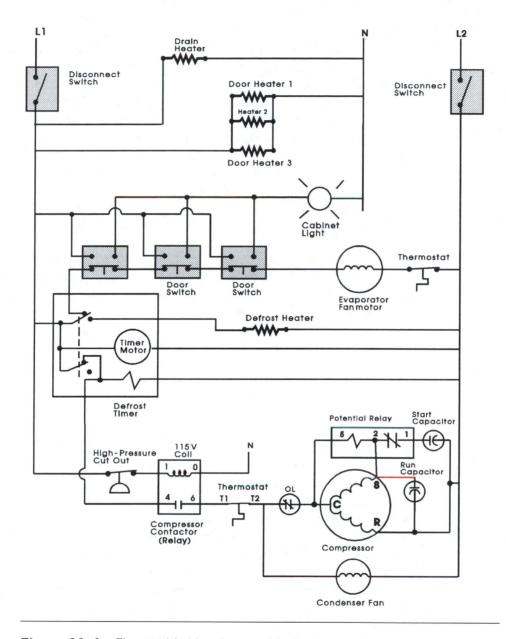

Figure 20–6   Electrical ladder diagram for the freezer case.

and test this wire for continuity. If you have voltage at test point 6 but not at test point 7, you should suspect the thermal overload on the compressor is open. Feel the area where the compressor overload is mounted to see if the compressor is warm to the touch. If the compressor has tried to start several times and has not started, the thermal overload will be opened due to excessive current. Allow the thermal overload to cool down and try to start the compressor again. If the contacts of the thermal overload will not close, replace the thermal overload. If you have voltage at terminal 7 but not at terminal 8, you should suspect voltage coming through the start capacitor called line 2. You can turn off power and test the wires connected to the start capacitor for continuity, and you can test the capacitor for normal operation.

Fig. 20–7 shows the voltage path for tests 1–7. In this diagram you can see that voltage must make its way from L1 through the contacts of the defrost switch, the contacts of the compressor relay, and then through the overload device until it reaches the C terminal on the compressor. If you do not have voltage at test point 1, you have lost incoming voltage and you should check to see that the disconnect is turned on and that the fuses are good. If you have voltage at test point 1 but not at test point 2, you have lost voltage in the contacts of the defrost switch or in the wires that connect to the defrost switch and to the compressor relay. You will need to verify if the system is in a defrost cycle. If the system is in a defrost cycle, the compressor is supposed to be deenergized. You can advance the defrost timer to bring the system out of defrost so that you can continue to make tests. If you do not have voltage through the defrost contacts, you can turn off power and test the timer contacts for continuity and the defrost timer for proper operation. You can also test the wires between L1 and the contacts of the defrost timer and the wire between the contacts of the defrost timer and the compressor contactor for continuity.

If you have voltage at test point 2 but not at test point 3, you should suspect the contacts of the compressor contactor are not closed. You should test to see if the coil of the contactor has 230 VAC across its terminals. If voltage is present, the contactor may be faulty and you can tests its coil for continuity. If voltage is not present, you can use Fig. 20–7 to trace the path of the voltage that energizes the compressor contactor coil. You can see from this diagram that voltage comes from L1, through the contacts of the high-pressure cut-out switch, to the coil of the contactor and back to L2. Leave one meter probe on L2 terminal and test for voltage at the input of the high-pressure switch. If you do not have voltage at this point, the wire between L1 and the switch is open. If you have voltage at the input of the high-pressure switch but not at the output, you should suspect the high-pressure switch is open. Be sure to test for high refrigerant pressure and correct its causes if necessary. If you have voltage at the output of the high-pressure switch but not at the coil, you have an open in the wire between the

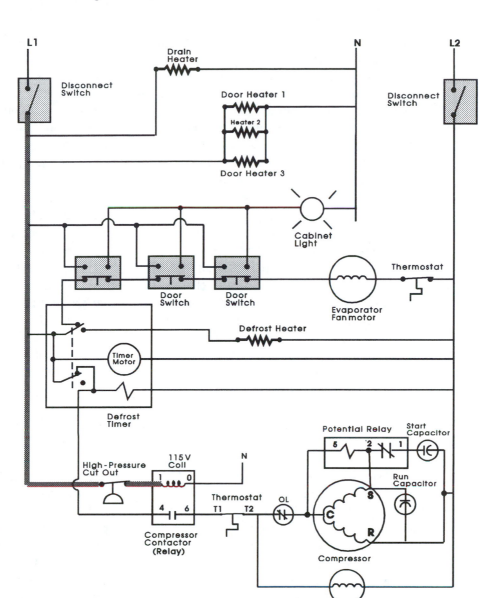

Figure 20–7 Electrical diagram of voltage path from L1 to coil of the compressor contactor highlighted.

switch and the coil. When you find the cause of the open circuit, make repairs as necessary and try to start the system.

If you have voltage at test point 3 but not at test point 4, you should suspect the wire between the compressor contactor and the thermostat. If you have voltage at test point 4 but not at test point 5, you should suspect the thermostat is open. Be sure it is set to its lowest position. If it still does not pass voltage, you can turn off power and test it for continuity. If you have voltage at test point 5 but not at test point 6, you should suspect the wire between the thermostat and the compressor terminal 2. If you have voltage at terminal 2 but do not have voltage at terminal C on the compressor, you should suspect the thermal overload on the compressor is open. You should remember the overload will protect the compressor if it tries to start and draws excessive current because one of its windings is open or if one of its capacitors is faulty. Sometimes the compressor continues to try and start with faulty conditions and it will stress the overload device until it is damaged. If you change the overload, be sure and continue to test the compressor circuit until you are certain that the compressor will start and draw normal current. Fig. 20–8 shows the circuit where L1 voltage is provided to the condenser fan, evaporator fan, and door heaters.

If you have voltage at terminal C of the compressor, it means that the path for the L1 side of the circuit is good. You can leave one meter probe on terminal L1 as a reference for tests 8–10. Fig. 20–9 shows the path of voltage from L2 to the compressor motor. From this diagram you can see that voltage on the L2 side goes directly to terminal R of the compressor. If you have voltage at L1-L2 but not at L1 to R, you can suspect the wire from L2 to R. Turn off power and test it for continuity.

If you have voltage at terminal R to C on the compressor, it should hum when it tries to start, since this is a complete circuit through the run winding. If the compressor does not hum and try to start, you can test this circuit for current. If no current is being drawn, you can suspect the compressor has an open in its run winding. Turn off power and remove all wires from the compressor terminals and test the compressor run and start windings for continuity. If the compressor hums and tries to start, you can test for a voltage loss in the start winding circuit, which includes the start capacitor and potential relay. If you have voltage at test point 9 but not at test point 10, you can suspect the start capacitor or the normally closed contacts of the potential relay. Remember the contacts of the potential relay will open after the compressor is running at approximately 75% and sufficient counter EMF is developed to energize the potential relay coil. You can turn off power and test these components for continuity, and you can test the capacitor for proper operation.

The condenser fan motors, evaporator fan motors, and heaters can be tested using test points 11–14. If you suspect a condenser fan or evaporator fan is not

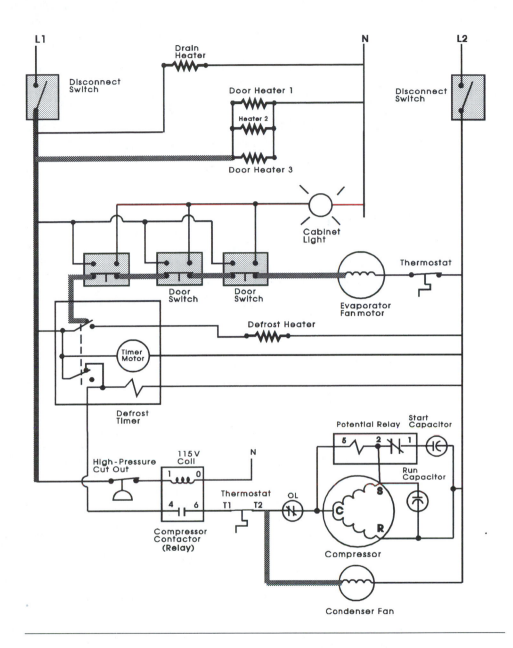

Figure 20–8   Electrical diagram of voltage path for condenser fan, evaporator fan, and door heaters highlighted.

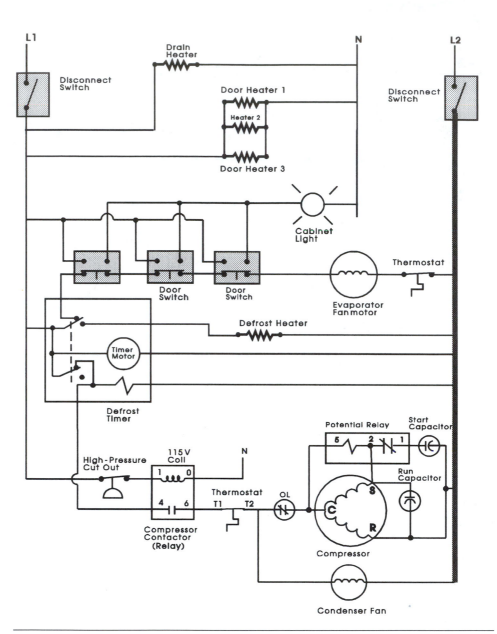

Figure 20–9 Electrical diagram of the path of voltage on the L2 side of the compressor highlighted.

running, you can physically test to see if their fan blades are turning. You should remember the evaporator fan should run anytime the doors are closed, and the condenser fan should run anytime the compressor is running. Since these are shaded-pole motors, you can test for the presence of voltage. If you have voltage and the motor does not run, you can test it for an open. If you do not have voltage, you can use voltage tests or continuity tests to locate the open in the wiring.

20.6   Troubleshooting the Defrost Cycle

Since the freezer case has evaporator temperatures below 32°F, frost can build up on the fins of the evaporator. The system can be sent into a defrost cycle by time or if the temperature in the coil is too low. The frequency and the duration of the defrost cycle can be set in the timer. When the system goes into a defrost cycle, 230 VAC is sent to the defrost heating coil. The defrost coil is mounted near the evaporator so that heat from it can melt frost from the evaporator coil. When the coil is covered with frost, it will activate the defrost thermostat to move its contact from the normally closed position to the normally open position. This interrupts voltage to the evaporator fan, while the system is in a defrost cycle. When sufficient frost has been removed, the thermostat returns to its normally closed position, which energizes the evaporator fan again. It also ends the defrost cycle.

The coil of the timer motor receives voltage anytime the system disconnect switch is closed. Fig. 20–9 shows the electrical diagram of the defrost cycle that includes the defrost timer and the defrost heater coils. You should notice that the defrost heater is a simple electric heating coil that is located on the evaporator coil so that the heat it produces will melt the frost. The main problem with the defrost cycle is that it will not start or it will not end. If the defrost cycle does not start, you should test for 230 VAC across the timer motor at terminals N and X. When the defrost control is in its normal condition, it should have continuity between terminals N-3 and from terminals 2–4. When the control goes to defrost, it opens terminals N-3 and 2–4, and closes 3–1.

20.7   Ice Makers

Ice makers can be either flake type or cube type and they are found in hotels, motels, commercial kitchens, and cafeterias. They provide a continuous source of ice cubes for the food industry. Fig. 20–10 shows typical cube type ice machines. Fig. 20–11 shows an electrical diagram of a typical cube type ice machine. From this diagram you can see that the main components of the cube type ice machine include the compressor, condenser fan, evaporator plates which make the ice cubes, a water pump, an inlet water valve, and a hot gas solenoid valve.

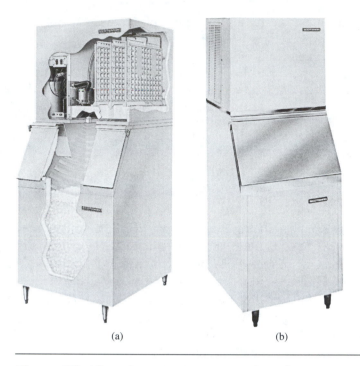

(a)   (b)

Figure 20–10a   A cut-away picture of a cube type ice machine. **b** A cube type ice machine. *(Courtesy of Scotsman)*

The typical controls for a cube type ice machine include an ice size sensor, a low-pressure cut-out, a high-pressure cut-out, and a bin control to turn the machine on and off. Control systems vary greatly from brand to brand.

20.8   Normal Operation of an Ice Maker

During normal operation the ice maker is connected to 230 volts at line 1 and line 2. When the system is making ice the master switch, high temperature cut-out switch, the high pressure cut-out switch and the bin thermostat are closed which provides power to coil of the compressor contactor. The compressor is wired as a capacitor-start, capacitor-run motor and it gets voltage for its start winding through a *potential relay*. The ice making part of this system has a water pump motor that is energized and a water solenoid valve that is energized to the open position to allow water to flood the evaporator plate. The ice bin sensor control turns the compressor contactor relay coil on or off, and the contracts of this relay open to de-energize L1 voltage to the compressor motor when the level of ice in the bin is full. The ice sensor also regulates the amount of time the water pump and solenoid stay energized to provide the correct amount of water

Air Cooled, 60 Hz, Single Phase Schematic Diagram

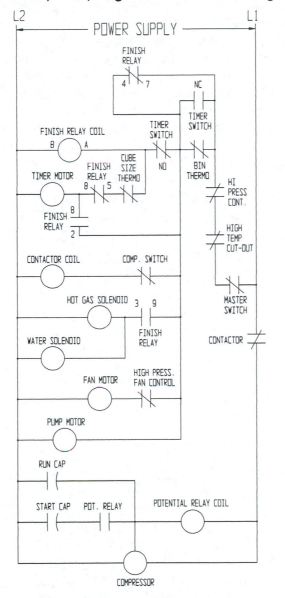

Figure 20–11   Electrical diagram of ice maker. *(Courtesy of Scotsman)*

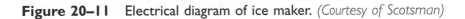

to ensure the ice cubes are full size. As water flows into the evaporator, it will begin to freeze on the evaporator plate and form ice cubes.

When the ice has frozen solid, the sensor starts the *harvest cycle*. During the harvest cycle, the hot gas solenoid and finish relay are energized to redirect the hot gas from the compressor to the evaporator plate to allow the ice to melt slightly so that it becomes free of the evaporator plate and falls into the storage bin. When the ice falls free of the evaporator, the harvest cycle is complete, and the freezing cycle is started again and the water pump and solenoid are energized to allow water to flood the evaporator again. If the bin that contains the ice cubes becomes to full, the bin level sensor will open and de-energize the primary voltage to the transformer, which prevents water from flowing to the evaporator plate. The refrigeration cycle will continue for several seconds after this point until the low pressure switch opens. The pressure in the evaporator will drop until it reaches its preset pressure since no water is available to freeze.

20.8.1   Troubleshooting the Ice Maker

When you must troubleshoot the ice maker you can use the following list to make voltage measurements at specific points in the circuit. After you have made the measurements, you can analyze them to determine where the problem is. If the compressor is not running you should use tests 1-11 to find the problem. If the compressor is running, but the condenser fan is not running, you should use tests 12 and 13 to determine the problem. If the compressor and condenser are running but the machine is not making ice, you should use tests 14 and 15 to determine the problem.

1. L1-L2 for 230 VAC _____
2. Master Switch Contacts to L2 for 230 VAC _____
3. High Temp. Cut-out Contacts to L2 for 230 VAC _____
4. Hi. Press. Contacts to L2 for 230 VAC _____
5. Bin Thermo. Contacts to L2 for 230 VAC _____
6. Contactor Coil to L2 for 230 VAC _____
7. Compressor Switch Contacts to L2 for 230 VAC _____
8. Start capacitor to L1 for 230 VAC _____
9. Potential Relay Contacts to L1 for 230 VAC _____
10. Terminal S on compressor to L1 for 230 VAC _____
11. L1 to L2 on the compressor for 230 VAC _____
12. High Press. Fan Control to L2 for 230 VAC _____
13. Across the condenser fan terminals for 230 VAC _____
14. Across water pump for 230 VAC _____
15. Across water solenoid for 230 VAC _____

If you do not have 230 VAC at test point 1, you should check the service box to determine if you have blown a fuse or a circuit breaker is open. If you have voltage at test point 1, but not at test point 2, you should suspect the master switch contacts are open or faulty. If you have voltage at test point 2 but not at test point 3, you should suspect the high-temperature cut-out switch is open. The problem may be that the temperature is too warm. If you have voltage at test point 3, but not at test point 4 you should suspect the high pressure contacts are open. Be sure to determine if the pressure in the system is high from a problem such as the condenser fan is inoperative, or if the condenser coil is covered with dirt so that air can not get through the coils to help remove heat. If the refrigerant pressure is at the correct level, and the high-pressure switch is still open, you should suspect the switch itself is faulty.

If you have voltage at test point 4, but not at test point 5, you should suspect the bin thermostat switch is open. This switch may be open because the bin is full. If you have a problem with any of these switches you must be sure to check the condition that may cause the switch to open. You must remember that you may have to correct the condition that is causing the problem rather than changing the switch. For example if the pressure is not at the correct level you may have to correct the pressure problem before the system will operate correctly. If the pressure is at the correct level and the switch is still open, you should then suspect the switch itself.

If you have voltage at test point 5, but not at test point 6, you should suspect the coil that closes the compressor contacts is not energized. If voltage is present at the coil but the contacts are not closed, you may need to test the relay coil to ensure that it has continuity. If you have voltage at test point 6, and the contactor coil is energized you should move to test point 7, and test to see if voltage is passing through the contacts of the contactor. If the contacts are closed and the compressor is trying to start but it can not run, it will cause excessive current which will open the overload after a short period of time. Other problems that may keep the compressor from running and allow it to draw excessive current include high refrigerant pressure which causes the compressor to work too hard, or low refrigerant pressure, which cause the compressor to run hot since not enough refrigerant is coming back to the compressor in liquid droplets to keep it cool. If all of the conditions in the compressor are normal and the overload is still open, you should suspect the overload itself.

If you have voltage at test point 8, but not at test point 9, you should suspect the wire between terminal C on the compressor and terminal 3 on the overload is open. If you have voltage at test point 9, but not at test point 10, you should suspect the wire that provides L2 voltage to the circuit through the start capacitor is open. If you have voltage at test point 10, but not at test point 11, you should suspect the contacts of the potential relay are faulty. If you have voltage at test point 11, the circuit is functioning properly and you should suspect the start winding or run winding of the compressor motor and you can test each winding for continuity. Be sure to turn off power before you make the test for continuity.

Another problem that causes the ice maker to fail is when the compressor is running, but the condenser fan is not running. When this occurs you should use test 12. This test is for voltage across the contacts high pressure fan control. If you have voltage at this point, you should use test 13 which test for voltage directly across the terminals of the fan motor. If you have voltage and the motor does not turn, you should suspect the motor is faulty and test it for continuity. The condenser fan motor is a shaded pole motor. If you do not have voltage at the motor terminals, you should suspect the wires that connect the motor directly to the voltage.

If the compressor and condenser fan are both running but the system is not making ice, you should suspect the water and ice control circuit. You can start with test 14. If you do not have voltage at test point 14 you can suspect the bin switch is open. Be sure to determine if the level of ice in the bin is keeping the switch open, or if the switch is faulty or not adjusted correctly. If the pump is running but you do not have water flowing, you should suspect the water solenoid and check for voltage directly at the terminals of the water solenoid. If you have voltage at these terminals, but the solenoid is not energized you should turn off power and test the solenoid coil for continuity. You should also remember that sometimes the water solenoid valve corrodes and can not operate correctly even though the electrical control part of the valve is functioning right.

If the machine is making ice but will not harvest it correctly, you should suspect the ice sensor control which is also known as the bridge thickness sensor, or the hot gas solenoid. If it have voltage but do not operate, you should suspect the valve coil and test it for continuity. You will find that troubleshooting a system is a process of determining what is operating correctly and what is not. You must trace voltage to determine how far it is moving through a circuit, and know where to find information about motors and controls in prior chapters for more detailed troubleshooting. The biggest mistake that most starting technicians make when troubleshooting is finding a control such as a high pressure control that is open an thinking the control is faulty, when in fact the control is simply open because it is doing its job correctly by detecting a high pressure condition. You must determine the cause of the high pressure condition and correct it rather than change the high pressure switch.

Questions for This Chapter

1. Identify the main parts of a reach-in cooler and explain their function.
2. Compare the parts of a cooler to the parts of a freezer and explain the basic differences.
3. Identify the controls for a freezer or cooler and explain their operation.
4. Explain why a freezer requires a defrost cycle and a reach-in cooler does not.
5. Identify the main parts of an ice maker and explain which parts control the refrigeration cycle and which parts control the water flow.

True or False

1. A cooler or freezer can use a low-pressure control as a thermostat to control the temperature in the display case.
2. The door heaters on a reach-in cooler are needed because the door handles are too cold when people touch them.
3. In most reach-in coolers the evaporator fan and condenser fans are three-phase motors.
4. Freezers require a defrost heater because frost will build on evaporator fins and block air flow.
5. The ice maker shown in the electrical diagram in Fig. 20–11 uses a hot gas solenoid to warm the evaporator plate during the harvest cycle.

Multiple Choice

1. If you are troubleshooting a reach-in cooler because its compressor is not running, you should _____
   a. test the door heaters because an open in a heater will prohibit voltage from reaching the compressor.
   b. test the supply voltage, compressor relay, and thermostat to see that voltage is passing through these components and reaching the compressor.
   c. test the door switches to ensure voltage is reaching the compressor.
2. The compressor in the electrical diagram for the reach-in cooler in Fig. 20–2 is connected as a _____
   a. capacitor-start, induction-run (CSIR) compressor.
   b. permanent split-capacitor (PSC) compressor.
   c. split-phase compressor with a current relay.

3. The condenser fan motor in the electrical diagram for the reach-in cooler in Fig. 20–2 is connected as a _____

   a. shaded-pole motor.
   b. capacitor-start, induction-run motor (CSIR).
   c. permanent split capacitor (PSC).

4. The compressor and condenser fan for the reach-in cooler in Fig. 20–1 is physically located _____

   a. outside since this is a split-system type unit.
   b. below the cooler case toward the back of the unit.
   c. on top of the cooler case.

5. If the defrost cycle for the freezer in Fig. 20–9 will not activate, you should suspect _____

   a. the defrost timer motor has stopped.
   b. one of the door switches is faulty.
   c. the drain line heater has failed and the drain is frozen shut.

Problems

1. Use the diagram in Fig. 20–2 and sketch the electrical diagram of the circuit for the evaporator fan.

2. Use the diagram in Fig. 20–2 and sketch the electrical diagram of the circuit for the compressor. Be sure to include the control circuit and all controls for the compressor relay.

3. Use the diagram in Fig. 20–2 and sketch the electrical diagram of the circuit for the condenser fan.

4. Use the diagram in Fig. 20–11 of the ice maker and sketch the electrical diagram of the circuit for the defrost hot gas solenoid and all of the controls for defrost.

5. Use the diagram in Fig. 20–11 of the ice maker and sketch the electrical diagram of the circuit for the compressor.

A Developing a Ladder Diagram from a Wiring Diagram

Many times you will troubleshoot a system and a wiring diagram of the electrical circuit will be provided, but a ladder diagram will not be available. You can develop your own ladder diagram by tracing the circuits in the wiring diagram. The material in this section will provide an example of converting a wiring diagram for a package air-conditioning system to a ladder diagram. Fig. A–1 shows a wiring diagram of a typical package air-conditioning system that is connected to 230 VAC. You can see in this diagram that the major components in this system are the compressor motor, the condenser fan motor, the evaporator fan motor, the compressor contactor, the evaporator fan relay, the transformer, and the thermostat.

Fig. A–2 shows the skeleton or the outline of the ladder diagram. Typically the compressor and condenser fan loads will be shown at the top of this diagram, and the transformer and thermostat will be shown at the bottom of this diagram. You can start developing the ladder diagram by placing the compressor motor, the condenser fan motor, and evaporator fan motor between the two rails of the ladder diagram. This is shown in Fig. A–3.

The next step in the procedure is to find the source of power. In the wiring diagram, the electrical disconnect is the source of L1 and L2 voltage. You should begin the process by tracing the wire that connects L1 to the contacts of the com-

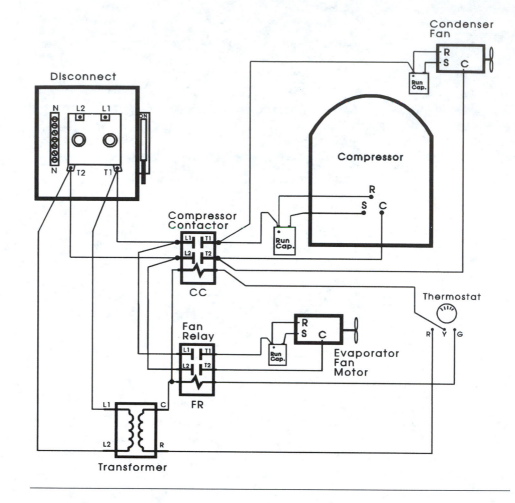

Figure A–1   Wiring diagram of a package air conditioner.

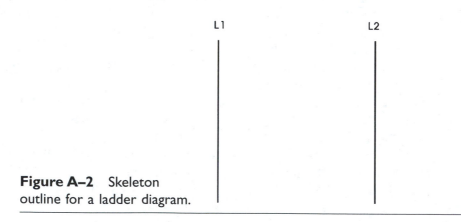

Figure A–2   Skeleton outline for a ladder diagram.

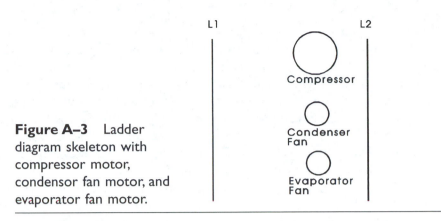

Figure A–3 Ladder diagram skeleton with compressor motor, condensor fan motor, and evaporator fan motor.

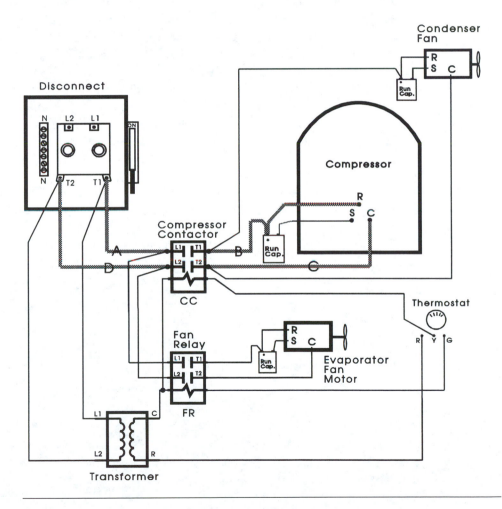

Figure A–4 Wires A, B, C, and D are highlighted in the wiring diagram.

pressor contactor. You can see that a wire identified as wire A runs terminal L1 of the disconnect to terminal L1 of the compressor contactor. A wire identified as wire B runs from terminal T1 of the compressor contactor to terminal R of the compressor. On the other side of the compressor a wire identified as wire C is connected to terminal C on the compressor and terminal T2 of the compressor contactor. You can follow a wire identified as wire D from terminal L2 of the compressor contactor to terminal L2 of the disconnect. These wires are highlighted in the wiring diagram shown in Fig. A–4, and they are placed in the ladder diagram you are developing in Fig. A–5.

The next component you can add to the ladder diagram is the condenser fan motor. You should notice in the wiring diagram in Fig. A–6 that a wire identified as wire E is highlighted and it runs from the compressor contactor terminal T1 to the R terminal on the condenser fan. You should also notice a wire identified as wire F that runs from terminal T2 of the compressor contactor to the C terminal of the condenser fan. Since the condenser fan is wired as a PSC-type motor, the run capacitor is connected between the R and S terminals of the motor. Fig. A–7 shows the condenser fan motor added to the ladder diagram you are developing.

The next component you will add to your diagram is the evaporator fan. In Fig. A–8 you will notice that the evaporator fan is controlled by the fan relay. Wire G runs between terminal L1 of the compressor contactor and terminal L1 of the fan relay. Wire H runs between terminal T1 of the fan relay and terminal R of the fan. Wire I runs between terminal C of the fan and terminal T2 of the fan relay. Wire J connects terminal L2 of the fan relay to terminal L2 of the compressor contactor, where it picks up supply voltage. Fig. A–9 shows the evaporator fan added to the ladder diagram.

The primary side of the transformer is connected with wire K at terminal L1 of the compressor contactor. The second wire of the primary side of the transformer is identified as wire L and it is connected to terminal L2 of the compres-

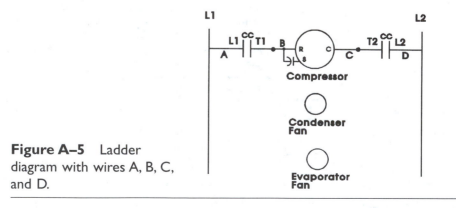

Figure A–5 Ladder diagram with wires A, B, C, and D.

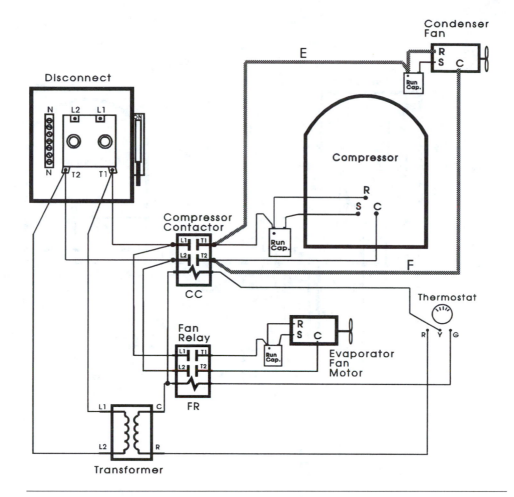

Figure A–6 The wires connected to the condenser fan are shown highlighted in the wiring diagram.

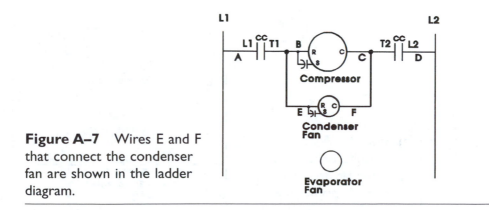

Figure A–7 Wires E and F that connect the condenser fan are shown in the ladder diagram.

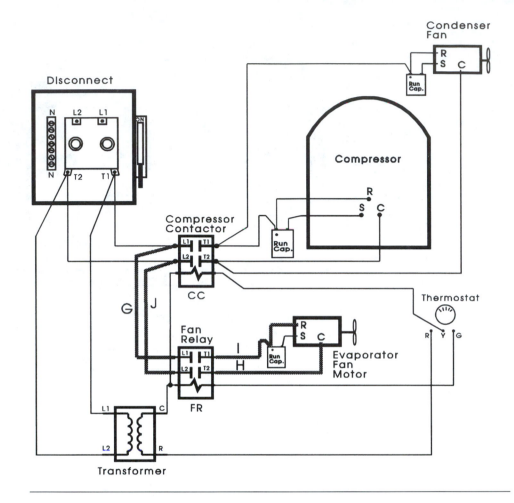

Figure A–8   The wires to the evaporator fan are highlighted in the wiring diagram.

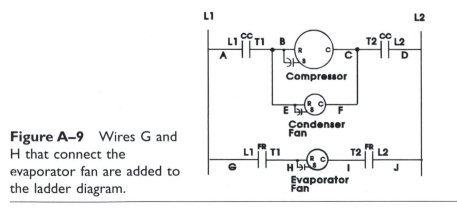

Figure A–9   Wires G and H that connect the evaporator fan are added to the ladder diagram.

sor contactor. The primary side of the transformer is connected to a 230 VAC power source anytime the disconnect switch is turned on. The secondary side of the transformer is identified as terminals R and C. Fig. A–10 shows the transformer highlighted in the wiring diagram. Fig. A–11 shows the transformer primary and secondary added to the ladder diagram.

When you are ready to draw the low-voltage control side of the circuit, you can place the secondary transformer winding, the coil of the compressor contactor, and the coil of the fan relay between the power wires of the ladder diagram skeleton. Fig. A–12 shows this.

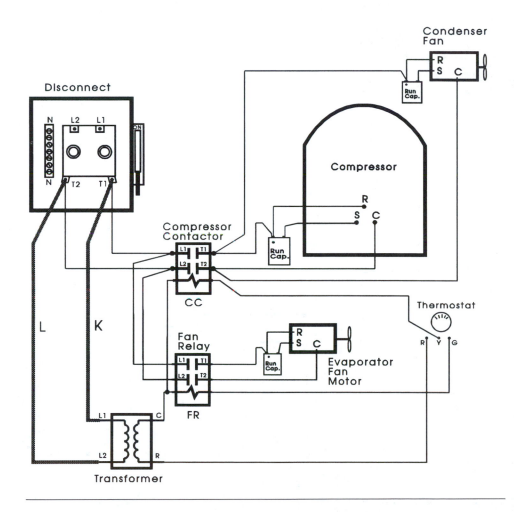

Figure A–10 The wires connected to the transformer are highlighted in the wiring diagram.

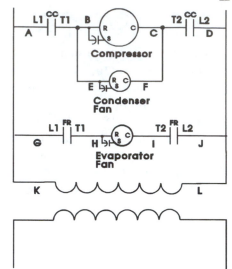

Figure A–11 The transformer and the wires connected to it are shown in the ladder diagram.

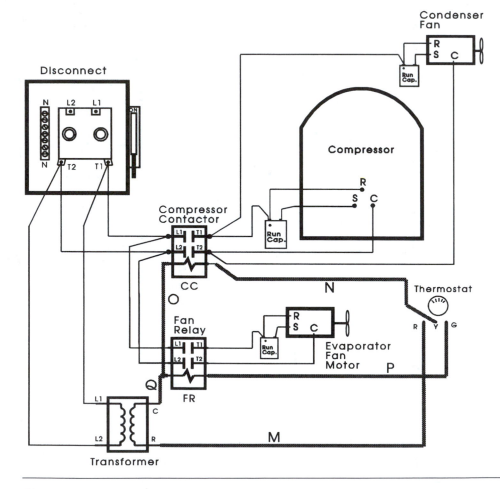

Figure A–12 The wires connected to the thermostat and the compressor contactor coil are shown highlighted in the wiring diagram.

A wire identified as wire L connects the thermostat terminal R to terminal R of the secondary of the transformer. Terminal Y on the thermostat is connected to terminal C1 of the compressor contactor with wire M. Wire N is connected to wire C2 of the compressor contactor coil and terminal C2 of the of the fan relay coil. Wire O is an internal connection between the fan switch and terminal R inside the thermostat. Wire P is connected between terminal G of the thermostat and terminal C1 of the fan relay coil, and wire Q is connected between terminal C2 of the fan relay coil and terminal C of the transformer. Fig. A–13 shows these wires in the wiring diagram, and Fig. A–14 shows these wires in the ladder diagram.

The connection between the cooling thermostat contact and the fan switch is inside the thermostat. The fan switch can be set to auto where it receives power from the Y terminal of the thermostat whenever the thermostat is calling for cooling, or it can be set for on where it receives power from the R terminal of the thermostat constantly. When the fan switch is in the on position, it will energize the fan relay coil constantly, which causes the fan to run continually.

Fig. A–15 shows a completed ladder diagram that you have made from the wiring diagram. You can see the high-voltage components at the top of the dia-

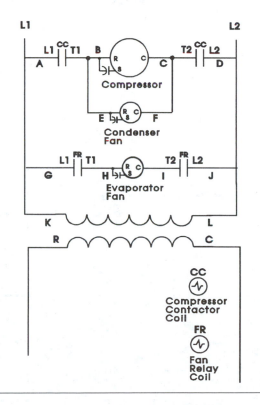

Figure A–13 The coil of the compressor contactor and the coil of the fan relay are added to the ladder diagram.

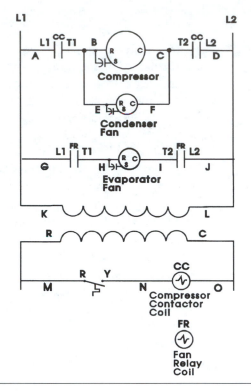

Figure A–14 The skeleton for the low-voltage side of the ladder diagram.

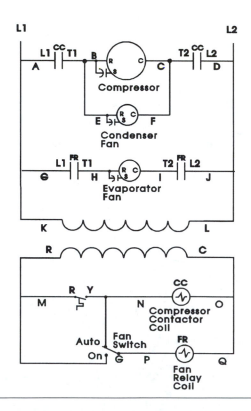

Figure A–15 The completed ladder diagram.

gram, and the low-voltage components at the bottom of the diagram. You should also notice that it is much easier to see the switches that control each of the loads in the diagram on the ladder diagram than on the wiring diagram. You should notice that the wiring diagram shows some of the wires crossing each other, which makes it difficult to use during troubleshooting. With practice, you will find it fairly easy to develop the ladder diagram from a wiring diagram, and the time it takes to change the diagram will be saved during troubleshooting.

Index